**GENETIC RESOURCES
CONSERVATION PROGRAM
(GRCP)**

Environmental Issues in Latin America and the Caribbean

Environmental Issues in Latin America and the Caribbean

Edited by

Aldemaro Romero
*Arkansas State University,
State University, AR, U.S.A.*

and

Sarah E. West
*Macalester College,
Saint Paul, MN, U.S.A.*

 Springer

A C.I.P. Catalogue record for this book is available from the Library of Congress.

ISBN-10 1-4020-3773-2 (HB)
ISBN-13 978-1-4020-3773-3 (HB)
ISBN-10 1-4020-3774-0 (e-book)
ISBN-13 978-1-4020-3774-0 (e-book)

Published by Springer,
P.O. Box 17, 3300 AA Dordrecht, The Netherlands.

www.springeronline.com

Printed on acid-free paper

Cover image of tree courtesy of A. Romero.
All other small cover images © 2005 JupiterImages Corporation.

Springer has the 'Right to Publish' Chapter 6.
All other chapters are copyright Springer.
All Rights Reserved
© 2005 Springer
No part of this work may be reproduced, stored in a retrieval system, or transmitted
in any form or by any means, electronic, mechanical, photocopying, microfilming, recording
or otherwise, without written permission from the Publisher, with the exception
of any material supplied specifically for the purpose of being entered
and executed on a computer system, for exclusive use by the purchaser of the work.

Printed in the Netherlands.

Dedication

To our families.

Contents

Dedication	v
Contributing Authors	xi
Preface	xv
Acknowledgments	xxi

PART 1: PAST AND PRESENT CONSERVATION CHALLENGES

1. In the Land of the Mermaid: How Culture, not Ecology, Influenced Marine Mammal Exploitation in the Southeastern Caribbean
   Aldemaro Romero and Joel Creswell — 3

2. Conserving the Pines of Guadalupe and Cedros Islands, Mexico: An International Collaboration
   Deborah L. Rogers, J. Jesús Vargas Hernández, A. Colin Matheson, and Jesús J. Guerra Santos — 31

3. Biodiversity Conservation in Bolivia: History, Trends and Challenges
   Pierre L. Ibisch — 55

PART 2: NATIONAL POLICIES, LOCAL COMMUNITIES, AND RURAL DEVELOPMENT

4. Peasant, Environment and Maize "Modernization" in Zacapoaxtla, Mexico, 1974-1982
   Bert Kreitlow — 75

5. Planting Trees, Building Democracy: Sustainable Community Forestry in Mexico
   Ross E. Mitchell — 95

PART 3: GETTING THE PRICES RIGHT: MECHANISMS FOR PROTECTING PUBLIC GOODS

6. Market-Based Policies for Pollution Control in Latin America
   Sarah E. West and Ann Wolverton — 121

7. A Deeper Solution for the International Commons: Building an Effort Control Regime for the Eastern Tropical Pacific Tuna Fishery
   Brian Potter — 149

8. Eco-Labeling in Latin America: Providing a Scientific Foundation for Consumer Confidence in Market-Based Conservation Strategies
   Thomas V. Dietsch — 175

PART 4: PUBLIC PARTICIPATION AND JUSTICE SYSTEMS

9. Public Prosecutors and Environmental Protection in Brazil
   Lesley K. McAllister — 207

10. Democracy by Proxy: Environmental NGOs and Policy Change in Mexico
    Raul Pacheco-Vega — 231

PART 5: THE EFFECTS OF TRADE AND DEVELOPMENT POLICIES ON THE ENVIRONMENT

11. Economic Progress in the Countryside, Forests, and Public Policy: Some Lessons from Ecuador and Chile
    Douglas Southgate, Boris Bravo-Ureta, and Morris Whitaker — 253

12. Environmental Implications of Cuba's Development Strategy
during the Special Period
Sergio Díaz-Briquets and Jorge Pérez-López 267

Index 295

Contributing Authors

Boris Bravo-Ureta
Office of International Affairs and Department of Agricultural and Resource Economics, University of Connecticut, Storrs, Connecticut

Joel Creswell
Yale School of Forestry and Environmental Studies, New Haven, Connecticut

Sergio Díaz-Briquets
Casals and Associates, Alexandria, Virginia

Thomas V. Dietsch
University of Michigan, School of Natural Resources and Environment, Ann Arbor, Michigan and Smithsonian Migratory Bird Center, National Zoological Park, Washington, DC

Jesús J. Guerra Santos
División Académica de Ciencias Naturales y Exactas, Universidad Autónoma del Carmen, Ciudad del Carmen, Campeche, México

Pierre L. Ibisch
Science Department, Fundación Amigos de la Naturaleza (FAN, Bolivia), former Integrated Expert, Center for International Migration and Development (CIM, Germany), Faculty of Forestry, University of Applied Sciences Eberswalde, Germany

Bert Kreitlow
Department of History, Carroll College, Waukesha, Wisconsin

A. Colin Matheson
Forestry and Forest Products, CSIRO, Canberra, ACT, Australia

Lesley K. McAllister
Energy and Resources Group, University of California at Berkeley, Berkeley, California

Ross E. Mitchell
Sustainable Ecosystems, Integrated Resource Management, Alberta Research Council, Edmonton, Alberta, Canada

Raul Pacheco-Vega
Institute for Resources, Environment and Sustainability, The University of British Columbia, Vancouver, British Columbia, Canada

Jorge Pérez-López
International Economist, Falls Church, Virginia

Brian Potter
Department of Political Science, The College of New Jersey, Ewing, New Jersey

Deborah L. Rogers
Genetic Resources Conservation Program, University of California, Davis, California

Aldemaro Romero
Department of Biological Sciences, Arkansas State University, State University, Arkansas

Douglas Southgate
Department of Agricultural, Environmental and Development Economics, Ohio State University, Columbus, Ohio

J. Jesús Vargas Hernández
Programa Forestal, Instituto de Recursos Naturales, Colegio de Postgraduados, Montecillo, Estado de México, México

Sarah E. West
Department of Economics, Macalester College, Saint Paul, Minnesota

Morris Whitaker
Department of Economics and Office of the Provost, Utah State University, Logan, Utah

Ann Wolverton
National Center for Environmental Economics, U.S. Environmental Protection Agency, Washington DC

Preface

We began this book with a simple goal— to assemble a collection of readings for an undergraduate interdisciplinary course taught by one of us (AR) at Macalester College. This course explored environmental problems and solutions in Latin America and the Caribbean using both natural science and social science methods. After a literature search failed to produce an anthology of interdisciplinary readings appropriate for the course, we set out to compile one.

We sought papers that dealt with the most salient environmental problems in the region, that were written by experts, and that were appropriate for undergraduate students. Most importantly, we sought papers that clearly demonstrate the contributions that experts from one discipline can make to analysis in another discipline. We sought papers that, for example, show how biological species assessments can be used to inform the politics of biodiversity conservation. We sought papers that show how economic analysis can be used to predict the likely effects of human behavior on ecosystems. We sought papers that pay close attention to how institutions, both national and international, affect the outcome of environmental initiatives.

To find essays that fit our needs, we sent out a world-wide call for papers, chose the most promising submissions, and subjected these submissions to peer review. The twelve approved essays represent the work of researchers from Latin America and the Caribbean, the United States, Canada, Europe, and Australia. All the authors have direct experience with Latin America and the Caribbean and the region's problems.

What distinguishes this book from others is its interdisciplinary nature. While each essay relies on a well-developed methodology from within one discipline, the book is a collection of essays from both the natural and social sciences. These essays are written so that readers outside of the authors' disciplines can understand them and, more importantly, can see how their methods are relevant to their own research. A biology student can use this book to learn how to present technical data on the status of flora and fauna in Latin America and the Caribbean while also learning how the politics of international institutions affect species' survival. An economics student can learn about how to use biologists' research on the benefits for birds of shade-grown coffee to develop optimal coffee-pricing schemes.

This book is not technical. It is appropriate for students without extensive training; it could be used, for example, in a course that has an introductory social or natural science course as its only prerequisite. Professors in either of these fields should find this book a source of both case studies from within their discipline and outside material that complements it. Professors in interdisciplinary fields will find this book integrates disciplines while maintaining methodological rigor. Researchers and policymakers looking for an overview of the region's more pressing and intriguing environmental issues should also find this book of interest.

Mindful of the great diversity of environmental problems in the region, we have assembled essays that analyze a representative set of problems in depth rather than skim the majority of the region's environmental issues. While they do not cover every country in the region, the essays examine problems in the country for which many environmental studies have been published (Mexico) and in the most enigmatic one in the region (Cuba). The essays also sample a group of countries from South America, Central America, and the southeastern Caribbean. They consider problems at international, regional, national, and local levels and deal with environmental policy and practice now and in years past.

We organize the twelve essays according to theme and approach into five parts: Past and Present Conservation Challenges; National Policies, Local Communities, and Rural Development; Getting the Prices Right: Mechanisms for Protecting Public Goods; Public Participation and Environmental Justice; and The Effects of Development Policies on the Environment.

Chapters in Part 1 are case studies of resource exploitation and conservation written by biologists. They also demonstrate how politics and economics affect the likelihood of conservation initiatives' success.

In the first chapter, Romero and Creswell provide a detailed history of marine mammal exploitation in five contiguous countries of the southeastern Caribbean. To examine the evolution of exploitation practices, the authors use information from archaeological findings, archival material, and interviews with people who had direct experience with the fauna. Romero and Creswell find that while Hispanic countries engaged essentially in dolphin fishing, those under British influence practiced mostly shore-whaling— despite the fact that all these countries share the same marine mammal fauna. And, while dolphins were captured by local fishers, shore-whaling was financed and operated by local elites who transferred technology and manpower from their agricultural business into the seasonal activity. The authors conclude that culture, economic forces, and social conditions, more than ecological conditions, determine the pattern of natural resource exploitation.

In Chapter 2, Rogers and her collaborators present a rich description of a unique conservation project involving the Monterey pine (*Pinus radiata*), a species whose natural populations are restricted to three along the central coast of California and two on the Mexican islands of Guadalupe and Cedros, off Baja California. This essay uses methods from conservation biology to describe the role of the pine in the islands' ecosystems. It also details how the United States, Mexico, and Australia cooperated in research, education, and fundraising to save the populations. Without the biological assessments that show the species' genetic uniqueness, those countries would not have lent their support. And without international coordination, the biologists could not have determined how to protect the valuable trees.

Chapter 3 examines the history of biodiversity conservation in Bolivia, one of the most biologically diverse countries in the world, with more than half of its territory in good or excellent conservation status. Ibisch emphasizes poverty's competing effects on biodiversity. When the government had no funds with which to build access roads into forested areas, the country's poverty acted to preserve biodiversity. Once access roads were finally built, however, poverty in existing settlements pushed humans deeper into the forest. This chapter provides valuable lessons about how economic considerations affect conservation efforts in the poorest countries.

In Part 2, chapters demonstrate the role played by local agricultural and forest communities in the evolution of environmental policy and practice.

In Chapter 4, Kreitlow analyzes farm modernizing efforts implemented by the Mexican government in the Zacapoaxtla region of the Northern Sierra. The author, a historian, uses fieldwork interviews, archival work, and technical reports written by agricultural technicians to explore the relationship between environmental and political change and the role that peasants play in that change. Beginning in the 1970s, the Mexican government attempted to introduce scientific technology such as hybrid seed, synthetic pesticides, and synthetic fertilizer, all standards in Green Revolution modernization programs. However, peasants rejected most of these proposed farm technologies, partly because the methods were environmentally and economically inappropriate. Kreitlow's chapter shows how the environment conditioned the state project and therefore participated in political change.

In Chapter 5, Mitchell examines the intersection of forestry management, forest trade, and local democracy in Mexican communities. He traces the historical development of environmental policy and the Mexican forest industry that eventually led to community control of forest resources. Mitchell also discusses implications of the North American Free Trade Agreement (NAFTA) and the forest certification process on trade and local people. He conducts case studies to explore the hypothesis that local control and democracy are necessary for environmental sustainability, especially in forest-based communities of Mexico. He concludes that communal forestry management offers new hopes for environmental and democratic sustainability.

Part 3 makes a strong case for using economic incentives and regulations to protect the environment. The essays emphasize the role that institutions, international pressures, and conservation biology can play in effective implementation of policies that "get the prices right" by increasing the relative cost of behavior associated with environmental degradation.

Economists tout market-based incentives such as pollution taxes and tradable permits as the most cost effective methods for addressing a wide variety of environmental problems. In Chapter 6, West and Wolverton examine these incentives and their applicability to Latin America. They find that the institutions so necessary for successful implementation of these policies in the United States and Europe are weak or nonexistent in Latin America. They argue that since weak government agencies make enforcement difficult and costly, pollution control policies should attempt to minimize the need for large amounts of direct monitoring. In addition, they suggest that preference should be given to environmental policies that are revenue-increasing on net, so that weak institutions can become self-sustaining.

The parable of the tragedy of the commons tells us that resources held under open access conditions are prone to over-exploitation. One might therefore conclude that regulations should limit the amount of resource extracted. In Chapter 7, Potter explains while regulations to limit aggregate catch improve fisheries' use of resources, they also promote over-investment or a "race for fish." An optimal regulatory regime would therefore limit the investments and labor dedicated to harvesting, thereby increasing the costs of resource depletion borne by the fishery. Achievement of this objective in global forums is quite rare. Potter examines the reasons why the member-states of the Inter-American Tropical Tuna Commission have attempted this difficult task. The negotiation process and likely success of international regulation of harvesting in the Eastern Tropical Pacific yellowfin fishery provide an example of effective management of global common property.

Market-based conservation strategies can encourage sustainable and environmentally-sensitive management practices in richly biodiverse areas. By paying a price premium, consumers provide a market signal that encourages producers to use less damaging management practices. However, as Dietsch explains in Chapter 8, without reliable certification programs, consumers cannot know whether the product they are buying was produced using environmentally-friendly techniques. Using methods from conservation biology and the example of benefits conveyed to birds by shade-grown coffee, he proposes a general approach for monitoring and evaluating certification programs. Dietsch's approach enables consumers to reliably distinguish environmentally-friendly products and to understand the effect their purchase has on tropical ecosystems.

Chapters in Part 4 examine how actors from within and outside Latin American governments have arisen to push for increased enforcement of environmental law.

Like many Latin American countries, Brazil has extensive and detailed environmental laws that are largely not enforced. In Chapter 9, McAllister describes and analyzes the involvement of the Brazilian Ministério Público in environmental protection. She uses legal analysis to explain the instruments that prosecutors use to defend environmental interests, describes and provides an explanation for the legal and institutional changes in the 1980s that allowed the Ministério Público to become a significant player, and assesses the effectiveness of prosecutorial enforcement of environmental laws. Her chapter provides an example of how a government entity can increase the enforcement of environmental law.

In Chapter 10, Pacheco-Vega examines how Mexican and international environmental non-governmental organizations (ENGOs) have exerted influence on the design and implementation of Mexican environmental policy. He uses an interdisciplinary framework to analyze how ENGOs successfully pressured the Mexican government to require polluters to report their toxic releases in the *Registro de Emisiones y Transferencia de Contaminantes*. This study shows how, by building coalitions and engaging in educational campaigns, ENGOs changed environmental laws in a country in which the state has traditionally dominated civil society.

The chapters in Part 5 examine the effects of economic development policies on the environment. Chapter 11 focuses on how increases in agricultural productivity have affected deforestation while Chapter 12 examines the effect of opening to international investment.

Recent literature on the causes of tropical deforestation indicates that, under certain circumstances, raising agricultural yields can accelerate farmers and ranchers' encroachment on tree-covered land. The authors of Chapter 11, however, contend that the linkage between intensification and habitat conservation is generally positive. Southgate, Bravo, and Whitaker use economic methods to contrast the experiences of Chile and Ecuador during the 1980s and 1990s. They find that because of productivity improvements, agricultural land use in Chile fell. In contrast, in Ecuador, where productivity-enhancing investment was low, large tracts of forests have been converted into cropland and pasture.

In Chapter 12, Díaz-Briquets and Pérez-López examine the effect of Cuban development policy on the environment. In particular, they examine the effects of policies instituted in the 1990s, during what has been called a "special period in time of peace" (*período especial en tiempos de paz*), when a severe economic crisis was triggered by disruptions in imports of oil and other raw materials from socialist countries. It focuses on case studies of joint ventures in the tourism, nickel mining, and oil production sectors. Their central argument is that Cuba's socialist development model is responsible for widespread environmental degradation on the island, but to the extent that it has accelerated the growth of extractive industries, recent foreign direct investment from non-socialist countries has "compounded environmental stresses associated with socialist development."

Acknowledgments

For their thoughtful essays and their patience during the publication process, we thank the book's contributors. For their helpful suggestions and keen judgment, we thank our peer reviewers. We thank Karine Moe for providing a good model for a book of readings for undergraduates. For managing the online submission process, we thank Ann Esson. For her attention to detail during the formatting process, we thank Erika Molina. We thank Kate West for helping us make the Preface clear and concise. We also thank the editors at Springer and especially Mrs. Betty van Herk for shepherding us quickly through the publication process. And, as always, for their love and support we thank our spouses and children.

PART 1: PAST AND PRESENT CONSERVATION CHALLENGES

Chapter 1

IN THE LAND OF THE MERMAID: HOW CULTURE, NOT ECOLOGY, INFLUENCED MARINE MAMMAL EXPLOITATION IN THE SOUTHEASTERN CARIBBEAN

Aldemaro Romero[1] and Joel Creswell[2]
[1] *Environmental Studies Program, Macalester College, 1600 Grand Ave., St, Paul, MN 55105-1899. Current Address: Department of Biological Sciences, Arkansas State University, P.O. Box 599, State University, AR 72467;* [2] *Environmental Studies Program, Macalester College, 1600 Grand Ave., St, Paul, MN 55105-1899. Current Address: Yale School of Forestry and Environmental Studies, 205 Prospect Street, New Haven, CT 06511*

Abstract: Although some progress has been made toward a better understanding of marine mammal utilization in the Southeastern Caribbean, no comparative analysis has been carried out to see how such practices originated, developed, and finally impacted the marine mammal populations in that region. We conducted field and archival studies for Venezuela, Trinidad and Tobago, Grenada, Barbados, and St. Vincent and the Grenadines. We analyzed records of whaling, dolphin fisheries, and manatee exploitation for those countries, interviewed local fishers, and explored the remains of whaling stations in each area. Our results show that each of these countries developed a different pattern of whale and dolphin exploitation, but similar patterns of utilizing manatees. We conclude that these five neighboring countries, although sharing essentially the same marine mammal species, developed different exploitation practices in terms of species targeted, capture techniques, and time periods in which that exploitation took place, due to different cultural circumstances.

Key words: whaling; dolphin fisheries; whales; dolphins; manatees

1. INTRODUCTION

Patterns of species exploitation by humans can be influenced by both environmental conditions and culture (e.g., religion) (Adeola, 1992; Richerson et al., 1996). Anthropogenic species extinctions (neoextinctions)

and/or depletion can provide a wealth of information regarding the biological and cultural aspects of interactions between humans and those species. In this chapter we intend to investigate the relative importance of both culture and ecology in determining how wildlife is exploited. To that end, we have chosen to study the history of exploitation of the same resource (marine mammals) among a group of five neighboring countries (Venezuela, Trinidad and Tobago, Grenada, Barbados and St. Vincent and the Grenadines) in the southeastern Caribbean.

Intentional captures of small cetaceans (whales and dolphins) throughout Latin America have been widely reported (see Romero et al., 1997 and references therein) as have incidental captures by gillnets in Mesoamerica and the wider Caribbean (Vidal et al., 1994). Organized commercial whaling of some kind has also been reported for the southern Caribbean (Caldwell and Caldwell, 1971; Romero et al., 1997; Romero and Hayford, 2001; Romero et al., 2002; Creswell, 2002). Dolphin fisheries have been reported in the same area (Romero et al., 1997 and references therein). It is also well known that manatees, *Trichechus manatus manatus*, have been overexploited in the Caribbean basin and that most, if not all populations, have become either extinct or severely depleted (Lefebvre et al., 2001). Despite the increasing amount of data in this field, this information has yet to be placed within a wider cultural perspective to ascertain how factors other than ecology can explain differences in modes of exploitation of the same resources among contiguous (by nature of their maritime borders) but culturally different countries.

In this chapter we examine marine mammal exploitation in the southeastern Caribbean and its cultural circumstances. We provide information that is consistent with our main argument, i.e., that local cultural, historical, economic, and political circumstances, are the determining factor in how those species have been exploited.

2. MATERIALS AND METHODS

Given the diversity in time and nature of many of the sources used for this research, we followed the basic principles of research synthesis (for details see specifics in Romero et al. 1997; Romero and Hayford, 2000; Romero et al. 2001; Romero 2002a,b and Creswell 2002). For St. Vincent and the Grenadines we relied on previously published reports since marine mammal exploitation activities have been widely documented in the past (e.g., Caldwell and Caldwell, 1975). Therefore, we summarize here both field and archival work carried throughout the study area.

2.1   The study area

For comparative purposes, we chose five neighboring countries whose exploitation of marine mammal resources has been well documented but that differ in history and culture. They are Venezuela, Trinidad and Tobago, Grenada, Barbados, and St. Vincent and the Grenadines.

These nations offer a unique opportunity to study faunal depletion/extinction because: (1) they have been subject to occupation by two highly different human cultures -Amerindians and Europeans- at different times and in multiple waves for which archaeological and historical data are available, and (2) they were among the first lands colonized and exploited by Europeans in the American continent, which allows us to understand, from historical records, how the social, economic, and political aspects of the colonization process affected species. Recent studies have shown that by combining paleontological, archaeological, ecological, historical, and economic data, one can reconstruct fairly precisely the historical ecology for some marine species of the Caribbean. This has been shown for coral reefs (Jackson, 2001), mollusks (Romero, 2003), whale sharks (Romero et al., 2000), whales and dolphins (Romero et al., 1997; Romero and Hayford, 2000; Creswell, 2002; Romero et al., 2002) and manatees (Lefebvre et al., 2001).

2.2   Historical setting

The early inhabitants of the Caribbean basin were the Paleo-Indians. They started colonizing the Caribbean basin as early as 7,000 YBP (years before the present). They traveled from the mainland to the islands by means of rafts and subsisted by fishing (Rouse and Cruxent, 1963). On the mainland, they were characterized as being collectors and gatherers of food as well as hunters of big game, and did not practice agriculture or used stone tools. It is widely held that these peoples probably migrated in response to the extinction of the large game animals and the subsequent decline of big game hunting culture on the mainland. In the Caribbean, Paleo-Indians were followed by Meso-Indians (about 3,000 YBP). They were more technologically sophisticated and reached the Greater Antilles from what is today Venezuela. They also invaded Trinidad in about 2,800 YBP but like the Paleo-Indians, they did not settle in the Lesser Antilles (with the exception of St. Thomas). They too lacked agricultural skills, being mostly gatherers and fishers. Only a few of them survived until the arrival of the Europeans (Watts, 1987).

Meso-Indians were supplanted in the Caribbean basin by the Neo-Indians. Two distinct groups composed the Neo-Indians: the Arawaks and the Caribs. The Arawaks originated in South America. They began colonizing the Greater Antilles around 2,100 YBP (Rouse and Cruxent, 1963). They were the first Amerindians encountered by Columbus. Arawaks were mostly farmers who sometimes lived in stable villages. They also consumed fish, shellfish, turtles, and manatees. Land animals played a minor role in their diet (Watts, 1987). The Caribs came from the Orinoco region and followed the Arawaks path of colonization. They colonized the Lesser Antilles in about 1,000 YBP. By the time of Columbus, the Caribs could be found in what is today northern Brazil, the Guaianas, Venezuela, and the Lesser Antilles while the Arawaks dominated in the islands north of Venezuela, Trinidad, and the Greater Antilles. Like the Arawaks, the Caribs also practiced agriculture, but because they tended to move more in pursuit of aggressive expansion, they depended on hunting, fishing, and collecting more than the Arawaks. It is difficult to determine the population size of these indigenous people (and, thus, their potential effect on natural resources); past attempts to estimate population figures have been highly controversial (Henige, 1998). A widely held figure for humans inhabiting the area considered for this study at the time Columbus' arrival is 50,000 (Lockhart and Schwartz, 1983). In general, remains of marine mammals have been found in archaeological sites associated with all of these cultures throughout the Caribbean (Wing and Reitz, 1982).

2.3 Species composition

The marine mammals of the southeastern Caribbean are poorly known. Beginning in the 1990's, some more systematic efforts have been carried out. These include Romero et al. (1997) and Romero et al. (2001) for Venezuela, Romero and Hayford (2000) and Romero et al. (2002) for Grenada, Creswell (2002) for Barbados, and Romero et al. (2002) for Trinidad and Tobago. Based on these efforts, we now have a more complete picture of the marine mammal composition of this region. Marine mammals in this part of the world can be divided into two major groups: sirenians (the manatee) and cetaceans (dolphins and whales). The species of marine mammals whose presence has been confirmed for the study area are listed in Table 1-1. This list excludes the boto (*Inia geoffrensis*), a dolphin found exclusively in freshwaters.

3. COMPARATIVE ANALYSIS

3.1 Venezuela

1. In the Land of the Mermaid

Venezuela is a continental, Hispanic country, with highly mixed ethnicity, whose economy is largely based on oil. It became formally independent from Spain in 1824. Fishing is largely a marginal activity (Romero, 1990). Legends about marine mammals occur among many peoples that traditionally inhabit the Orinoco Delta area. Human characteristics are attributed to whales and dolphins and most other wildlife. Tales about some animals portray them to be good while others clearly consider them to be bad. Some indigenous people kill some species without any cultural consideration while others species are respected and beloved. There is no uniform set of characteristics attributed to all marine mammals, thus traditional beliefs only influence the exploitation of certain species, and not marine mammals as a group. What follows is summarized from Romero et al. (1997, 2001) unless otherwise noted.

3.1.1 Manatee exploitation

Manatees have been exploited in Venezuela since pre-Columbian times. Indigenous peoples of the lower Orinoco River (the Waraunos, an Arawak tribe) believed that manatees had special powers that were released during the eclipse of the moon. They also called the Milky Way 'the road of the manatee.' Manatee meat has been used as food, their oil for cooking, and their skin for the manufacture of whips. These products, as well as ear bones, have also been used for medicinal purposes in different forms. In addition, ear bones were used as amulets. Tribes in the Amazon, however, believed that people that drown or ate manatees became manatees themselves. There is little question, however, that manatees have been heavily exploited in the past and that illegal hunting occurs to the present day leading to a severe depletion of the fragmented populations of this species (Lefebvre et al., 2001).

Table 1-1. List of marine mammal species whose presence has been confirmed in the countries of the study area.

Species	Venezuela	Trinidad and Tobago	Grenada	Barbados	St. Vincent/ Grenadines
T. manatus	X[1]	X[2]	X[3]		
B. borealis	X				
B. edeni	X	X[4]	X[5]		
B. physalus	X				
M. novaeangliae	X	X[2]	X[3]	X[6]	X
P. macrocephalus	X		X[3]	X[6]	X
K. breviceps				X[6]	X
K. sima					X
M. europaeus		X			
S. bredanensis	X	X[5]		X	X
S. fluviatilis	X	X			
G. griseus	X				X
T. truncatus	X	X[2]	X[3]	X[6]	X
S. frontalis	X	X	X[6]	X[6]	X
S. attenuata	X	X[5]	X		X
S. longirostris	X		X[6]		X
S. clymene	X				X
S. coeruleoalba	X				X
D. capensis	X		X[6]		
L. hosei	X[7]				X
P. electra	X[8]		X[6]		X
F. attenuata	X				X
P. crassidens	X	X	X[6]		X
4O. orca	X	X[2]	X[6]	X[6]	X
G. macrorhynchus	X	X[2]	X[6]	X[6]	X
Z. cavirostris	X			X	X
Total	23	12	13	9	19

Numbers refer to most recent source(s). For all records prior to 2001 (without superscript) Romero et al., 2001 and references therein.
1 Lefebvre et al. (2001)
2 Romero et al. (2002a)
3 Romero and Hayford (2000), Romero et al. (2002b)
4 Swartz (et al. 2001)
5 Romero et al. (2002b)
6 Creswell (2002)
7 García et al. 2001
8 Bolaños et al. 2001

3.1.2 Whaling

Venezuelan fishers have never been involved in whaling. Yankee whalers visited the Gulf of Paria, between Venezuela and Trinidad, between 1837 and 1871 but may also have visited at other times. They hunted predominantly for humpbacks, but occasionally they killed sperm whales and 'blackfish,' *G. macrorhynchus*. At least nine whaling voyages were carried out by Yankee whalers, capturing at least 25 whales (Reeves et al., 2001). There was very little, if any, interaction between the whaling crews and Venezuelans. Therefore, there is no evidence that they ever influenced any marine mammal exploitation practice in Venezuela. There has never been shore whaling in Venezuela.

The ecological impact of whaling in Venezuelan waters is not known, except for the fact that whales are no longer found in the Gulf of Paria (see Trinidad and Tobago section below). Humpbacks are still found in other Venezuelan waters.

3.1.3 Dolphin fisheries

Venezuelan fishers usually employ small boats with a crew of at least three people (the captain, the harpooner, and his assistant) to hunt for dolphins. Indigenous peoples used harpoons for marine and freshwater captures of cetaceans and manatees. Although all other fisheries tools used in Venezuela today can be traced to Mediterranean origins, harpoon points sharpened from bone date back to Meso-Indian times (7,000 - 3,000 YBP). The ancient harpoon heads are remarkably similar to those made today along the Upper Orinoco. Hand-thrown harpoons are still used today for most intentional catches of marine mammals in Venezuela. Throughout the country, all harpoons are very similar in design and structure. The local names used for the different parts of the harpoons are also similar, despite some of them deriving from many different Arawak and Carib languages. This structural and linguistic consistency suggests that use of harpoons is of ancient origin and is widespread. Today harpooning is the preferred method for dolphin fisheries in Venezuela. Dolphins that are accidentally netted are usually consumed.

Cetacean oil was widely used in the nineteenth century for lamps, as a lubricant, and for medicinal purposes. Even today some fishers employ the blubber of the boto, *Inia geoffrensis*, as asthma remedy, and fishers at Maracaibo Lake rub the fat of the tucuxi, *Sotalia fluviatilis*, on the chests of sufferers of coughs, flu, and asthma. However, today dolphins are captured almost exclusively for the purpose of obtaining meat as bait for shark (and sometimes crab) fishing. The liver of the dolphin is commonly consumed directly as a delicacy and dolphin teeth are sometimes used to make necklaces. Occasionally, some freshwater dolphins have been captured for exhibition in aquaria both in and outside Venezuela; however, recent attempts to do so have encountered stiff criticism (Romero, 2000).

There is evidence that at least 12 of the 23 marine mammal species found in Venezuela have been taken by fishers. The species most frequently mentioned are the common dolphin, *Delphinus capensis* (25%), the bottle-nosed dolphin, *Tursiops truncatus* (23%), and the boto *Inia geoffrensis* (16%). Most of the animals taken inhabit coastal areas.

Capture occurs throughout the year. There has been an increase in the last few decades in reports of dolphin hunting. That probably reflects an intensification of fisheries in general that followed the Venezuelan government's 1960's policy of granting fishing licenses, docking rights, and Venezuelan flags to numerous long-line fishing boats of Japanese, South Korean, and Taiwanese origin.

There is no formal system for monitoring or reporting cetacean catches in Venezuela. Fishers are aware that the activity is illegal. Catches thus go unreported. When interviewed, fishers say that they capture several dolphins per sortie, a minimum of two or three, sometimes as many as 12. Some remote beaches are used to butcher dolphins out of sight of authorities, and contain numerous remains of dolphins in different states of decomposition (Fig. 1-1).

The Venezuelan government estimates that a total of 200 to 300 dolphins are killed every year, but all other sources put the figure 25 to 70 times higher. Fishers of the eastern part of the country, where there is an abundance of dolphins, are the ones that most commonly hunt dolphins. Nothing is known about the population status of the species involved in cetacean fisheries; therefore, there is no way to quantitatively ascertain the impact of the fisheries on their stocks.

Venezuela has no legislation that specifically addresses the exploitation of marine mammals. There are two pieces of legislation designed to protect wildlife in Venezuela: the Wildlife Protection Law (a civil statute enacted in 1970) and the Environmental Criminal Law (a penal statute enacted in 1992); but enforcement is rare.

Figure 1-1. Remains of *Delphinus capensis* in La Francesa Beach, Sucre State, Venezuela (Photo by A. Romero).

3.2   Trinidad and Tobago

This is basically a two-island country, originally inhabited by the Arawaks. It was under Spanish rule until 1797, when the local Spanish government capitulated to a British force, and was formally ceded to Great Britain in 1802. Slavery was abolished in 1833 and between 1845 and 1917 more than 150,000 Muslim and Hindu Indians were brought to the island of Trinidad by the British to replace plantation slaves. Tobago, originally inhabited by the Caribs, was successively a Spanish, British, Dutch, and French possession until 1814, when France ceded it to Britain. Tobago formed a part of the Windward Islands Colony until 1889, when it was joined to Trinidad. Today Trinidad and Tobago, which became independent in 1962, is a multicultural nation with a mixed industrial-agricultural economy in which fishing plays a minor role. What follows has been summarized from Romero et al. (2002a) unless otherwise noted.

3.2.1 Manatee exploitation

Manatees have been harpooned for their meat, oil, and hide from Colonial times until relatively recently. Today the only remaining area in Trinidad still inhabited by manatees is the Nariva Swamp. A 1997 survey revealed the presence of at least 18 individuals, down from an estimate of 25-30 in 1991. If manatees were accidentally caught in nets, they were butchered. Despite legal statutes aimed at protecting both the species and its habitat, the manatee is still locally threatened with extinction. The enforcement arm of Trinidad's Wildlife Section does not have sufficient staff to enforce the laws. Therefore, poaching, squatting, harmful agriculture practices, indiscriminate harvesting of mangroves, and mining continue to this day.

3.2.2 Whaling

All available data indicate that there was never much interaction between Yankee and shore whalers. Yankee whaling in the area did not start until the 1830s, when Trinidadian shore whaling was already in full swing.

Shore whaling by Trinidadians, on the other hand, was widely practiced as an opportunistic endeavor. When humpback whales (the only species they regularly captured) were spotted from shore, the whalers launched a small boat called 'pirougue' (wooden skiff) led by a captain at the stern, six stalwart oarsmen, and a harpooner in the bow. If a cow and calf were encountered together, the whaler attempted to wound the calf with the least possible injury in order to ensure that the mother could be easily approached and harpooned, due to the strong mother-calf bond exhibited by the target species. Once struck, a flag was stuck in the whale, the mouth was sewn up so the whale would not take in water and sink, and the carcass was towed to the station. Local laborers worked for up to 24 hours at a time flensing the animal, as near to the shore as possible. The slices of blubber were placed in sugar coppers (copper kettles formerly used to boil sugarcane) and boiled to extract the oil. During this period, numerous sharks showed up to take bites out of the remains of the whales. Apparently they were so numerous that the whaling company employed men to kill them with harpoons and hatchets. A Bermudan whaler known as 'Old Abraham' may have been instrumental in the introduction of some whaling techniques. In 1834 a professional harpooner from Germany was brought in.

By the early 1830's there were already four whaling stations in operation in the 'Bocas' (passages between nearshore islands) area in northeastern Trinidad. The whaling stations were very primitive consisting essentially of either of shacks or one-story buildings (Fig. 1-2).

1. In the Land of the Mermaid

Figure 1-2. Whaling Station, Copper's Hole, Monos Island, Trinidad Ca. 1900 (Photo by an anonymous photographer; picture found at the Library of the University of the West Indies in St. Augustine, Trinidad).

Today there is little left of these whaling operations except for two coppers and one container submerged just a few meters from the location of one of the former stations.

The oil was taken to Port of Spain and was mostly consumed locally as lamp oil or medicine-whale oil (mixed with honey, a flu remedy). Oil that was exported was sent mostly to British colonies. The meat was consumed locally. Whalebone was sent to London for use as manure.

Although there are some unconfirmed reports of shore whaling taking place in Trinidad at the end of the eighteenth century, commercial whaling most likely began around 1826 initially employing slave labor. Whaling operations ceased at the end of the 1870's due to the depletion of the local whale population. Whaling took place between January and May every year, when humpbacks visited Trinidadian waters.

The number of whales caught annually was usually between 20 and 35 totaling least 500 whales killed. Baleen whales no longer frequent Trinidadian waters and are only seen in very small numbers around Tobago.

The establishment of whaling in Trinidad was authorized in 1827. Japan requested that the government of Trinidad and Tobago join the IWC and oppose the ban on commercial whaling. To this day, the Trinidadian government has done neither. Legal protection for cetaceans in Trinidad is ambiguous.

Shore whaling in Trinidad must have required important local investment, given that from the beginning we find the names of upper-class Trinidadian families involved in this business. All of these were merchant and planter families. Some had been involved in coconut oil production and sales; therefore, they saw in whale oil an opportunity to expand their business. The development of commercial whale fisheries in Trinidad coincided with bad economic times for the island and with the decline in the local population between 1827 and 1833.

3.2.3 Dolphin fisheries

There is only one piece of evidence of possible utilization of cetaceans by pre-Columbian inhabitants of Trinidad: an unidentified bony remain of a cetacean at St. Joseph (the first Spanish capital of Trinidad) on a branch of the Caroni River on the south side of the North Range. There is no indication of the type of capture.

Dolphin fisheries have always been rare in Trinidad and Tobago waters. Reports of these activities are scant and most are related to accidental nettings. When captured, the animals are butchered and usually sold. They are sometimes labeled as 'shark.' The species involved in these incidental takings are *Stenella* spp. and *Tursiops truncatus*. The largest animal ever taken in this way was an orca, *Orcinus orca*. The only other current utilization of marine mammals is an occasional dolphin watch operation in Tobago.

3.3 Grenada

The Caribs originally inhabited this small country. The first Europeans to establish themselves permanently here were the French in 1650. In 1783, the island was ceded to the British who immediately imported large numbers of slaves from Africa and established sugar plantations. The colony gained independence in 1974. It maintains a strong mixture of French and British culture as evidenced by names of people and places, and by the overwhelming predominance of Catholicism. It has a small-scale agricultural economy with a fledgling tourism industry. Fishing is a marginal activity. The information below has been summarized from Romero and Hayford (2000), and Romero et al. 2002b) unless otherwise indicated.

3.3.1 Manatee exploitation

Archaeological and historical records indicate that manatees were hunted for their meat, using harpoons, by pre-Columbian people as well as Europeans up to the seventeenth century).

3.3.2 Whaling

Some Yankee whaling took place in these waters from as early as 1857 until 1888. Whaling ships primarily hunted humpbacks, but occasionally landed sperm whales. The ships provided whale meat to the local market of Grenada and the neighboring southern Grenadines (Reeves et al., 2001).

However, it was shore whaling the activity that most impacted marine mammals populations in the waters of Grenada. Between 1920 and 1923 shore whaling was purely opportunistic. Local fishers harpooned whales from small boats that had previously been used for the same purpose in Barbados (Creswell, 2002). Fishers from Bequia were known to whale in Grenadian waters. In 1925 and 1926, Norwegian whalers brought two and three, respectively, modern steam-driven whaling vessels from Norway. Each vessel had a crew of eleven men and employed harpoons with explosive heads. Whaling always took place between January and April.

The Norwegians built a modern, 2-story whaling station on Glover Island (in the south of Grenada) in 1924 (Fig. 1-3) and directed the entire operation from that point onward.

Figure 1-3. Only known picture of the whaling station at Glover Island, Grenada (picture by unknown photographer at the Grenada Archives).

About seventy people (mostly locals) were employed at Glover Island. They almost exclusively captured humpback whales although there was a landing of a Bryde's whale, *Balaenoptera edeni*.

Between 1920 and 1923 whale oil was exported to Trinidad, Barbados, and the United Kingdom. Under the Norwegians, whale oil was exported to Norway, the Netherlands, the United Kingdom, Barbados, and Demerara (today's Guyana). Some of the whale meat was sold for human consumption. Manure was exported to Trinidad.

We have very precise figures on the number of whales taken in Grenadian waters by shore-whaling operations: 10, 1, and 5 whales were taken in 1920, 1921, and 1923, respectively. In 1925 and 1926, the Norwegians captured 105 and 72 whales respectively. After that, the operation was abandoned due to the scarcity of whales. No operations were carried out in Grenada during the 1927 season. Whales in the area are very rare today.

With the establishment of the whaling operation by the Norwegians, the colonial government of Grenada required that every piece of the animal had to be utilized and that strict measures be taken to avoid the foul smell of these operations spreading to the main island of Grenada.

The establishment of the Norwegian operation required extensive investment. After the 1926 season, where no whales could be seen and/or captured, the operation collapsed. All attempts to refinance the company failed. This coincided also with a more general depletion of whale stocks in the Northern Hemisphere. In addition, beginning in 1925, factory whaling ships were equipped with ramps that allowed the catch to be processed on board, making land stations like the one at Glover Island unnecessary. In 1928, the whaling industry at Glover Island was finally abandoned and the factory was dismantled in 1929. The only current utilization of marine mammals is by two whale watching operations.

The Grenadian government has supported Japan's position in favor of commercial exploitation of whales in the most recent meetings of the International Whaling Commission (IWC). During the 1999 meeting of the IWC that took place in Grenada, abundant literature in English, printed in Japan, promoted the development of 'traditional' whaling. One week before the aforementioned meeting, a humpback whale and its calf were killed in the Grenadines. The news was highly publicized in the Grenadian media, which was critical of their government's pro-whaling policies, leaving Grenadians wondering if the practice of whale killing will reach their island.

1. In the Land of the Mermaid

3.3.3 Dolphin fisheries

There are no records of intentional dolphin exploitation in Grenada. We heard of an instance of a stranded dolphin which was butchered and its meat later consumed by the locals.

3.4 Barbados

From the time of its discovery by the Spaniards, Barbados was raided for slaves, and by 1500, was entirely depopulated. After that, the Portuguese visited occasionally. The British established it as a colony in 1627. Slavery was abolished in 1833, leading to a substantial increase in agricultural production, particularly sugar. It gained independence in 1966. It is the most densely populated of the Lesser Antilles and its economy is dependent on tourism, agriculture, and offshore banking. Fishing is a secondary industry but, comparatively speaking, is slightly more important than in the other countries covered in the present study. The information below is summarized from Creswell (2002).

3.4.1 Manatee exploitation

There is no evidence that manatees ever inhabited Barbados.

3.4.2 Whaling

Under British rule, Bridgetown became a busy port and Barbados a regular stop on the routes of many cargo, passenger, and whale ships. Despite early descriptions of an abundance of humpbacks in Barbadian waters, it seems that Bridgetown was mostly used as a port for supplies replenishment and crew recruitment. Curiously there was relatively little Yankee whaling in Barbadian waters. Whaling campaigns were recorded between 1859 and 1866.

Shore whaling was a different story. There are archaeological remains of a sperm whale and dolphins in Barbados, but none of humpback whales. These remains most likely correspond to stranded animals. The first historical record of a whale (probably a stranded humpback calf) utilization in Barbados dates to 1813. The meat was consumed by the local black population.

Shore whaling was an opportunistic operation based on humpback whale sightings from the shore. Whalers employed boats powered by both sails and oars that were operated by a relatively small crew. They used harpoons to capture the whale and an explosive lance ('bomb lance'), to kill it (Fig. 1-4).

Figure 1-4. Aldemaro Romero holds a whaling gun, from which the bomb-lance was shot. Shotgun courtesy of Charles Jordan. Picture by J. Creswell.

The whalers jumped into the water and sew the whale's mouth shut, to prevent it from filling with water and sinking. The whale was then dragged ashore, alongside a jetty on the beach for flensing. Although there are reports of sharks attacking whale carcasses, the local operations never employed anyone to kill the sharks while the animal was being flensed. In fact, the whalers could reportedly walk through the water in the midst of the frenzied sharks and not be harmed. Barbadians were routinely recruited to join the crews of Yankee whaling ships. Some of these Barbadians returned home after the voyages, having gained the necessary skills to hunt whales, and started their own operations.

Beginning in 1867, three shore whaling stations were established. They consisted of shacks on the beach where the whaling gear was stored. The blubber was boiled in copper kettles, of the same design as those used to boil sugar cane juice. The oil was exported to England and Canada. The bones were ground and used as fertilizer. The meat was sold locally for consumption and eaten by the black populations. The baleen plates were used to make brooms.

Records of whaling operations span from 1879 to 1910, always between January and May. The average number of whales taken per year between

1889 and 1902 was 11.6 with a maximum of 37 in 1901. The total number of whales killed in the Barbadian shore whaling industry was at least 187. Shore whaling ceased due to the scarcity of whales. At that time, the two remaining whaling boats were taken to Grenada and operated there from 1920 to 1923.

There was usually significant competition between the stations over whales. Two of the stations were located next to each other and both saw whales at essentially the same time. In 1904, the government passed the Fisheries Regulation Act, updating all of Barbados' fishing regulations and consolidating them into a single bill. This included laws governing competition between whaling boats from different operations, probably as a response to quarrels between the two stations. These laws include provisions that establish ownership of a whale by the first boat that strikes it and the ownership of a mother by a boat that strikes her calf, and vice versa. They even detail how profits and expenses are to be split if two boats happen to strike the same whale.

Local business families owned and ran the whaling operations but whaling was not their primary source of income. Although Barbados is not a member of the IWC, it may soon join. It has received financial assistance from Japan to upgrade its disaster emergency mechanism, a move seen by many as Barbados accepting a bribe in exchange for supporting the Japanese pro-whaling agenda.

3.4.3 Dolphin fisheries

The rare dolphin captures in Barbados are limited to accidental nettings. The primary use of nets by Barbadian fishers is to catch flying fishes for which gill nets are used.

3.5 St. Vincent and the Grenadines

Caribs were the original inhabitants of these islands. They aggressively resisted European settlement until 1719, when the French settled St. Vincent. During the eighteenth century, African slaves from St. Lucia and Grenada, intermarried with the Caribs and became known as 'black Caribs.' The French established plantations for a variety of agricultural products and imported many slaves. In 1763 St. Vincent was ceded to Britain. Slavery was abolished in 1834. This resulted in shortages of labor, attracting Portuguese and East Indies immigrants. The colony gained independence in 1979. Today its economy depends heavily on agriculture (mostly banana) and tourism. Fishing is a minor industry. Bequia, an island in the northern

Grenadines, is home to most of the whaling and dolphin fisheries in the country.

3.5.1 Manatee Exploitation

There is no evidence that manatees ever existed in these islands.

3.5.2. Whaling

3.5.2.1 Yankee whaling

Yankee whalers visited the St. Vincent and the Grenadines area very consistently between the 1830's and the 1880's. Some of the ships whaled there for several months at a time.

Yankee whalers influenced the beginning of this activity at the local level by recruiting some seamen/fishers in the Grenadines to man their whaling boats. That was the case of William Thomas Wallace, Jr. ('Old Bill'), born in Bequia in 1840. Of Scottish ancestry, he was a planter who served as a seaman on numerous whaleships beginning in 1857. He went to Provincetown and New Bedford, major whale ports at the time, marrying in the former and learning ship building in the latter. He returned to Bequia where he established a shore whale fishery sometime between 1875 and 1876. This activity was carried out as a supplement to income from the agricultural industry, which, by this time, was in decay. The whaling season coincided with the time of year when less labor was needed for the cultivation of crops (Adams, 1970).

3.5.2.2 Shore Whaling/Dolphin Fisheries

Unlike the other countries covered in this study, St. Vincent and the Grenadines' shore-based marine mammal exploitation encompasses both large whales and dolphin fisheries. Some of those activities still take place today; hence, the use of present tense in some of our statements.

Most captures took/take place on the windward side of the islands in order to help bring the whales to shore after they had been harpooned. Whales were/are spotted from shore-based lookouts. The spotters use signal mirrors to direct boat crews toward whales. Whalers originally employed Nantucket-type sailing boats. Much later, some were outfitted with engines. Initially they used hand-held harpoons, but gun harpoons were introduced in 1958. Butchering took/takes place in the water near the beach. The flensed parts of the animals were/are placed in boats pulled alongside the carcass and transported to shore. Men in boats lance sharks scavenging in the area (Adams, 1970, 1971).

The number of whaling stations has varied through time. At one point there were seven whale-fishing operations in Bequia and Ile-de-Caille, with four boats, each employing six men. In Bequia the whaling stations have always been primitive and consist of a small shed for storing blubber and a small structure that supported their boiling kettles, but no buildings of substantial size. In 1931 a local resident started a whaling company with three boats to hunt pilot whales, sperm whales, and dolphins. Later others became interested and the fleet was expanded to 15 boats; today that number is somewhat reduced. All of these whaling initiatives were developed by local planters with the necessary capital to invest in the operations. Crewmen, however, were/are from the lower social classes (Adams, 1971; Beck, 1986).

Captured humpbacks generate meat and oil as primary products. A humpback yields between 400 and 1,500 gallons (ca. 1,500 and 5,700 liters respectively) of oil, and adults could yield more than 2,000 pounds (ca. 908 kilograms) of meat (Rack, 1952; Adams, 1973). In the past, most of the oil was sent to Kingston, St. Vincent, where it was subsequently routed to England and the U.S. Some (about 1,000 gallons – 3,785 liters- a year) was sent directly from Bequia to Trinidad and Barbados. Between 1893 and 1903 an average of 25,000 gallons (about 95,000 liters) was exported. Meat was very popular among blacks, but unpopular among whites (Adams, 1970). Because there is no market for whale oil today, meat is the main product obtained from modern whaling. The procedures for flensing the animal and preparing oil today are essentially the same as that used in the past (Fenger, 1958; Beck, 1986). The main usage of non-humpback marine mammals is for meat, although their oil can be used for cooking, medicinal purposes, and as a lubricant (Brown, 1945; Adams, 1970). An average pilot whale yields 25 gallons (ca. 95 liters) of oil.

In addition to humpback whales, the others species exploited in waters of St. Vincent and the Grenadines have been: pilot whales ('blackfish'), *G. macrorhynchus*; sperm whales ('sea-guaps'), *P. macrocephalus*; orcas ('Whitefish'), *O. orca*; and dolphins (*T. truncatus* and *Stenella* spp.). Other species, such as the false killer whale, *P. crassidens*, Cuvier's beaked whale, *Z. cavirostris*, the rough-toothed dolphin, *S. bredanensis,* the pygmy killer whale, *F. attenuata,* the dwarf sperm whale, *K. sima*, and Fraser's dolphin, *L. hosei*, may be taken occasionally, on an opportunistic basis (Caldwell and Caldwell, 1975; Caldwell et al., 1976).

Humpbacks were taken very actively between 1875 and 1925 and the whaling season was between January and May (Adams, 1970). Initial hunting of pilot whales took place from the nineteenth century to the present in a very occasional and intermittent way; only since 1931 it has been a major operation. This activity is a year-round one with lows between mid-December and mid-January and during holidays (e.g., Carnival) or special occasions (e.g., weddings, funerals) (Rack, 1952).

The total number of humpbacks taken in St. Vincent's and the Grenadines waters has yet to be fully ascertained. Most estimates suggest at least 550 whales in total killed through shore whaling alone between early 1860's to the present. It seems that early on about 115 humpbacks were taken per year. Around 1900, 50 were captured and from 1910 until the present no more than 10 have been landed per year. Between 1940 and 1957 no whales were taken. Between 1958 and 1984 between 52 and 70 humpbacks were landed. This decline is the result of overexploitation (Adams, 1971; Beck, 1986; Price, 1985; Agard and Gobin, 2000). Today the quota allowance granted by the IWC is three humpbacks per year.

The number of pilot whales captured between 1962 and 1979 was between 25 and 422 per year (average = 194 per year), while the number of killer whales was 0 to 12 per year, sperm whales 0 to 6 per year, and false killer whales 0 to 15 per year. The total number of dolphins captured has been between 200 and 500 per year for the same period. In all cases, the declining number of catches beginning in the 1970's has been attributed to overfishing (Adams, 1973; Caldwell and Caldwell 1975; Price, 1985).

There is a law in St. Vincent and the Grenadines, governing the whale fishery at Bequia. The 'Whalers Ordinance of 1887' lays out many of the same rules as the Fisheries Regulation Act of Barbados. The Whalers Ordinance was the result not only of quarreling, but also of hostile behavior between whaling companies, such as boats ramming each other and whalers scaring whales away to prevent their competitors from catching them (Adams, 1971; Beck, 1986).

In 1868 whaling ranked fourth in the value of exports from St. Vincent. Shore whaling was a prestigious activity among locals (Adams, 1970). St. Vincent and the Grenadines and St. Lucia have enough marine mammals present to support small whaling operations (Hoyt, 1994). These two countries are the only ones in the Caribbean countries that have whaling allowances (three and two, respectively) from the IWC on aboriginal whaling grounds (Agard and Gobin, 2000).

4. DISCUSSION

Archaeological evidence strongly suggests that pre-Columbian cultures in the Caribbean opportunistically exploited any and all marine mammal resources available to them. The little we do know about Amerindians of the Caribbean suggests that they did not have widespread cultural barriers to marine mammal consumption. In fact, on some islands, marine mammals made up a portion of the Carib and Arawak diets.

Manatees were present in Venezuela, Trinidad, and Grenada and were consumed by the indigenous people of the Lesser Antilles before the arrival of Columbus. Archaeological remains indicate that the West Indian manatee was distributed throughout most of the Lesser Antilles and that indigenous people inhabiting those islands consumed them (Ray, 1960; Wing et al., 1968; Watters et al., 1984; Lefebvre et al., 2001; Wing and Wing, 1995). A single manatee yielded an average of around 440 kg of meat, plus some fat, making it an excellent, easily exploitable food source. They were hunted from canoes, using harpoons with floats and ropes attached (Rouse, 1948; McKillop, 1985; Watts, 1987). With the arrival of the Europeans their exploitation accelerated to the point that they became extinct in Grenada and severely depleted in Venezuela and Trinidad. Historical accounts also support the idea that manatees were hunted with harpoons throughout Colonial times (Du Tertre, 1667; Dapper, 1673; Labat, 1742; Bullen, 1964; Wing and Wing, 1995; for additional citations on pre- and post-Columbus use of manatees in the Caribbean see McKillop, 1985).

The pattern of cetacean exploitation developed into two differentiated types in the region: Venezuela, a Hispanic country, did not engage in whaling but rather in dolphin fisheries. This is the same pattern followed by other Latin American countries (Romero et al. 1997 and references therein). On the other hand, the other four countries of our study, all of them at some point under British influence, developed shore whaling, with St. Vincent and the Grenadines also engaging in dolphin fisheries, an alternative developed after local whale population became depleted.

Despite the fact that Yankee whaling took place in all of the countries in our study area, it did not have any significant cultural influence either in Venezuela or in Trinidad. The reason the Yankee whalers had little influence on Trinidad was because shore whaling was already taking place by the time the first of them arrived. Although Yankee whalers visited Grenada, there is no evidence of their influence in the country, since shore whaling began in the twentieth century, i.e., well after Yankee whalers stopped visiting those waters. Grenada's whaling industry did really not take off until the Norwegians started to exploit the resource in 1925.

Barbados and St. Vincent and the Grenadines were different cases. Yankee whalers recruited natives of both countries as crews for their ships. Further, the last known whaling campaign by a Yankee whaler in Barbadian waters was in 1866 and the very next year, Barbadians started to develop shore whaling, for which they used Nantucket-type sailboats very similar to those employed by Yankee whale ships to pursue whales. The same can be said of St. Vincent and the Grenadines, where they still employ similar boats today. Whalers from Barbados and St. Vincent and the Grenadines used different boats than the ones from Trinidad and Tobago: the former used sailboats and the latter rowing boats. Therefore, Caldwell and Caldwell's (1971) statement that 'All of the formal whaling in the Lesser Antilles is patterned after the style of the New England whalemen' is not completely accurate.

Whaling stations were also different: the ones in Barbados consisted of a shack on the beach in which whaling gear was stored. There were no permanent buildings. This situation is similar to the one in Bequia, in which the whaling stations had a small shed for storing blubber and a small structure that supported their boiling kettles, but no buildings of substantial size. The stations in Trinidad ranged from the shack type to one-story buildings, but none had structures to assist with bringing whales onto the beach, as whales were flensed in the water. These examples stand in stark contrast to Grenada, whose Norwegian-built station on Glover Island was a two-story building with a concrete chute for dragging whales up to a flensing platform. In Venezuela, the flensing of dolphins takes place either on isolated beaches or on board long-line fishing boats, in order to hide this illegal activity.

Oil was the main reason for hunting whales. In the case of Barbados, the whale oil export market exhibited a normal economic relationship between supply and demand. The value of oil remained relatively constant at low levels of production, but when production increased dramatically, the price of oil dropped substantially. Other than the one-year peak in 1907, the price of whale oil in the twentieth century never returned to its previous high value. Indeed at the peak of production, in 1901, the price per barrel dropped below £1 (from a maximum of £4). The fact that the price per barrel remained relatively low after the peak suggests that the market was saturated in those years. It also suggests that the demand for whale oil was decreasing as people began to substitute alternatives for whale products (Creswell, 2002). The same factor contributed to the demise of Trinidad's whaling industry. From the 1870's onward, there was an overproduction of whale oil, and kerosene was being used as lamp fuel, instead of whale oil. Thus the price of whale oil plummeted (Romero et al., 2002a).

The level to which whale products were used varied from country to country. In Trinidad and St. Vincent and the Grenadines people use(d) oil for medicinal purposes while the meat was consumed locally. In Trinidad the bones were used for manure. A similar pattern was followed in Grenada, except that law mandated the full utilization of the carcass. In Barbados people of African descent were the main consumers of whale meat and the whale plates were used to make brooms. It is interesting to note that contemporary accounts tend to emphasize that whale meat was popular among people of African descent but not among those of European descent throughout the Caribbean.

Humpback whales were overwhelmingly the target species in the former British colonies, although in the case of St. Vincent and the Grenadines catches were progressively diversified to include all available species of marine mammals.

Another interesting fact among the countries involved in whaling was that they developed whaling at different points in history. The first to develop shore whaling was Trinidad (1830-1862), then Barbados (1879-1910), and then Grenada (1920-1926) (Fig. 1-5).

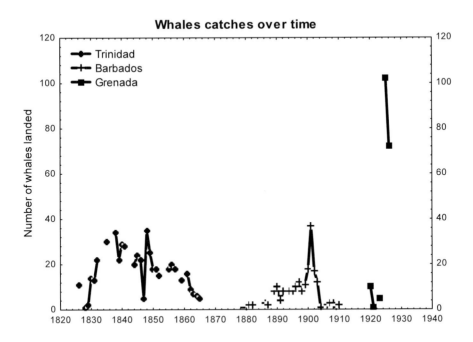

Figure 1-5. Whale catches for Trinidad, Barbados, and Grenada, not counting those from Yankee whalers (based on Romero et al., 2002; Romero and Hayford, 2000; and Creswell, 2002; respectively).

St. Vincent and the Grenadines had a more complicated history: They began shore whaling for humpbacks in 1875 and temporarily stopped doing so around 1920, which means that their industry operated concurrently with the one in Barbados; yet, unlike Barbados, humpbacking has continued in a small scale and very intermittently since then. Further, as humpbacking declined, the capture of other species, from sperm whales to pilot whales to small dolphins, surged dramatically from the 1930's to the 1970's. Whaling always took place between January and May (when humpbacks were present in the southeastern Caribbean for breeding and calving). On the other hand, dolphin fisheries in both Venezuela and St. Vincent and the Grenadines take place virtually year-round, since they target resident populations of these species.

There can be little doubt that the combined whaling efforts of Yankee whalers and shore whalers in the southeastern Caribbean must have had an important impact on the humpback population of the region. Data summarized in this paper show that if we add the number of whales taken in Trinidad (500+), the ones taken in Grenada (193), in Barbados (187), and in St. Vincent and the Grenadines (550+), we are talking about at least 1,400 whales having been captured through shore whaling. This figure is conservative since it does not include: (1) whales taken for which data are not available because the source (e.g., a Blue Book) for a specific year is not available; (2) whales whose capture may have gone unreported (see, for example, some cases in Romero et al., 1997); and (3) whales that were struck but not recovered and died shortly after as a consequence of the wounds.

Yankee whalers may have had an even higher impact in the area. According to Mitchell and Reeves' (1983) between 1866 and 1887, Yankee whalers killed 2,421 humpbacks in the Caribbean. Therefore, the figure of somewhere between 3,821 and 4,500 humpbacks killed in a little more than 100 years must have had a significant impact. The best supporting evidence that whaling had a large impact is that humpback whaling ceased in Trinidad, Barbados, and Grenada (and has been severely reduced in St. Vincent and the Grenadines) due to the scarcity of whales. Further, the population for those waters has yet to be recovered. A recent visual and acoustic survey (Swartz el al., 2001) showed very few humpbacks in the area. This is consistent with our own fieldwork that includes field surveys as well as extensive interviews with both fishers and whale-watching operators that indicate that humpbacks are very rare in the southeastern Caribbean. This is in sharp contrast with both contemporary descriptive narratives and whaling figures for the area from the 1830's to the 1920's.

All countries involved with whaling developed, in one way or another, legislation to specifically permit and/or regulate the industry. The least sophisticated of these was Trinidad's, probably because it was the first one, while Grenada had the most sophisticated and strict legislation, probably because it was the last one to enact such laws. Although there is some general legislation such as the Wildlife Protection Law and the Criminal Environmental Law, Venezuela does not have specific regulations regarding the protection of marine mammals. This, combined with the general lack of law enforcement, can explain why dolphin fisheries continue up to this day with very little control.

Despite the decimation of the humpback population in the southeastern Caribbean and the possible impact on other species as a result of uncontrolled marine mammal exploitation, this resource may become, again, the target of exploitation in these countries. The Venezuelan government has shown interest in studying dolphin populations in order to ascertain the feasibility of their 'harvesting' (Cohen, 1994), and it has persecuted those who opposed the exploitation of marine mammals in Venezuelan waters (Stone, 1995). St. Vincent and the Grenadines has transitioned into dolphin fisheries while continuing to humpback, despite the fact that its marine mammal resources seem depleted at the present time. Eastern Caribbean countries have been under heavy pressure to join the IWC and/or support the resumption of commercial whaling while opposing any attempt to put small marine mammals under the regulatory mandate of the IWC (Anonymous, 2001b).

Some have argued that instead of attempting to pursue whale and dolphin watching, it would be more economically feasible for southeastern Caribbean countries to resume whaling. The main proponents of this idea are Grenada, Dominica, Antigua, St. Kitts and Nevis, St. Lucia, and St. Vincent and the Grenadines. They argue that the IWC's international ban on whaling is not supported by sufficient scientific data, but is rather a form of 'Eco-Posturing.' They, along with, Japan, Norway, and China are attempting to form a pro-whaling voting bloc within the IWC to pass a resolution allowing what they call sustainable use of marine resources. Japan has been particularly active in the past few years in providing financial and technical assistance to eastern Caribbean countries, a policy that has been denounced as a way to buy their votes in the IWC (Alleyne, 2001; Ally and Peltier, 2001; Anonymous, 2001a,b).

5. CONCLUSIONS

Despite the fact that the southeastern Caribbean countries share the same marine mammal fauna, those animal species have been exploited differently in time and space, due to differences in the dominant cultures, and whether each culture had intrinsic (Amerindian, Latin, British) or extrinsic (Yankee whaling, Norwegian whaling) motivations to begin whaling. This is evidenced by variations in the animals targeted for capturing, specific methods used in capturing the animals (boat and whaling station design), and by the way the whaling products were/are used. The utilization of these resources has always been largely opportunistic and in almost all cases has led to the depletion and/or extinction of the resource in question at the local level. The legislation designed to regulate that exploitation (when enacted) was drawn up to regulate competition among whaling enterprises rather than to protect the resource in question.

The inclination of most of the nations within our study area to pursue legal avenues to resume marine mammal exploitation is a matter of concern not only because of their past history and present trends, but also because such intentions are not based on scientific knowledge of species diversity and population status of the fauna involved. Again, foreign influence may convince some of these countries to utilize these resources without a clear understanding of the ecological, economic, and political consequences of doing so.

ACKNOWLEDGMENTS

A number of individuals have been of great help during the years I (AR) have been working on marine mammals of the Caribbean. I want to thank the coauthors of papers I have published on the subject: A. Ignacio Agudo (Fundacetacea), Ruth Baker (Macalester College), Joel E. Creswell (Macalester College), Steven M. Green (University of Miami), Kyla T. Hayford (Macalester College), Michael Manna (Florida Atlantic University), Annabelle McKie (Florida Atlantic University), Giuseppe Notarbartolo di Sciara (Tethys Research Institute), Andrea Romero (Carleton College) and Jessica Romero (University of Chicago), and Anuradha Singh (University of West Indies, St. Augustine Campus). Other people that provided support are mentioned in the acknowledgement sections of Romero et al. (1997), Romero and Hayford (2000), Romero et al. (2001), Romero et al. (2002a,b), and Creswell (2002). Different portions of this research have been funded by Keck Student-Faculty Collaboration (Macalester College) grants, Wallace Travel Grants (Macalester College), and funds generously provided by Mr. and Ms. Gail Ordway. A. I. Agudo and S.M. Green read the MS and made valuable suggestions.

REFERENCES

Adams, J.E., 1970, *Marine industries of the St. Vincent Grenadines, West Indies*, Unpublished Ph.D. Diss., University of Minnesota – Twin Cities.
Adams, J.E., 1971, Historical geography of whaling in Bequia Island, West Indies. *Caribbean Studies* **11**:55-74.
Adams, J.E., 1973, Shore whaling in St, Vincent Island, West Indies. *Caribbean Quarterly* **19**: 42-50.
Adeola, M.O., 1992, Importance of wild animals and their parts in the culture, religious festivals, and traditional medicine, of Nigeria. *Environmental Conservation* **19**:125-134.
Alleyne, D. 2001, Eco-posturing or whale management? Barbados: *The Daily Nation* (19 July 2001); p. 2.
Anonymous, 2001a, Whaling body ready for battle. Barbados: *The Daily Nation* (19 July 2001); p. 20.
Anonymous, 2001b, Region likely to vote for pro-whaling. Barbados: *The Daily Nation* (18 July 2001), p. 12A.
Beck, H.P., 1986, 'Bleows' The whaling complex in Bequia. *Folklife Annual,* **1986**:42-61.
Bolaños, J., Quijada, A., and Villarroel, A., 2001, First record on the melonheaded whale (*Peponocephala electra* Gray 1846) for Venezuela and a review of its distribution in the Caribbean Sea, in: 14th. *Abstracts, BiennialConference on the Biology of Marine Mammals, Nov. 28 - Dec. 3, 2001*,pp. 28-29.
Brown, H.H., 1945, *The fisheries of the Windward and Leeward islands. Developmentand welfare in the West Indies*, Bulletin 20. Barbados: Advocate Co.

Bullen, R.P., 1964, The archaeology of Grenada, West Indies. *Florida State Museum, Social Sciences Contribution* **11**:1-67.

Caldwell, D.K., and Caldwell, M.C., 1971, The pygmy killer whale, *Feresa attenuata*, in the Western Atlantic, with a summary of world records. *Journal of Mammalogy* **52**:206-209.

Caldwell, D.K., and Caldwell, M.C., 1975, Dolphin and small whale fisheries of the Caribbean and West Indies: occurrence, history, and catch statistics -with special reference to the Lesser Antillean island of St. Vincent. *Journal of the Fisheries Research Board of Canada* **32**:1105-1110.

Caldwell, D.K., Caldwell, M.C., and Walker, R.V., 1976, First records for Fraser's dolphin (*Lagenodelphis hosei*) in the Atlantic and the melon-headed whale (*Peponocephala electra*) in the western Atlantic. *Cetology* **25**:1-4.

Cohen, E., 1994, Multaron con 50 mil bolívares a Aldemaro Romero e Ignacio Agudo. *El Nacional* (Caracas, Venezuela), (28 Abril 1994).

Creswell, J.E., 2002, The exploitative history and present status of marine mammals in Barbados, W.I., *Macalester Environmental Review* (28 May 2002); http://www.macalester.edu/environmentalstudies/MacEnvReview/rinemammalsbarbados.htm

Dapper, O., 1673, *Die unbekante neue Welt, oder Beschreibung des Welt-Teils Amerika, und des Sud-Landes.* Amsterdam: Bey Jacob von Meurs.

Du Tertre, P., 1667, Histoire générale des Antilles habitées par les François T. Iolly, Paris, 4 vols.

Fenger, F.A., 1958, *Alone in the Caribbean*. Belmont: Wellington Books.

García, L., Bolaños, J., and Gonzalez-Fernandez, M., 2001, A live stranding of the Fraser's dolphin (*Lagenodelphis hosei* Fraser, 1956) in the central coast of Venezuela: first record for the southern Caribbean Sea, in: *Abstracts, 14th. Biennial Conference on the Biology of Marine Mammals,* Vancouver: *November 28 – December 3, 2001,* p. 79.

Hoyt, E., 1994, *Discover whale and dolphin watching in the Caribbean*. Bath, U.K.: Whale and Dolphin Conservation Society.

Jackson, J.B.C., 2001, What was natural in the coastal oceans? *Proceedings of the National Academy of Sciences (USA)* **98**:5411-5418.

Labat, P., 1742, *Nouveau voyage aux Isles de l'Amerique*. 4 volumes, Paris: Chez Theodore le Gras.

Lefebvre, L.W., Marmontel, M., Reid, J.P., Rathbun, G.B., and Domning, D.P., 2001, Status and biogeography of the West Indian manatee, in: *Biogeography of the West Indies*, C.A. Woods and F.E. Sergile, eds. Boca Raton, Florida: CRC Press, pp. 425-474.

Lockhart, J., and Schwartz, S.B., 1983, *Early Latin America: a history of colonial Spanish America and Brazil*. Cambridge: Cambridge University Press.

McKillop, H.I., 1985, Prehistoric exploitation of the manatee in the Maya and circum-Caribbean areas. *World Archaeology* **16**:337-353.

Mitchell, E. and Reeves, R.R., 1983, Catch history, abundance, and present status of Northwest Atlantic humpback whale. *Reports of the International Whaling Commission,* Special Issue 5, SC/33/PS14, pp. 153-209.

Price, W.S., 1985, Whaling in the Caribbean: historical perspective and update. *Reports of the International Whaling Commission* **35**:413-420.

Rack, R.S., 1952, *Fisheries in the Caribbean. Report of the Fisheries Conference held at Kent House, Trinidad, March 24-28, 1952.* Port-of Spain, Trinidad.

Reeves, R.R., Swartz, S.L., Wetmore, S., and Clapham, P.J., 2001, Historical occurrence and distribution of humpback whales in the eastern and southern Caribbean Sea, based on data from American whaling logbooks. *Journal of Cetacean Research and Management* **3**:117-129.

Richerson, P.J., Mulder, M.B., and Vila, B.J., 1996, *Principles of Human Ecology*. Needham Heights: Simon & Schuster.

Romero, A., 1990, Iniciativas de manejo costero en Venezuela, in: *El manejo de ambientes y recursos costeros de América Latina y el Caribe. Vol. 1*, Buenos Aires, Argentina: Senado de la Nación Argentina and Organization of American States, pp. 203-224.

Romero, A., 2000, Should Venezuelan botos be imported into the U.S.? *Whalewatcher* **32**:13-15.

Romero, A., 2003, Death and taxes: the case of the pearl oyster beds depletion in sixteenth century Venezuela. *Conservation Biology* **17**:1013-1023.

Romero, A., Agudo, A.I., and Green, S.M., 1997, Exploitation of cetaceans in Venezuela. *Reports of the International Whaling Commission* **47**:735-746.

Romero, A., Agudo, A.I., Green, S.M., and Notarbartolo di Sciara, G., 2001, Cetaceans of Venezuela: their distribution and conservation status. *NOAA Technical Reports NMFS* **151**:1-60.

Romero, A., Agudo, A.I., and Salazar, C., 2000, Whale shark records and conservation status in Venezuela. *Biodiversity: The Journal of Life on Earth* **3**:11-15.

Romero, A., Baker, R., Creswell, J.E., Singh, A., McKie, A. and Manna, M., 2002a, Environmental history of marine mammal exploitation in Trinidad and Tobago, W.I., and its ecological impact. *Environment and History* **8**:255-274.

Romero, A. and Hayford, K.T., 2000, Past and present utilisation of marine mammals in Grenada, West Indies. *Journal of Cetacean Research and Management* **2**:223-226.

Romero, A., Hayford, K.T., Romero, A., and Romero, J., 2002b, The marine mammals of Grenada, W.I., and their conservation status. *Mammalia* **66**:479-494.

Rouse, I., 1948, The Arawak, in: *Handbook of South American Indians*, J.H. Steward, ed., Washington: Government Printing Office.

Rouse, I. and Cruxent, J.M., 1963, *Venezuelan Archeology*. Yale University Press, New Haven.

Stone, R., 1995, Portrait of a killing. *Science* **267**:4428.

Swartz, S.L., Martinez, A., Cole, T., Clapham, P.J., McDonald, M.A., Hildebrand, J.A., Oleson, E.M., Burks, C., and Barlow, J., 2001, Visual and acoustic survey of humpback whales (*Megaptera novaeangliae*) in the Eastern and Southern Caribbean Sea: Preliminary Findings. *NOAA Technical Memorandum NMFS-SEFSC* **456**:1-37.

Vidal, O., Van Waerebeek, K., and Findley, L.T., 1994, Cetaceans and gillnet fisheries in Mexico, Central America and the wider Caribbean: a preliminary review. *Reports of the International Whaling Commission,* Special Issue, **15**:221-233.

Watters, D.R., Reitz, E.J., Steadman, D.W., and Pregill, G.K., 1984, Vertebrates from archaeological sites on Barbuda, West Indies. *Annals of the Carnegie Museum* **53**:383-412.

Watts, D., 1987, *The West Indies: patterns of development, culture and environmental change since 1492*. Cambridge: Cambridge University Press.

Wing, E.S., Hoffman, C.A., and Ray, C.E., 1968, Vertebrate remains from Indian Sites on Antigua, West Indies. *Caribbean Journal of Science* **8**:123-139.

Wing, E.S. and Reitz, E.J., 1982, Prehistoric fishing economies of the Caribbean. *Journal of New World Archaeology* **5**:13-32.

Wing, E.S. and Wing, S.R., 1995, Prehistoric ceramic age adaptation to varying diversity of animal resources along the West Indian archipelago. *Journal of Ethnobotany* **15**:119-148.

Chapter 2

CONSERVING THE PINES OF GUADALUPE AND CEDROS ISLANDS, MEXICO: AN INTERNATIONAL COLLABORATION

[1]Deborah L. Rogers, [2]J. Jesús Vargas Hernández, [3]A. Colin Matheson, and [4]Jesús J. Guerra Santos

[1]*Genetic Resources Conservation Program, University of California, One Shields Avenue, Davis, CA 95616;* [2]*Programa Forestal, Instituto de Recursos Naturales, Colegio de Postgraduados, Montecillo, Estado de México, México;* [3]*Forestry and Forest Products, CSIRO, Canberra, ACT, Australia;* [4]*Depto. de Ciencias Agrícolas, Facultad de Estudios Superiores Cuautitlán, Universidad Nac. Autónoma de México, México; Current Address: División Académica de Ciencias Naturales y Exactas, Universidad Autónoma del Carmen, Ciudad del Carmen, Campeche, México*

Abstract: Monterey pine (*Pinus radiata* D. Don) is an enigmatic species. Native to only Mexico and USA, it is restricted to three populations along the central coast of California and one on each of two Mexican islands off Baja California—Guadalupe and Cedros Islands. Commercially, it is grown in exotic plantations worldwide on over 4 million hectares with high economic value, yet there is little value in its countries of origin. Overall, the species has lost perhaps 50% of its natural habitat and is threatened by various human-related influences. The two insular pine populations are well differentiated genetically and have different ecological associations both from each other and the mainland populationsAlthough Guadalupe Island has protected status under the Mexican Ministry of the Environment and Natural Resources (SEMARNAT), the pine population may be headed towards extinction because ofgrazing pressure from introduced goats. On neither island are the pines protected from current threats or do they have dedicated funding or a specific conservation plan. Effective longterm conservation of the pines requires a consistent, institutionalized strategy and dedicated funding. International interest in the insular pine populations can contribute to their conservation through research, providing information to support conservation plans and public education materials, providing the technical justification and a proposal to include the island populations of Monterey pine on the threatened and endangered species list in Mexico, maintaining a backup seed collection for restoration, publicizing the value and vulnerability of these populations, and providing funds, as possible. A multinational expedition to both islands in 2001 to collect seeds and information for conservation purposes is an example of the feasibility and value of international collaboration in protecting the Mexican island pines.

Key words: Monterey pine, radiata pine, *Pinus radiata*, conservation genetics, *in situ* conservation, *ex situ* conservation, forest genetic resources, genetic restoration.

1. INTRODUCTION

Monterey pine (*Pinus radiata* D. Don) is an enigmatic species. Native to only Mexico and USA, it has a narrow range and is currently restricted to three populations along the central coast of California and one on each of two Mexican islands off Baja California—Guadalupe and Cedros Islands. Commercially, it is grown in exotic plantations worldwide on over 4 million ha, primarily in Australia, New Zealand, and Chile. Seed collections from the native populations have formed the basis of breeding programs in these, and several other, countries. The domesticated and commercially improved progeny from these programs are now of significant economic value. Indeed, the species' international presence shows an upward trend but it has lost perhaps 50 percent of its natural habitat in the last century. In California, Monterey pine has modest commercial value as a Christmas tree and horticultural species. However, its greater value in California is best measured with other currencies—adding aesthetic value to coastal landscapes, defining a forest ecosystem with a diverse array of understory species, and harboring a gene pool that constitutes the species' evolutionary potential and traits possibly of future economic interest. In México, there have been some field tests of Monterey pine in plantation culture, but the species is not (commercially) competitive with other pines, and has little economic value here. The two native populations on Guadalupe and Cedros Islands suffer from various threats and do not have specific or well-funded conservation plans. Therefore, conservation of the remaining natural habitat and native gene pool of Monterey pine is a challenge that is most effectively pursued through international collaboration—both between conservation interests in Mexico and the USA, and with the international community that has derived economic benefit from and continues to value the native gene pools of this species.

The current natural distribution of Monterey pine and the forest ecosystems it defines consist of five disjunct populations (Figure 2-1). Three of these are located along the central California coast, in a narrow fogbelt region, near Año Nuevo, the Monterey peninsula, and Cambria. Each of the occurrences has a unique assemblage of associated species. For example, near Año Nuevo, Monterey pine grows in close association with knobcone pine (*Pinus attenuata*) and even forms natural interspecific hybrids with this species. On the Monterey peninsula, it grows variously with bishop pine (*Pinus muricata*) and other coniferous species such as Monterey cypress (*Cupressus macrocarpa*), as well as several herbaceous species that are

2. Conserving the Pines

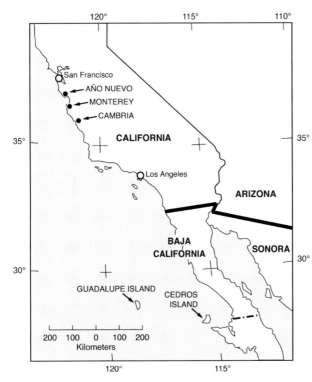

Figure 2-1. Locations of current natural populations of Monterey pine: Año Nuevo, Monterey, and Cambria in California and Guadalupe and Cedros Islands in Mexico. Figure modified and reprinted from the *Journal of Forestry*, 1998, Volume 96, Number 1, page 36, (Ledig, F.T., J. Jesús Vargas-Hernández, and K.H. Johnsen. The conservation of forest genetic resources: Case histories from Canada, Mexico, and the United States). Copyright 1998 by the Society of American Foresters. Reproduced with permission of the Society of American Foresters via Copyright Clearance Center.

endangered or threatened. At its southernmost US occurrence, near Cambria, California, Monterey pine is the only coniferous species within its local ecosystem. In this population it intermixes with California live oak (*Quercus agrifolia*) towards the inland, drier extent of its range. In each of its California occurrences, Monterey pine is a defining environmental feature: influencing the microclimate, condensing the fog to precipitation beneath its canopy, and significantly determining the types of other plants and animals with which it coexists. Ownership of the Monterey pine forests in California is diverse, with perhaps 75% of the total area privately owned, and the remainder owned and managed by State government and nongovernmental organizations such as universities and land trusts.

In México, Monterey pine grows on two islands off the Pacific coast of Baja California. It is the only pine species, and one of the few tree species, that naturally inhabits these islands. Guadalupe Island, located approximately 280 km off the coast, is located almost at the limits of the ocean economic zone of exclusivity assigned to Mexico. It was discovered in 1602 by a Spanish expedition under the command of Sebastián Vizcaíno (Berzunza, 1950). The history of naming the island is not conclusively documented but seems to have originated after the 1837 visit of the French Admiral Abel du Petit Thouars to the west coast of North America.

The island is about 30 km long and 11 km wide (Rico C., 1983). It is part of an archipelago of volcanic origin, approximately seven million years old, and has never been in contact with the continent. At the northern end, Mount Augusta (elevation 1,298 m) marks the crest of a volcanic mountain that slopes 3,660 m into the floor of the Pacific Ocean (Bostic, 1975). Conditions are dry to arid with annual rainfall of approximately 150 mm. Accounts from several botanists who explored the island in the late 19th and early 20th century reveal many endemic plant species (Howell, 1942; Rico C., 1997a). There is a total known flora of 216 species: not all of them concurrent, and perhaps 45 are relative newcomers (weeds), leaving 171 species that are possibly native (Moran, 1996). Over 30 species have probably gone extinct. Monterey pine occurs only at the north end of the island, mostly on crests of the ridges. Three other tree species currently grow on the island: island live oak (*Quercus tomentella*), Guadalupe Island fan palm (*Brahea edulis*), and Guadalupe cypress (*Cupressus guadalupensis* ssp. *guadalupensis*).

The island has never been inhabited by humans on a permanent basis. However, in the past few centuries, it has occasionally been a temporary home for pirates, whalers, and fishermen who have used the island both as a refuge and as a 'hunting' field, particularly for marine mammals. Goats were introduced to the island in the nineteenth century as a source of fresh meat and quickly surpassed the carrying capacity of the island, affecting its native flora (Ravest S., 1983). Currently, the island is under the protection of the Ministry of the Environment and Natural Resources (SEMARNAT, Mexican Secretaría de Medio Ambiente y Recursos Naturales) through the National Commission for Natural Protected Areas (CONANP). The only human inhabitants on the island now are a Garrison of Marines at the naval post at the southern tip of the island who guard the island and monitor a small meteorological station, and about 30 fishermen who occupy a temporary camp on the southwest side of the island during the fishing season for abalone and lobster. They belong to a fishing cooperative based at Ensenada, Baja California.

Cedros Island lies about 30 km northwest of Guerrero Negro, Baja California, and about 300 km southeast of Guadalupe Island. Varying in width from 4 to 14 km wide (Rico C., 1997b) and with a length of about 32 km, it resembles Guadalupe Island in both size and shape. Cedros Island has a continental origin, geologically similar to Sierra Vizcaíno which lies on the continent, southeast of the island (Osorio T., 1948). Drawing from this similarity, it is assumed that Cedros Island is geologically old, formed from Jurassic sediments. Despite this, most soils on the island are quite shallow, almost skeletal.

In contrast to Guadalupe Island, Cedros Island has been inhabited by humans since pre- Colombian times (Ravest S., 1983). Spanish explorers discovered the island in 1540, giving it the name 'Cedros' in reference to the trees located on the mountain ridges which can be seen from the ocean (Osorio T., 1948). On the sheltered east coast, on the southern tip of the island, there is a small village of about 4,500 permanent residents (CONABIO, 2000). The island is otherwise uninhabited, except during the fishing season for abalone and lobster when fishermen camp near the lighthouse at the northern tip and on the west coast. The main economic activity on the island is the fishing industry, with a cannery that can process over 200 tonnes of fish per day (Osorio T., 1948). Cedros Island also serves as distribution point for salt produced on the continent, in the town of Guerrero Negro, where the world's largest salt producing industry is located. From Cedros Island, the salt is taken all over the world (Ravest S., 1983). During the late nineteenth and early twentieth centuries, after the gold rush in California, mining activity was common on Cedros Island, as in many other islands in the California region (Osorio T., 1948). There is still some evidence of mining on the northern part of the island; however it never was a very important economic activity.

The total area of pine forest is approximately 140 ha. The pines occur in two main populations: inland towards the center of the island and at the northern end of the island, separated by approximately 14 km. The vegetation on the island is diverse, even in the dry southern areas, and includes perhaps 15 endemic species and varieties. The desert canyons contain such shrubs as the silver-leaved sunflower (*Viguiera lanata*), elephant trees or torote (*Pachycormus discolor*), bursage (*Ambrosia chenopodiifolia*), and the Cedros sage (*Salvia cedrosensis*) (Oberbauer, 1986). Like Guadalupe Island, this island is owned by the federal government of México and is considered an important area for bird conservation by the National Commission for the Knowledge and Use of Biological Diversity (CONABIO, Comisión Nacional para el Conocimiento y Uso de la Biodiversidad) (CONABIO, 2000).

Among the five native populations of Monterey pine in Mexico and USA, there are notable differences in biological attributes, ownership and management entities, local values and importance, and influences on their conservation status. This chapter will focus on the two pine populations on Guadalupe and CedrosIslands,Mexico,referring to the California populations only as appropriate to provide context or contrast. Within its native range, *Pinus radiata* is known commonly as Monterey pine and thus will be referred to by that name here. However, in many southern hemispheric countries where it is grown in plantation culture, it is commonly referred to as radiata pine.

2. STATUS OF THE PINES

Monterey pine is on the World List of Threatened Trees and the five populations have been classified according to the International Union of Conservation of Nature and Natural Resources' (IUCN) Red List Categories. In the IUCN system, the Guadalupe and Cedros Island populations (designated in that system as *Pinus radiata* var. *binata* Engelm.) are listed as endangered (IUCN, 2000). IUCN also has classified Guadalupe Island as a scientific reserve because of the large number of endemic species (IUCN, 1992).

Each of the five populations of Monterey pine has some unique genetic features as well as a different ecosystem context. Field and laboratory studies have revealed considerable genetic diversity within Monterey pine and strong genetic differentiation among the five populations that is somewhat unusual for an outcrossing coniferous species (Moran et al., 1988; Burdon, 1992; Wu et al., 1998). Some of these differences are readily apparent. Whereas the mainland populations are largely three-needled, the pines on the islands generally have their needles in bundles of two. The cones of the island pines are smaller, on average, than those of the mainland populations. Trees from both island populations have shown significantly greater resistance to wind-induced toppling than those from the mainland populations (Burdon et al., 1992). The island pines have also shown superiority in basic wood density relative to the mainland populations (Burdon, 1992). The among-population differences are such that two, and sometimes three, varieties of Monterey pine are recognized, depending on whether or not the island populations are grouped together as one variety (see next section).

Although Guadalupe Island is a protected area under SEMARNAT, the Monterey pine forest has been declining over the last century. The number of trees has dropped to approximately 220 (+/- 20) in 2001 (Rogers et al., 2004) and this population appears to be following a trajectory towards extinction (Ledig et al., 1998). All of the remaining trees are large and presumably very old, there having been no effective regeneration of the pines for decades. Feral goats—introduced in the 19th century to serve as a source of fresh meat for the sailors, fishermen, whalers, or others who might visit the island—quickly grew in population and exerted tremendous grazing pressure on the island's plant species. Therefore, for decades, seedlings have been eaten very quickly after germination. A few pine seedlings were seen in May, 2001, but were expected to be eaten by goats in the near future (Rogers et al., 2004). In addition to the goats, other problems include proliferation of introduced plants that displace the native species. Species that are particularly invasive on Guadalupe Island include Tocalote or Malta starthistle

2. Conserving the Pines

(*Centaurea melitensis*), tree tobacco (*Nicotiana glauca*), and slender wild oat (*Avena barbata*) (Moran, 1996).

The pines occur in a scattered progression along the ridgetop and down steep slopes at the north end of the island at elevations between approximately 500 and 1200 m (Figure 2-2). Sometimes in clusters of 2 to 10 trees, they are also often found as individual matriarchs separated from the next nearest tree by 0.5 km or more (Figure 2-3; Rogers et al., 2004). The seeds from such trees may well be inbred and show loss of vigor if they ever had the opportunity to germinate and grow.

It is clear that the pines on Guadalupe Island play critical ecosystem roles and that local extinction of the pines would likely be followed by extinction or extirpation of other species. In particular, the pines condense the fog that is common throughout the year particularly along the northern ridges of this volcanic-origin island. In an otherwise arid environment, water can be seen streaming down the trunks of the pines, collecting in the furrows of fallen branches and moistening the ground beneath the trees (Rogers et al., 2004). Birds use this water source and find habitat in the branches. Through moisture condensation, shade, organic matter provision, and physical stature, the pines contribute an essential and irreplaceable suite of ecosystem services.

Figure 2-2. The Monterey pines on Guadalupe Island are all located at the northern end of the island, typically scattered along fog-swept ridgetops as in this May 16, 2001, scene. (Photo credit: A.C. Matheson).

Figure 2-3. Some members of the multinational expedition in May, 2001, collect pine cones from one of the majestic Monterey pines on Guadalupe Island. (Photo credit: A.C. Matheson).

Conservation of the pines on Guadalupe Island is challenging because of the stressful environmental conditions, continued heavy predation pressure from feral goats, remote location of the island, and physically challenging access to the pines—on mountain ridges and steep slopes with no direct vehicle access. Key to the maintenance of remaining genetic diversity of the pines, the success of any natural regeneration or restoration efforts, and the longterm survival of this population is the effective control or removal of the introduced goats. Several recent efforts have focused on this goal. In the last two years, several thousand goats have been removed by Mexican ranchers and a binational nonprofit conservation organization (Grupo de Ecologia y Conservacion de Islas A.C., *Island Conservation*) has organized fence construction in some critical areas, including three exclosures around some of the pines. Seedlings were noticed within these exclosures early in 2002 (Sanchez-Pacheco, pers. comm..). However, fence exclosures are not absolute protection from goat predation over time, and the only longterm effective strategy for pine conservation is the complete elimination of the goats.

A census has not been taken of the Monterey pines on Cedros Island—their numbers being far greater than those on Guadalupe Island—nor has a comprehensive map of their distribution been made. The pines on this island are generally much smaller and younger than those on Guadalupe Island. They are found at a lower elevation than on Guadalupe Island, between 285 and 690 m. Also in contrast with Guadalupe Island, there is evidence of recent and ongoing reproduction of the pines, with fire influencing reproduction and the age structure of pine stands on Cedros Island. Fire scars on some of the older trees, notes from grey literature, conversations with local fishermen, and the even-age of some stands, all suggest both recent and frequent fires on this island (Rogers et al., 2004). The cones of Monterey pine are semi-serotinous—that is, being tightly closed and opening under intense dry heat. Thus, depending on the intensity and frequency of fire, seed dispersal and natural regeneration can actually be assisted by fires. The climate here seems to be drier than on Guadalupe Island, with lower relative humidity and less fog, particularly in the southern stand of Monterey pine. As on Guadalupe Island, the presence of fog seems to be critical for the growth of Monterey pine, the pines being restricted to the ridges and moist canyon slopes where fog occurs frequently (Figure 2-4).

The two main populations—one central and the other northern—are primarily distributed along the windward side of the main ridge. Both are composed of several narrow stands, but the northern population occupies an area about twice as large as the southern population. Within each population, stands differ greatly in the size and age structure of trees and the underlying soil type. For example, some stunted trees grow on very shallow soils. Some stands are fairly young and even-aged; others have a varied age structure

Figure 2-4. The Monterey pines on Cedros Island are smaller, younger, and more numerous than those on Guadalupe Island. Small stands are scattered along ridges and in moist canyon slopes (as in this May, 2001 scene) in the central and northern portion of the island. (Photo credit: D.L. Rogers).

with some much older trees. The presence of many patches of young trees in the northern population, particularly on disturbed soils from abandoned mining sites, suggests that this population may even be expanding in range (Rogers et al., 2004).

Cedros Island is not a protected area, nor are there any specific reserves or conservation measures for Monterey pine. Reports from several expeditions to the island in the last 40 years concur in noting the presence of small seedlings and patchy forested areas of varying ages (Libby et al., 1968; Rico C., 1997b; Rogers et al., 2004). Indeed, the abundant natural reproduction of Monterey pine here, particularly following fires, prompted one scientist to conclude that "Cedros Island remains the least endangered of the five native radiata [pine] populations" (Libby, 1978). Although natural regeneration does not seem to be a problem, there is little known about the pines. The effect of frequent fires on the genetic structure and level of inbreeding is not known. There are numerous introduced plants and animals on this island and, while not as severe in their impact as on Guadalupe Island, there could be some influences. Although there was probably some harvesting of trees during the brief period of gold and copper mining in the late 1800s, there seems to be little interest in the pines at present. Only a few topped trees, noticed in 2001, suggest that the current residents might make occasional use of the pines as Christmas trees (Rogers et al., 2004). In summary, currently this population of Monterey pine does not seem to have any severe threats to its longevity. But it is largely unstudied in its genetic structure and ecology, and vulnerable to introduced diseases or other biota.

Fortunately, based on findings from a recent multinational expedition, no evidence has been found that the fungus (*Fusarium circinatum*) associated with the pitch canker disease is on either island (Rogers et al., 2004). This disease has caused significant mortality in all three California populations of Monterey pine.

3. SCIENTIFIC AND NATURAL HISTORY INTERESTS

The volcanic origin of Guadalupe Island provides a natural laboratory to study species evolution and adaptation, with interesting contrasts between island flora and fauna and their continental relatives. Although Cedros Island was initially connected with and is not completely isolated from the mainland, a similar process of genetic differentiation has occurred there with both plant and animal species (Osorio T., 1948). Indeed, scientists with the Biodiversity Research Center of the Californias at the San Diego Natural History Museum have as their geographic research focus the region of southern California and the Peninsula of Baja California, including Guadalupe and Cedros Islands. Although the island pine populations are certainly not pristine, their remote location and freedom from most urban influences provide some research advantages. For some research questions, they may provide a model and simplified system by which to better understand general principles and the ecosystem dynamics of the mainland Monterey pine populations.

The relationship of the pines of Cedros and Guadalupe Islands to each other and to the mainland populations have long challenged and intrigued taxonomists. Indeed, the Cedros Island pines once were described as a variety of Bishop pine (*Pinus muricata* var. *cedrosensis*) (Newcomb, 1959). Even today there is some disagreement among taxonomists. Although there is general agreement that the two island populations are Monterey pine and a different variety from the mainland populations (*Pinus radiata* var. *radiata*), some consider them a single variety, *P. radiata* var. *binata* (Perry, 1991), while others describe Cedros Island as a different variety (*P. radiata* var. *cedrosensis*) (Fielding, 1961; Axelrod, 1983; Bannister and McDonald, 1983). Recent studies have shown that the island populations retain characteristics of an ancestral pine from which the continental populations of Monterey pine (*Pinus radiata* var. *radiata*) descended (Axelrod, 1980; Millar et al., 1988). Thus, both island populations can be regarded as relictual variants of Monterey pine, with the Cedros Island population being considered the most ancient of the five current populations.

4. POLITICAL, ECONOMIC, AND SOCIAL INTERESTS: MEXICO

Mexico has an intrinsic political interest in conserving the pine populations on the islands. This interest is related to maintaining control of the Mexican insular territory in the area and associated natural resources, beginning with the explicit incorporation of Guadalupe and Cedros Islands to the national territory in 1917. Conserving the islands and their associated natural resources is a matter of pride and heritage, honor and nationalism, but also of economic policy, particularly for Guadalupe Island. Because of its strategic location, maintaining this possession allows Mexico to increase substantially its oceanic economic zone of exclusivity in an area rich in natural resources[1].

In 1922, Guadalupe Island was declared a 'National Reserve' by Mexican President Alvaro Obregon. This designation, however, was primarily

---

[1] Julio Berdegué, unpublished data, 1957.

focused on protecting the marine mammals that use the island as a natural refuge (SAF, 1922). In 1928, it was declared a 'Protected Area' by President Plutarco Elias Calles (SAF, 1928). In 1980, the federal government established a specific program to promote the "integral development of Guadalupe Island" (Ravest S., 1983). No similar program has ever been developed for Cedros Island. The main objective of the program was to create a basic infrastructure on the island to allow colonization and maintenance of a permanent human population there to utilize the natural resources available in an 'orderly manner', assuming this would be enough to protect and conserve those resources. As a result of this program, about 60 km of dirt roads were built on the island, from the southern tip to the northern part of the island where the springs are located (Ravest S., 1983). In addition, a small airstrip was built to allow landing for small aircraft and fuel-based, electricity generators were installed at the navy base and the fishing camp.

In 1982, efforts to incorporate Guadalupe Island into the 'National System of Protected Natural Areas' were initiated, but a federal decree to make this designation official has never been published. The draft decree assigned the island's technical management to the Ministry of Urban Development and Ecology (SEDUE). The management program included plans to eradicate the goats from the island; unfortunately, for several reasons, these plans have not been implemented (Rico C., 1983). As part of the same management program, actions to fence and protect the cypress forest from the goats were started. However, the fence was never finished and after a few years the protection program was abandoned (Ravest S., 1983). Moreover, the Monterey pine population was not included in that fencing program.

Currently, Guadalupe Island is considered broadly as a 'natural, protected area', but this designation is under review (CONANP, 2002). Consequently, it might even be withdrawn from conservation status. Under the current category, protection and technical management of the natural resources, including the Monterey pine population, is the responsibility of SEMARNAT. CONABIO considers both islands as important areas for bird conservation (CONABIO, 2000). Together, CONABIO (created in 1992) and SEMARNAT (originally created as SEDUE in 1982, then evolved to SEMARNAP in 1994, and to SEMARNAT in 2000) have raised the political profile of environmental conservation in Mexico, setting priorities and securing some funds for conservation of biodiversity.

Despite the political actions taken to protect Guadalupe Island, the small size of the natural insular populations of Monterey pine, and the value of these populations as an international genetic resource, the species is not officially considered threatened, endangered, or even at risk in Mexico (INE, 1994). For this reason, no specific management plans for conservation or restoration of the Monterey pine populations have been developed for both protected areas. Certainly, a revision of the threatened and endangered species list is necessary to incorporate both insular populations of Monterey pine.

The Monterey pine populations do not have a direct economic or social interest either locally or nationally in Mexico. With the exception of sporadic use of Monterey pine trees on Cedros Island by local residents for firewood, Christmas trees, or fencing material, no other commercial product is obtained from these pine populations. Despite the huge economic importance of Monterey pine internationally as plantation species, this is not the case for Mexico (Arteaga M., and Etchevers B., 1988). Several attempts have been made to introduce and establish Monterey pine plantations throughout Mexico (Salinas Q., and Gómez N., 1975; Arteaga M., and Etchevers, B., 1988)[2,3] and even as street trees in urban areas of Mexico City[4]. All those attempts failed for various reasons ranging from climatic and edaphic factors, to air pollution, insects, and disease problems (Calderón Flores, 1967; Eguiluz-Piedra and Cibrián-Tovar, 1976; Arteaga M., et al., 1985; Salinas Q. and Gómez N., 1975). The only exceptions are a few small experimental plantings (under 100 ha, total) established in Ayotoxtla, Guerrero on the west-central region of Mexico (Arteagea M., et al., 1985). At this location, Monterey pine has shown good survival and a higher growth rate as compared to other non-native pine species. However, because of the severe insect and disease problems that Monterey pine has experienced in other areas of Mexico (Salinas Q. and Gómez N., 1975; Eguiluz-Piedra and Cibrián Tovar, 1976; Peña B. and Cibrián T., 1980.), the species is no longer planted in Mexico.

A possible new venue of economic interest is export of germplasm to other countries where Monterey pine has economic value as a plantation species. This possibility, however, is strongly limited by three factors. First, the amount of seed production in both pine populations is relatively low, and the collection costs are extremely high. Second, international phytosanitary regulations affect and sometimes limit the movement of germplasm across country borders. Finally, the international demand for seed is relatively low, both because only small samples of seed are required to evaluate its potential for introduction to new regions and to start new domestication programs, and because most of the countries with the best potential for plantation industries with Monterey pine already have sufficient genetic resources.

An economic activity that is growing over the entire Baja California region is ecotourism. This activity might increase the economic interest in conserving the populations in their natural habitat. Despite the accessibility

---

[2] Rafael Ojeda, unpublished data, 1977.
[3] Abraham Escárpita Herrera et al., unpublished data, 1986.
[4] Othon Yáñez Márquez, unpublished data, 1981.

difficulties, the dramatic scenic value of the island pine populations could generate sufficient interest to justify ecotrips. Currently, the ecotourism market in the Baja California region is dominated by US companies, particularly charter boat companies, that conduct whale-watching or other natural history trips. As such, the main economic value to Mexico is only in the permit fees and port charges that these businesses are charged by the Mexican government. However, one such company does contribute a percentage of profit towards conservation activities in the Baja California region[5].

5. POLITICAL, ECONOMIC, OR SOCIAL INTERESTS: USA

Although harvested in its native forests particularly in the 18th and 19th centuries for residential and commercial purposes, Monterey pine has only modest commercial value today in California or the rest of the United States. For the latter half of the 20th century, it was often used in revegetation projects along major coastal transportation corridors in California, with stock usually derived from New Zealand commercial sources. However, since the advent of pitch canker disease, this use of Monterey pine has been discontinued in the state. It is still grown for Christmas tree and landscaping purposes in California and Oregon (Rogers, 2002). However, except for some progressive urban tree projects in Cambria and on the Monterey peninsula where native seed is collected and genetically local seedlings are transplanted into parks and street-side settings, there is little direct use of the native populations.

Today, Monterey pine is valued in California mainly for aesthetic reasons and for ecological roles within the geographically restricted ecosystems that it dominates. Extrapolating from research conducted on coast redwood (*Sequoia sempervirens*) within the coastal fog belt of California (Dawson, 1998), it is likely that Monterey pine plays a significant role in condensing fog and delivering moisture to the plants and soil within its vicinity. Indeed, on Guadalupe Island the authors often noted water streaming down the trunks of pines, being absorbed into the soil and pooling in bits of woody debris beneath the trees Although largely unquantified, Monterey pine unquestionably affects other microenvironmental attributes such as light penetration, wind disturbance,soil stability, and soil pH. Thus, this pine defines habitats forotherspecies—bothwidespreadandrestricted. Several plant species that are threatened (T) or endangered (E) at the State (S) or federal (F) leveloccurlargelyorentirely within native Monterey pine forests. These

---

[5] Searcher Natural History Tours. URL: http://www.bajawhale.com

include Yadon's rein orchid (*Piperia yadonii*, FE), Gowen cypress (*Cupressus goveniana* var. *goveniana*, FT), Hickman's cinquefoil (*Potentilla hickmanii*, SE, FE) and Monterey clover (*Trifolium trichocalyx*, SE, FE). One well-known ecological function of these forests is as overwintering sites for Monarch butterfly (*Danaus plexippus*) populations that breed east of the Rocky Mountains and migrate to the mountains of central Mexico (Weiss et al., 1991).

Aesthetically, Monterey pine contributes profoundly to the tourism appeal, property values, and general caché of the Californian communities within its range. The internationally known resorts and golf courses of the Pebble Beach Company, for example, are nestled next to (in some cases, carved out of) native Monterey pine forests. The aesthetic value, however, is frequently dwarfed by the value of this habitat for other land uses. With the majority of its extant natural range in California in an increasingly populated and urbanized area, Monterey pine has lost considerable habitat and its genetic integrity may be compromised. Figures describing its loss of habitat in California vary, but most estimates converge on 50%. And with the majority of the native Monterey pine stands occurring on private property, land conversion continues. Other influences associated with urbanization—including genetic contamination from planted pines, habitat fragmentation, natural regeneration problems, and introduced invasive species—continue to stress and reduce the native pine forests.

The major values of the Guadalupe and Cedros Island pine populations, from a US perspective, could be classified as research and conservation. Research on the island populations provides an evolutionary and biological context in which to better understand and more appropriately interpret similar kinds of information on the California populations. Numerous US studies of Monterey pine—genetic and evolutionary—have included one or both of the island populations to provide a species- or near-species-wide perspective (Murphy, 1981; Wu et al., 1999; Dvorak et al., 2000).

The pines of Guadalupe and Cedros Islands have been of conservation interest to US scientists and conservationists for decades (see section below on international collaborations). Several nonprofit organizations in California are devoted, exclusively or predominantly, to the protection of native Monterey pine forests. These include *Greenspace—The Cambria Land Trust* (Cambria, California) and *Monterey Pine Forest Watch* (Carmel, California). In addition, a project initiated under the auspices of the University of California, the *Monterey Pine Forest Ecology Cooperative*, provides a nonpolitical educational forum for encouraging the use of science in conservation of native Monterey pine forests. Although most of these organizations and programs focus on the California populations, their interests and concerns, and sometimes their activities, extend and apply to the Monterey pine populations on Guadalupe and Cedros Islands.

6. POLITICAL, ECONOMIC, OR SOCIAL INTERESTS: AUSTRALIA

Juxtaposed with the declining ecosystem health, lack of commercial interest, increasing urbanization pressure, and general lack of populace support for Monterey pine within its native range in Mexico and USA, there is enormous support for and investment in the species in its exotic plantation context, particularly in Australia, New Zealand, and Chile. We focus on the Australian involvement in this chapter because of the recent and ongoing collaborations among Australia, Mexico, and USA in conservation of the island pines. Monterey pine's success as a plantation species is largely attributable to its capacity to respond to both genetic improvement and management. The planted-forest technology for Monterey pine is probably the most advanced of any tree species (Sutton, 1999). In Australia, Monterey pine plantations account for 75% of the total pine plantations currently established. Current value of the sawn lumber produced from the (total) pine plantations is over $1 billion (Australian dollars) per annum. Furthermore, research investments have paid off handsomely, with first-generation breeding efforts for Monterey pine recently evaluated as representing a benefit/cost ratio of approximately 15 (CSIRO, 1999).

Although most of the trees now in use in Australian Monterey pine plantations have their genetic origin in the mainland California populations and there are seeds from both island populations in Australian seed banks (Eldridge, 1997), Australian scientists remain interested in the conservation of the native gene pools of both the island and mainland populations. Genetic resources in nature are usually better buffered from catastrophic loss than individual seed banks. The native forest context is the only one in which genetic diversity continues to be shaped by natural selection over time. There could be new genes or allelic combinations that may have value in Australian plantation culture in the future either because this is a new form of genetic diversity or because environmental or market conditions have changed to make existing genetic diversity more valuable. In particular, the apparent drought resistance of Guadalupe Island pines is very relevant to Australian conditions. Other traits that could have commercial application in the future include the higher level of resistance of the Guadalupe Island pines to red band needle blight (caused by *Scirrhia pini*) and western gall rust (caused by *Endocronartium harknessii*) (Cobb and Libby, 1968; Old et al., 1986) and the thinner bark, higher wood density (Nicholls and Eldridge, 1980), and greater frost resistance of both island populations relative to mainland populations (Alzard and Destremau, 1982). This interest is currently academic, however, as there is a moratorium on Monterey pine seed importation to Australia from the USA or Mexico because of the risk of importing the pitch canker fungus. Nevertheless, other types of collaboration are possible and a 1982 agreement between Mexico and Australia[6] could provide an appropriate framework.

The popularity and support for Monterey pine in plantation-growing countries such as Australia and New Zealand is largely, but not entirely, because of the economic value of the species. For example, highly productive and concentrated forest plantations can make it possible to allocate native forests to parks and reserves, and at the same time generate positive environmental effects as plantations replace degraded marginal agricultural lands (Gladstone and Ledig, 1990; Sedjo, 1999). Thus, Monterey pine plays a role in conserving the native forests of Australia.

The commercial interest in Monterey pine has also stimulated considerable investment in research, much of which is highly applicable to understanding the biology—primarily genetics—of this species and can inform conservation efforts of the island populations (Matheson, 1980; Moran et al., 1988).

7. INTERNATIONAL COLLABORATIONS: ACCOMPLISHMENTS AND PROSPECTS TOWARDS CONSERVATION OF THE PINES

Effective conservation of the ecology and genetic diversity of the island pine populations requires both *in situ* (i.e., protection of native forests) and *ex situ* (i.e., collecting and preserving seeds that reflect the genetic diversity of those populations and could be used in restoration efforts) approaches. *In situ* conservation of these populations would allow maintenance of the evolutionary trajectory of Monterey pine within these specific island environments. In addition, this would help protect all the other species associated with the pines. There have been several expeditions by Mexicans to the is lands to provide *ex situ* conservation for the pines. Between 1975 and1982, there were at least four expeditions to Guadalupe and Cedros Islands by faculty and students from the Universidad Autónoma de Chapingo to collect Monterey pine seed (Ravest S., 1983). In 1980 on Guadalupe Island, a temporary nursery was fenced to raise seedlings, but because of insufficient funding and long-term research commitment, the seedlings were never

---

[6] Basic Agreement for Scientific and Technical Cooperation between Mexico and Australia.

planted out among the mature pine trees. Today, only a few seeds from those collections remain in storage, most probably with very low viability.

Interest in the pines—that could perhaps increase domestic commitment—and funding can be enhanced through international collaborations. With the exception of one seed collection expedition in 1978 that involved Australian and American, but no Mexican, scientists, there is scant evidence of international assistance for conservation of the pines in the 20th century. However, in 2001, a multinational collaborative effort led to an expedition to the islands in support of Monterey pine conservation. The purposes of the expedition were to collect pine seeds primarily for conservation and research purposes, to determine the status of the pines including any evidence of pitch canker disease, and to increase public awareness of the pines to improve the likelihood and success of conservation efforts (Figure 2-3).

The expedition was collaborative at every stage, from planning to fundraising to execution of the trip. It was also a highly multinational effort, with Mexican, American, Australian, and Canadian participants. The principal investigators who organized and led the expedition are this chapter's authors, and represent Mexico, USA, and Australia. Financial support came from Australian (Department of Industry, Science, and Technology; and the Commonwealth Scientific and Industrial Research Organization), American (the University of California's (UC) Genetic Resources Conservation Program, the University of California Institute for Mexico and the United States (UC MEXUS), and personal and research contributions from some participants), and international (Food and Agriculture Organization of the United Nations) sources. All 14 participants, many of them volunteers including a professional climber, contributed their time and skills and worked diligently during the physically challenging expedition. The Mexican consular offices in San Francisco, California and in Canberra, Australia, provided research visas for non-Mexican expedition participants. Patricia Beller (San Diego Natural History Museum), Dr. Exequiel Ezcurra (President of the National Institute of Ecology, Mexico), and Dr. Kenneth Eldridge (Senior Fellow, CSIRO, Canberra, Australia) provided helpful advice in preparing for the expedition. The Mexican Ministry of the Interior (Secretaría de Gobernación, Subdirección de Territorio Insular) provided permission to land on the islands, and the National Institute of Ecology (Dirección General de Vida Silvestre and Comisión de Áreas Naturales Protegidas) provided permission to collect cones and stem cores from the pines. On Guadalupe Island, staff of the Mexican navy helped to carry the bags of cones on foot over difficult terrain. Two Mexicans who were working on Guadalupe Island, Francisco Javier Maytorena and Mario Urias, contributed their time and vehicles towards transporting participants, equipment, and bags of pine cones between the camp site and the navy station, some 30 km apart. The owner-operators of *Searcher Natural History Tours*, Art Taylor and Celia Condit, provided valuable assistance in trip preparation and the *Searcher*'s captain (Art Taylor) and crew provided excellent support during the expedition. In every aspect and at every stage, this expedition was a reflection of multinational goodwill and passion for conserving the pines on these islands.

The accomplishments of this expedition can be measured in various currencies. Seeds were collected from almost 200 trees and will be used for longterm conservation (i.e., for restoration of the native populations, if required) and for research. The seeds were distributed to two locations—one in the US and one in Mexico—so as to minimize the risk associated with keeping all seeds at one facility. Research is underway that will provide information on the pattern of genetic diversity in those populations. No evidence of pitch canker disease was found on pines on either island. However, there was evidence of insects on Cedros Island that are associated with pitch canker disease as infection vectors. And whether vectored by insects or other means, the pines are vulnerable to this disease if the associated fungus is introduced. Two participants from the Área de Protección de la Flora y Fauna in Ensenada, Mexico (Celerino Montes and Ana Ma. Padilla Villavicencio)—one of them from the division (Islas del Golfo de California) directly responsible for protecting the islands' biota—developed a much better appreciation of the status of the pines, while contributing significantly to the expedition. Information gained from this expedition was useful to a nongovernmental organization, *Island Conservation*, that subsequently built some fences around several ecologically sensitive areas on Guadalupe Island, including three Monterey pine areas. Perhaps most importantly, the subsequent presentations, publications, and sense of mission generated among participants will continue to improve the international visibility of the pines and build a community of concern.

There are several challenges for effective, longterm conservation of the pines on Guadalupe and Cedros Islands. Although Guadalupe Island has protected status, it may not retain this status as it is currently under review. Furthermore, even having protected status is not sufficient to protect the ecological and genetic attributes of the pines. That pine population requires recognition and perhaps even federal designation as threatened or endangered. The pine populations on both islands require specific conservation plans as Cedros Island does not have protected status and any general plans for protection of Guadalupe's biota do not contain sufficient detail for appropriately managing the pines.

Also needed is a longterm conservation commitment in the form of an institutionalized conservation strategy. This is a challenge in Mexico, as in other democratic countries, because changes in federal administration can have significant impacts on management policies for federally managed lands. Consequently, there is not enough stability or continuity in conservation plans for federally controlled lands, even nature reserves, because there could be a change in administration every six years.

There is a critical need for funding that is targeted specifically for protection, and restoration if appropriate, for the pine populations. Funds are required to support conservation staff, to develop a conservation plan, to effectively extinguish the goat problem on Guadalupe Island, to monitor regeneration and health indicators in both pine populations, and to prepare publications to improve public awareness of the significance of these populations and engender a domestic conservation constituency.

Two activities are urgent for conservation of the island pine populations: goat removal on Guadalupe Island and strict adherence to pitch canker prevention practices for both islands. Introduction of this, or perhaps any disease to which the pines are vulnerable, could be lethal for these populations. This is particularly true for Guadalupe Island with its relatively small number of remaining trees. Persons with clothing or equipment that has been used within zones of pitch canker infestation on the California or Mexican mainland could inadvertently vector the pitch canker fungus if they visited the island pines without proper prophylactic practices.

Realistically, longterm support and an effective delivery structure for protecting and monitoring the pines must be provided domestically. Funding is perhaps the critical issue. As mentioned, ecotourism could possibly provide some support for the pines but this would require new domestic policies to encourage investment and participation by Mexican companies. It would also require careful adherence to pitch canker prevention policies and management practices that do not contribute to further ecological degradation through heavy visitor traffic, introduction of new weeds or diseases, or other human-related disturbances. Germplasm exportation might yield some income in the future if pine seed production exceeds that needed for natural regeneration, if there is reason to believe that there are valuable (and not yet sequestered) genes in the native populations, if phytosanitary concerns are eased to allow international transfer, and if issues concerning property rights of indigenous germplasm do not present trade barriers. Minimally, international commercial interest in the genetic resources of the island pines could be used to subsidize conservation efforts, as was done with the 2001 expedition where seeds were collected for both conservation and research purposes using mostly nondomestic funding sources.

International support for conserving the pines is also promising. Several conditions and developments have conspired to produce a more positive environment for conserving the island pines than perhaps ever before. Several organizations created in the 1990s are now focused on conservation issues that either specifically target the Baja California region (e.g., *Island Conser-*

vation) or that include this region in their geographic area of concern (e.g., the Commission for Environmental Cooperation, an organization created under the North American Agreement for Environmental Cooperation). The San Diego Natural History Museum organized a binational expedition to Guadalupe Island in 2000 to survey and assess the status of some of the island's biota which continues to illuminate the need for conservation. The Food and Agriculture Organization (FAO) of the United Nations remains interested in and supportive of genetic conservation efforts for the pines, considering the global importance of the species in developing and developed countries alike.

The collaborators in the 2001 expedition remain interested in assisting in conservation efforts for the pines. Specific activities that this multinational group could undertake include providing support for maintaining the protected status of Guadalupe Island, encouraging the adoption of protected status for Cedros Island, providing the technical justification and a proposal to include the island populations of Monterey pine on the threatened and endangered species list in Mexico, developing a specific conservation plan for the pines on these islands, developing guidelines to reduce the possibility of introducing pitch canker disease or other diseases that would threaten the native biota, and conducting research on the pines. The advanced state of genetic research and seed storage techniques and technologies for Monterey pine in Australia and New Zealand could contribute to genetic conservation in the native populations. In summary, conservation contributions of the international community could be made through continuing research on the pines, providing information to support conservation plans and public education materials, maintaining a backup seed collection for restoration, publicizing the international value and vulnerability of these island populations, and providing funds, as possible.

Fortunately, the biological window of opportunity for conserving the pines is still open. In the more urgent situation on Guadalupe Island, despite the harsh environmental conditions, natural regeneration still seems possible if seedlings are protected from the goats. No evidence of pitch canker disease has been found in either island population. With immediate attention to the more urgent conservation activities, and by engaging international interest and support, a requiem for the Monterey pine populations on Guadalupe and Cedros Islands may not be needed.

ACKNOWLEDGMENTS

We thank Dr. Exequiel Ezcurra and one anonymous reviewer for insightful reviews and helpful comments on an earlier draft. Support for the senior author was provided, in part, by grant # 2000-14325 from the David and Lucile Packard Foundation and #2-T-PRRP-3-161 from the University of California's Pacific Rim Research Program. This material is based upon work supported by a grant from the University of California Institute for Mexico and the United States (UC MEXUS) and the Consejo Nacional de Ciencia y Tecnología de México (CONACYT). We are grateful to the Food and Agriculture Organization of the United Nations for a grant to J. Jesús Vargas Hernández. In addition to the individuals and organizations named earlier in the text, the success and richness in experience of this expedition were greatly enhanced by participants David Bates, Ernesto Franco, Richard Hawley, Carl Jackovich, Laurie Lippitt, Javier López Upton, Tadashi Moody, and Nicole Nedeff.

REFERENCES

Alazard, P., and Destremau, D.X., 1982, De l'expérimentation en France de *Pinus radiata*. Résultats préliminaires. *Ann. Rech. Sylvic.* **1981**:4–33.

Arteaga M., B., and Etchevers B., J.D., 1988, Pinus radiata *en México y el Mundo*, Colegio de Postgraduados, Chapingo, México.

Arteaga M., B., Etchevers B., J.D., and Volke H., V., 1985, Influencia de las características fisiográficas y edáficas en el crecimiento de *Pinus radiata* D. Don en Ayotoxtla, Guerrero. *Agrociencia* **60**:109–121.

Axelrod, D.I., 1983, *New Pleistocene Conifer Records, Coastal California*. University of California Publications in Geological Sciences. University of California Press, Berkeley, CA.

Axelrod, D.I., 1980, *History of the Maritime Closed-Cone Pines, Alta and Baja California*. University of California Press, Berkeley, CA.

Bannister, H.M. and McDonald, C.R.I.,1983, Turpentine composition of the pines of Guadalupe and Cedros Islands, Baja California. *N. Z. J. Bot.* **21**:373–377.

Berzunza, C.R., 1950, La Isla de Guadalupe. *Bol. Soc. Mex. Geogr. Estad.* **70**:7–62.

Bostic, D. A., 1975, *Natural History Guide to the Pacific Coast of North Central Baja California and Adjacent Islands*. Biological Educational Expeditions, Vista, CA.

Burdon, R. D., 1992, Genetic survey of *Pinus radiata*. 9: General discussion and implications for genetic management. *N. Z. J. For. Sci.* **22**:274–298.

Burdon, R. D., Bannister, M.H., and Low, C.B., 1992, Genetic survey of *Pinus radiata*. 2. Population comparisons for growth rate, disease resistance, and morphology. *N. Z. J. For. Sci.* **22**:138–159.

Calderón Flores, E., 1967. Experiencias con *Pinus radiata* en España y en Bosques de Chihuahua. *México y sus Bosques* **18**:25–27.

Cobb, E. W. Jr., and Libby, W. J., 1968, Susceptibility of Monterey, Guadalupe Island, Cedros Island, and Bishop Pines to *Scirrhia (Dothistroma) pini*, the cause of red band needle blight. *Phytopathology* **58**:88–90.

CONABIO (Comisión Nacional para el Conocimiento y uso de la Biodiversidad). 2000, http://conabio_web.conabio.gob.mx.

CONANP (Comisión Nacional de Áreas Naturales Protegidas, Secretaría del Medio Ambiente y Recursos Naturales), 2002, http://www.conanp.gob.mx/

CSIRO (Commonwealth Scientific and Industrial Research Organization), 1999. Benefits from CSIRO Research for the Forestry, Wood and Paper Industries Sector: Impact Analysis and Evaluation. CSIRO Forestry and Forest Products, Canberra, Australia.

Dawson, T. E., 1998, Fog in the California redwood forest: Ecosystem inputs and use by plants. *Oecologia* **117**: 476–485.

Dvorak, W. S., Jordan, A. P., Hodge, G. R., and Romero, J. L., 2000, Assessing evolutionary relationships of pines in the *Oocarpae* and *Australes* subsections using RAPD markers. *New Forests* **20**:163–192.

Eguiluz-Piedra, T., and Cibrián-Tovar, D., 1976, La roya en plantaciones de Pino. *Bosques y Fauna* (México) **13**:1–11.

Eldridge, K. G., 1997, Genetic resources of radiata pine in New Zealand and Australia, in: *IUFRO '97 Genetics of Radiata Pine. Proceedings of NZ FRI-IUFRO Conference and Workshop*, R. D. Burdon and J. M. Moore, eds., Forest Resources Institute Bulletin No. 203., Rotorua, New Zealand, pp. 26–41.

Fielding, J. M., 1961, The pines of Cedros Island, Mexico. *Aust. For.* **25**:62–65.

Gladstone, W. T., and Ledig, F. T., 1990, Reducing pressure on natural forests through high-yield forestry. *For. Ecol. Manage.* **35**:69–78.

Howell, J. T., 1942, A list of vascular plants from Guadalupe Island, Lower California. *Leaflets West. Bot.* **3**:145–155.

INE, Norma Oficial Mexicana NOM-059-ecol-1994, que Determina las Especies y Subespecies de Flora y Fauna Silvestres Terrestres y Acuáticas en Peligro de Extinción, Amenazadas, Raras y las Sujetas a Protección especial y que Establece Especificaciones para su Protección. Publicada por el Instituto Nacional de Ecología en el Diario Oficial de la Federación.

IUCN, (International Union for Conservation of Nature and Natural Resources), 2000, IUCN Red List. <http://www.redlist.org>.

IUCN, 1992. Protected Areas of the World: A Review of Natural Systems. Vol. 4. Neartic and Neotropical (Cambridge, UK.

Ledig, F. T., Vargas-Hernández, J. J., and Johnsen, K. H., 1998, The conservation of forest genetic resources. Case histories from Canada, México, and the United States. *J. Forest.* **96**:31–41.

Libby, W. J., 1978, The 1978 expedition to collect radiata seed from Cedros and Guadalupe Islands, *I.U.F.R.O. Working Party Newsletter No.* **2**: 9–12.

Libby, W. J., Bannister, M.H., and Linhart, Y. B., 1968, The pines of Cedros and Guadalupe Islands, *J. Forest.* **66**:846–853.

Matheson, A. C., 1980, Unexpectedly high frequencies of outcrossed seedlings among offspring from mixtures of self and cross pollen in *Pinus radiata* D. Don. *Aust. For. Res.* **10**:21–27.

Millar, C. I., Strauss, S. H., Conkle, M. T., and Westfall, R. D., 1988, Allozyme differentiation and biosystematics of the Californian closed-cone pines (*Pinus* subsect. *Oocarpae*). *Syst. Bot.* **13**:351–370.

Moran, R., 1996. *The Flora of Guadalupe Island, Mexico* (San Francisco, California: California Academy of Science, 190.

Moran, G. F., Bell, J. C., and Eldridge, K. G., 1988, The genetic structure and the conservation of the five natural populations of *Pinus radiata*, *Can. J. For. Res.* **18**:506–514.

Murphy, T. M., 1981, Immunochemical comparisons of seed proteins from populations of *Pinus radiata* (Pinaceae). *Am. J. Bot.* **68**:254–259.

Newcomb, G. B., 1959, The relationships of the pines of insular Baja California in: *Proceedings of the IX International Botanical Congress, Volume II, Abstracts* (Montreal, Quebec, Canada, 281.

Nicholls, J. W. P., and Eldridge, K. G., 1980, Variation in some wood and bark characteristics in provenances of *Pinus radiata* D. Don. *Aust. For. Res.* **10**:321–335.
Oberbauer, T., 1986, Baja California's Pacific Island jewels. *Fremontia* **14(1)**:3–5.
Old, K.M., Libby, W.J., Russell, J.H., and Eldridge, K.G., 1986, Genetic variability in susceptibility of *Pinus radiata* to western gall rust. *Silvae Genet.* **35**:145–149.
Osorio T., B., 1948, La Isla de Cedros, Baja California. Ensayo Monográfico. *Boletín Sociedad Mexicana Geografía y Estadística* **66**:319–402.
Peña B., V., and Cibrián T., D., 1980, Ciclo Biológico de *Dioryctria* Grupo *Baumhoferi* Heinrich y su Relación con *Cronartium sp.* en Plantaciones de *Pinus Radiata* D. Don. en Tequesquinahuac, México. *Revista Chapingo* **23–24**:3–10.
Perry, J. P. Jr., 1991, *The Pines of México and Central America*, Timber Press, Portland, Oregon.
Ravest S., G., 1983, Salvar Isla Guadalupe: Un Deber de Mexicanidad. *Revista Chapingo* **8**:5–45.
Rico C., J., 1983, Mapa de Vegetación de Isla Guadalupe, *Revista Chapingo* **8**:46–54.
Rico C., J., 1997a, La flora endémica de Isla Guadalupe, Baja California Norte, México: En peligro de extinción. *Revista Chapingo, Serie Ciencias Forestales* **1**:17–24.
Rico C., J., 1997b, Variación morfológica en *Pinus radiata* var. *binata* (Engelm.) Lemmon y *Pinus radiata* var. *cedrosensis* (Howell) Axelrod de isla Guadalupe e isla Cedros, B.C.N., México, Tesis Profesional, División de Ciencias Forestales, Universidad Autónoma Chapingo, México.
Salinas Q., R., and Gómez N., Ma. S., 1975, Enfermedades del *Pinus radiata* D. Don. INIF, SARH-SFF. Nota Técnica No. 8.
Rogers, D. L., 2002, *In situ* genetic conservation of Monterey pine (*Pinus radiata* D. Don): Information and recommendations, Report No. 26. Genetic Resources Conservation Program, University of California, Davis, CA.
Rogers, D. L., Matheson, A. C., Vargas-Hernández, J. J., and Guerra-Santos, J. J., 2004, Genetic conservation of insular populations of Monterey pine (*Pinus radiata* D. Don). *Biodiv. Cons.* In press.
Secretaría de Agricultura y Fomento (SAF), 1922., Acuerdo reservando la Isla Guadalupe, de la Baja California para el foemento y desarrollo de las riquezas naturales que contiene. Secretaria de Agricultura y Fomento, Diario Oficial de la Federación.
Secretaría de Agricultura y Fomento (SAF)., 1928., Acuerdo por el cual se declara zona de exclusión para la caza y pesca de especies animales y vegetales la Isla Guadalupe, Baja California y las aguas territoriales que la circundan (Secretaria de Agricultura y Fomento, Diario Oficial de la Federación.
Sedjo, R. A., 1999, The potential of high-yield plantation forestry for meeting timber needs. *New Forests* **17**:339–359.
Sutton, W. R. J., 1999, The need for planted forests and the example of radiata pine. *New Forests* **17**:95–109.
Weiss. S. B., Rich, P. M., Murphy, D. D., Calvert, W. H., and Ehrlich, P. R., 1991, Forest canopy structure at overwintering Monarch butterfly sites: Measurements with hemispherical photography. *Cons. Bio.* **5**:165–175.
Wu, J., Krutovskii, K. V., and Strauss, S. J., 1998, Abundant mitochondrial genome diversity, population differentiation, and evolution in pines. *Genetics* **150**:1605–1614.
Wu, J., Krutovskii, K. V., and Strauss, S. J., 1999, Nuclear DNA diversity, population differentiation, and phylogenetic relationships in the California closed-cone pines based on RAPD and allozyme markers. *Genome* **42**:893–908.

Chapter 3

BIODIVERSITY CONSERVATION IN BOLIVIA: HISTORY, TRENDS AND CHALLENGES

Pierre L. Ibisch[1]
[1] *Science Department, Fundación Amigos de la Naturaleza (FAN, Bolivia), former Integrated Expert, Center for International Migration and Development (CIM, Germany), Faculty of Forestry - University of Applied Sciences Eberswalde, Germany, Alfred-Möller-Str. 1, 16225 Eberswalde.*

Abstract: Bolivia is one of the most biological diverse countries of the world maintaining vast, intact humid and dry forest ecosystems; yet, it is the poorest country in South America where both poverty and development lead to biodiversity degradation and loss. Conservation efforts have evolved rapidly from the first species-protection-laws in the nineteenth century, to the creation of the first national park in the mid-twentieth century, to the implementation of the U.N. Convention on Biological Diversity, the formulation of a national biodiversity strategy, and a current national protected-area-coverage of 16%. However, there are severe conflicts with accelerating economic and development. Threats in the most sensible ecoregions (e.g. population shifts from the Andes to the forest lowlands, increasing agricultural activities, growing activities of the oil and gas sector, deforestation, climate change) represent important conservation problems. Based on the current institutional, social, economic and ecological situation, the paper describes the general challenges for future biodiversity conservation.

Key words: Bolivia; biodiversity; environmental policy; protected areas

1. INTRODUCTION: A CONSERVATION PROBLEM ARISING IN A MEGADIVERSITY COUNTRY

Bolivia is located in the center of South America, in a transition zone between humid tropical and dry subtropical climates including lowlands below 300 m above sea level (asl) as well as high mountains up to almost 7,000 m asl. It is one of the world's top biodiversity countries, containing virtually all tropical vegetation formations and a great variety of geologic formations (Ibisch et al., 2003a, compare map of ecoregions, Fig. 1). Bolivia is also one of the most forest-rich countries of the world (Ibisch, 1998) with more than 20,000 species of plants (Ibisch and Beck, 2003), 45% of all South American bird species (Herzog, 2003), 356 species of mammals (Salazar and Emmons, 2003), and more than 200 species of amphibians (Reichle, 2003).

More than 50% of the country is characterized by ecosystems with good or excellent conservation status (Fig. 2, Ibisch et al., 2003b) despite the fact that what is today Bolivia has been occupied since prehistoric times. Some (semi)arid Andean ecosystems have been used for thousands of years and are today severely degraded (Seibert, 1993).

Bolivia is today the poorest country of South America, and both poverty and development has led to biodiversity degradation and loss (Ibisch, 1998; UNFPA[1]). The current population density is rather low (8.25 million inhabitants in a territory of more than 1 million km^2)[2]; while population growth is rather high and geographically uneven. The highest population density is in the Andes but because of migration, growth rates are much higher in the lowlands. For example, in the last 50 years, population has multiplied by eight in the lowland Department of Santa Cruz, while it has lost about 40% in the Andean Department of Potosí. There is also an increasing urbanization: today, more than half of the people live in urban centers. Transportation and production infrastructure are still poorly developed.

Despite general economic stagnation isolated regions are opening up to development. The most booming sector is the oil and gas industry. In the last three years, Bolivia has become the country with the largest known gas reservoirs of the continent. Several pipelines have been built to export the gas. The pipelines cut across intact forest ecosystems without road access and therefore may increase colonization of those areas.

In Bolivia, as in other countries, the presence or absence of roads is the key issue for conservation of the last wilderness areas. In the lowlands, people traditionally settled along the natural access ways, the large rivers. Rather poor soils and lack of access impeded early colonization of larger territories (Jones, 1995). Bolivia was excluded from the international market and the internal demand for lowland products like rice or cotton was insignificant (Hindery, 1997). Bolivia was a true *Andean* country until the opening up of three roads towards the lowlands, in the 1930s. In 1959 a road from La Paz reached Caranavi in the Andean foothills (Monheim,

---

[1] Website UNFPA (United Nations Population Fund) 2001. The state of world population 2001. Footprints and milestones: population and environmental change. New York (Social, demographic and economic indicators: http://www.unfpa.org/swp/2001/english/indicators/indicators2.html).
[2] Compare results from the census of 2001. Webpage INE (Instituto Nacional de Estadística, La Paz; www.ine.gov.bo): Censo Nacional de Población y Vivienda de 2001.

1965), today an important colonization area. In 1939 a road reached Villa Tunari in the Cochabamba Andean foothills: this was the starting point for the development of one of the most important colonization and drug production areas of Bolivia, the Chapare region. Until the 1990s, immigration was rather high. An annual deforestation rate of about 3% between 1990 and 1998 was among the highest in the country (Bruckner et al., 2000).

A third road was constructed from Cochabamba to Santa Cruz (1954), initiating the development of the Santa Cruz area. The city of Santa Cruz grew from about 50,000 inhabitants in 1950 (Montes de Oca, 1989) to about 1.5 million today. From 1955 onwards, peasants from Japan and Mennonites from North and Central America were invited by the Bolivian government to promote the development of the lowlands, converting virgin forest areas into cash crop farms using modern technology (Demeure, 1999; Steininger et al., 2001a). In the 1990s the Cochabamba-Chapare road was finished, establishing an important transport corridor that facilitated a great deal of Andean-parallel rain forest destruction. A development and deforestation boom started with development projects in the 1980s in the Santa Cruz area. From the mid-1980s to the end of the 1990s, the deforestation rate increased by 200% (CUMAT, 1992; Pacheco, 1997; Steininger et al., 2001a, b). In the Department of Santa Cruz, about 10% of the original forest has been eliminated (Steininger et al., 2001a,b). The Chiquitano Dry Forest has the highest deforestation rates of the world. Deforestation accelerated from 160 km^2/year in the early 1980s to 1,200 km^2/year in the late 1990s (Steininger et al., 2001a). According Camacho et al. (2001), at least three originally forest-covered municipalities in the western part of the ecoregion were completely deforested and from 1993 to 2000 deforestation increased by 80%, and 45% of the deforestation happened in those years (= 1,424,035 ha). 42% of the deforestation corresponds to areas not suitable for agriculture where, according to the governmental land-use plan for the Department of Santa Cruz, forest was to be preserved for protection or production purposes. Although the Bolivian deforestation rate was overestimated for a time, currently it is the country with the highest deforestation rate per capita of the world (Ibisch, 1998).

The uneven distribution of population is the main reason why many remote areas are still virgin. In the past, the lack of technology and access as a consequence of the human and economic poverty, has contributed more to biodiversity conservation than active conservation efforts (Ibisch, 1998, 2002). Nevertheless, the country has reached a turning point where active conservation is urgently required if healthy wilderness areas are to be maintained. Thus, it is time to review the history of biodiversity conservation in Bolivia and sketch the future trends and challenges.

2. HISTORY AND TRENDS OF BIODIVERSITY CONSERVATION IN BOLIVIA[3]

The first ever reported measures toward conservation in Bolivia date back to the fifteenth century. According to Wilkes and Lowdermilk (1938) an Inca governor commanded that forest relicts be protected because of soil degradation caused by the intensification of agriculture around the Lake Titicaca. The Incas were known for their sustainable agricultural techniques such as the implementation of large terrace systems on steep slopes.

Early Andean cultures had pantheistic beliefs regarding soils, plants, and animals (e.g., *pachamama* = mother earth), but the Incas were agrocentric. Thus, even before the Spanish *Conquista*, which accelerated deforestation due to demand for firewood and the massive introduction of European domesticated animals many Bolivian natural ecosystems were converted into agroecosystems.

2.1 From nineteenth century to beginning of the twentieth century

In 1826 and 1832 the first conservation laws were enacted and aimed at protecting the cascarilla-tree (*Cinchona officinalis* used for malaria treatment) and the chinchilla (*Chinchilla brevicaudata*) an animal highly prized for its fur. In 1906, the hunting of chinchillas was prohibited (Marconi, 1991) and in 1918, the export of chinchilla and vicuña fur (*Vicugna vicugna*) was forbidden (in the case of the chinchilla, without any success, Suárez 1985, being the only Bolivian mammal extinct in the wild).

2.2 1930s

In 1939, the first Bolivian protected area was created: the Sajama National Park around the Sajama volcano (Marconi, 1991), the highest mountain of Bolivia where the world's highest tree formations have evolved. The creation of the protected area was motivated by the need to protect firewood. In the same year, an area in Nor Lípez, Potosí, was declared a protection area for the chinchillas.

---

[3] Partially reproduced in Ibisch (2003a).

2.3   1950s

The German ecologist Heinz Ellenberg (1958) began discussion about the naturalness of the high Andean forest-free areas that continues today (e.g., Seibert, 1993; Kessler and Driesch, 1994; Kessler, 1995). From the 1950s onwards, forestry projects were implemented in Bolivia, especially in order to protect water and soils (e.g. La Paz, Cochabamba), but only exotic species like *Eucalpyptus globules* were used (Stolz et al., 1986). Native tree species were not utilized before the 1990s.

2.4   1960s

In 1960/1961 a decree was issued to protect lowland animal resources like crocodiles and in 1965 the first lowland protected area was created: the large Isiboro Secure National Park (1.2 million ha; Suárez, 1985). However, during decades this park existed just as a 'paper park' without management and protection.

2.5   1970s

The notion of sustainable development was introduced into international development cooperation. German cooperation turned from classical afforestation projects to more integrated erosion protection ones and institution building: a forestry school was established in Cochabamba and an ecology institute in La Paz (1978; Morales, 1982; Stolz et al., 1986). The latter was mainly proposed and promoted by Ellenberg whose ecological research made clear the necessity of planning a more sustainable development (Ellenberg, 1981). The creation of the National Ecology Institute marks the beginning of the systematic research of Bolivian flora and fauna that soon was supported by other international institutions. Many of the leading Bolivian biologists were trained there. In 1979 Bolivia ratified CITES.

2.6   1980s

Nature conservation became more concrete and intensive. In 1982 the Beni Biological Station was created. For some time it was considered as a model protected area funded, in part, by the world's first debt-for-nature-swap. Four years later it was declared as a *Man and the Biosphere*-UNESCO reserve (Salinas, 1995). In 1982, the first ecological journal of Bolivia was established: *Ecología en Bolivia*.

The importance of lowland natural resources was recognized and the protected area concept was discussed more intensively. In the mid-80s, about 6.4% of the country was legally protected, but only three national parks existed: Isiboro Secure, Huanchaca (today, Noel Kempff Mercado), and Amboró. These protected areas were established without applying scientific criteria, planning methods, or appropriate management until 1985 (Stolz et al., 1986; Baudoin, 1995). In 1986, the first strategic analysis of Bolivian protected areas is carried out in order to prioritize areas under threats (Hanagarth and Arce, 1986); however, it failed to have any effect on the development of the national protected area system.

Up to this point, almost exclusively, international institutions had promoted the conservation idea in Bolivia (especially IUCN). Since the national government could not carry out biodiversity conservation adequately, several NGOs were established, such as the CDC in La Paz, FAN in Santa Cruz, and PROMETA in Tarija. Initially they were mainly dedicated to the promotion of protected areas, but soon became players in the development of the national environmental agenda. In 1985 the national environmental umbrella organization LIDEMA (*Liga para la Defensa del Medio Ambiente*) was founded (Marconi, 1992b).

2.7   1990s

The NGO Conservation International promoted knowledge of remote areas by organizing its rapid assessment program (RAP) (e.g., Killeen and Schulenberg, 1998). These studies were instrumental in pushing forward protected area proposals or were even the starting point of new areas. After the World Conservation Congress of 1992, the CDC edited the pioneer publication "Conservation of the biological diversity of Bolivia" (Marconi, 1992a). It was the first Bolivian publication that used the terms biodiversity and conservation and for the first time, a Geographic Information System was used to produce thematic maps. Barthlott and Winiger (1998) was the first international publication that gives an overview of several aspects of the biodiversity of Bolivia.

In 1990 the government banned hunting and collecting of wild animals and plants, modifying a similar 1979 law (Marconi, 1991). This action led to a significant decrease in hunting, especially of mammals; the feline populations, for example, recovered visibly. In 1992, the first environment law was passed, and a biodiversity law draft was proposed to the national congress (not yet enacted) (Marconi, 1992b). In the same year the first specialized governmental office for biodiversity was established, the National Directorate for the Conservation of Biodiversity (DNCB 1997). Before this, the only environmental authority was the Center for Forest Development (CDF) which was, of course, more oriented towards timber production. The first government of Sánchez de Lozada (1993-1997) created a Ministry for Sustainable Development and Environment. A forestry law with strict environmental regulations was passed, and a National Institute for Agrarian Reform (INRA) was established. In 1994 Bolivia ratified the biodiversity convention.

For the first time regional land-use planning was done in Bolivia. The first land-use plan was prepared for the Department of Santa Cruz. Although its main purpose was to regulate rural development taking into account land-use potential, biodiversity conservation was promoted by proposing new protected areas (Chaco National Park, Otuquis National Park, and San Matías reserve).[4] They were decreed in 1994 and land-use plans were prepared for other lowland departments.

In 1995 two of the largest national parks of Bolivia were established (Madidi and Chaco, 1.9 million ha and 3.4 million ha respectively). The Chaco National Park is the largest dry forest reserve of the world (Baudoin, 1995). In the same year, the government allowed the participation of the civil society in protected area management when FAN was asked to co-manage the Noel Kempff Mercado National Park (Justiniano, 1998; Sánchez de Lozada, 1998). The co-management was clearly successful, especially related to the development of funding mechanisms. The park was enlarged and supported significantly through a project of the UN-Framework Convention on Climatic Change. In the context of this project and other initiatives, the park gained international attention. In 2000 it was a model project of the World Exposition in Hanover and was declared the first Bolivian UNESCO Natural Heritage Site. Despite these successes, a strong mistrust towards national NGOs persists in governmental circles.

The most immediate consequences of the biodiversity convention were decrees on the access to genetic resources and on biosafety in 1997 (yet to be enforced). 1997 marks the beginning of the latest phase of conservation policy in Bolivia. The change in government led to a loss of prominence of environmental issues as part of the national political agenda: for example, the above-mentioned environmental ministry was renamed the Ministry for Sustainable Development and Planning. Also, several donors began to give priority to other projects.

The biodiversity directorate, the former DNCB, was divided into two authorities: the General Directorate for Biodiversity (*Dirección General de Biodiversidad*, DGB) and the National Service for Protected Areas (*Servicio Nacional de Areas Protegidas*, SERNAP). As a consequence the national

---

[4] Compare, e.g., unpublished report: Navarro, G. 1992. Estudio de parques nacionales y otras áreas protegidas. Informe final. Proyecto de Protección de los Recursos Naturales en el Departamento de Santa Cruz (Componente Proyecto Tierras Bajas). Cooperación financiera del Gobierno Alemán. CORDECRUZ – KFW – Consorcio IP/SCG/KWC, Santa Cruz.

biodiversity authority as artificially split, impeding integral approaches. This was especially counterproductive given the increasing need to establish a regional biodiversity vision and to coordinate actions involving protected areas and non-protected areas. In the mid-1990s international cooperation started projects and funding programs supporting the national protected area system (e.g. GTZ: MAPZA bufferzone project). Support from international donors strengthened the SERNAP while the DGB was weakened. Nevertheless, the DGB promoted the implementation of the Biodiversity Convention. In 1998, at the Conference of the Parties in Bratislava, the first national report on convention implementation was presented with the participation of non-governmental organizations. From 1999 onwards, a national biodiversity strategy was elaborated (MDSP, 2002). A regional biodiversity strategy among Andean countries that was elaborated shortly afterwards did not have a similar effect and participation.

2.8 2000 onwards

Independently of governmental efforts, the discussion of regional and transnational ecosystem conservation became significant. From the late 1990s onwards there was a coordination of bi-national protection activities (e.g. La Paz: Madidi, Peruvian border; Oruro: Sajama, Chilean border; Tarija: Tariquia, Argentinean border; Santa Cruz: Noel Kempff Mercado, Brazilian border). The World Wildlife Fund and others introduced ecoregional and corridor planning (Ibisch et al., 2001).

At the same time the original monopoly of institutions from La Paz is fading while strong players are emerging in Santa Cruz and Tarija. Some indigenous peoples have become actual or potential allies because they are the legal authorities of large territories (TCO: *Tierras Comunitarias de Origen*). In some cases they promote protected area management (e.g. *Capitania Bajo y Alto Izozog*, CABI, managing the Chaco National Park Kaa Iya). The 'municipalization' of the country is an important process, giving more political and economic power to smaller and more operational administration units that are more accountable. They will be responsible for the small-scale land-use planning. Recently, the first municipal protected area was created: the Tucavaca Valley and Santiago Mountains area (Robison et al., 2002). The Bolivian protected area system (16% of the national territory) is characterized by integrating several very large reserves. Only 6% of the protected areas of the world are larger than 1 million ha (Green and Paine, 1999). In Bolivia there are six protected sites belonging to this category: Gran Chaco (about 3.5 million ha), San Matías (almost 3 million ha), Madidi-Apolobamba and Pilón Lajas complex (almost 2.8 million ha), Noel Kempff Mercado (about 1.5 Mio. ha), and Amboró/Carrasco complex (about 1.25 million ha). The best represented ecological regions are the subandean Amazon rain forests (belonging to the southwest Amazon ecoregion) with 58% covered by protected areas. Other well represented regions are Pantanal savannas (40.7%), Cerrado (39%), Yungas (39.2%), Yungas Páramo (33%), and Gran Chaco (30.8%). Low protected area coverage, e.g. in degraded Andean puna regions, are not necessarily cause for concern. However, there are some regions where representation should be improved, especially to guarantee long-term viability of taxa and ecological functionality (e.g. Amazon gallery forests, Ibisch, 2003b).

The creation of the Tucavaca Valley and Santiago Mountains Municipal Protected Area was made economically possible by a foundation created by environmental organizations and energy companies (Ibisch et al., 2002b). Especially since the oil and gas industry has become so significant, affecting valuable and intact ecosystems, first experiences have shown how these companies can participate in biodiversity conservation or mitigation projects. Industrial activities and the involvement of industrial sectors in conservation are polemic. In recent years the conservationist movement in Bolivia has split into a fundamentalist faction on one side and a pragmatic one on the other. The Chaco National Park was the first protected area to be affected by a pipeline and receive compensation funding. Plans to explore oil reservoirs by The Andina corporation within their legally assigned concession in the Amboró National Park were rejected (after strong public outcry). Simultaneously, concessions were awarded and exploited in other protected areas. Unfortunately, legal provisions for protection of high biodiversity areas with intact ecosystems are superseded by many oil and gas concessions (Fig. 3).

Biodiversity conservation in Bolivia has been initially relatively conflict-free and several large protected areas could be established in areas with a low population density (Fig. 4) while managers trying to involve local communities to minimize conflicts. Technical assistance is given to communities in protected area buffer zones and strategies are developed to give alternative income to the local peasants. But now conservation faces a growing problem: on the one hand, landless peasants seek free areas to colonize; on the other, economic activities increasingly threaten protected areas or last wilderness regions. The clash of economic interests with conservation is now a national issue.

The conservation sector has also been divided on the main objective of the National Biodiversity Strategy. The government proposed that the strategic goal should be to develop the economic potential of biodiversity to realize its economically sustainable conservation goals. The philosophy of the governmental proposal was that the traditional sectors (many of them threatening biodiversity) would recognize the importance of biodiversity only if it is made clear that its conservation means economic growth. However, some of the more fundamentalist conservationists rejected this approach, preferring to stress the intrinsic value of biodiversity and proposed pure conservation for the sake of it. After a rather strong dispute, an all-inclusive compromise was reached. The strategic objective now reads "Develop the economic potential of the country's biological diversity, ensuring the conservation and sustainable use of ecosystems, species, and genetic resources. This will be achieved through the strengthening of the productive capacity of distinct actors and realization of a fair distribution of the benefits generated, with the goal of contributing towards national development through improving the quality of life in the population" (MDSP, 2002).

Although it is important to achieve this objective, it may be financially unfeasible. Another strategy is to force the development sectors that cause biodiversity damages or loss to internalize in their activities adequate mitigation and/or compensation costs. An important precedent for the recognition of additional environmental costs not identified by the national legislation has initiated a model process in the area of the Chiquitano Dry Forest ecoregion. There, in the context of the construction of a gas pipeline an environmental NGO showed that the most serious impact of the pipeline would be the access road created by it promoting rural development and deforestation of the intact dry forest. Because oil companies in Bolivia are not required to mitigate or compensate secondary impacts, the oil companies together with the environmental NGO, made a voluntary commitment in addition to environmental and social compensation and mitigation measures that are legally required in the framework of the construction of the pipeline. Following the governmental approval of the gas pipeline construction, a private foundation was created in order to address ecoregional conservation problems arising in the context of the infrastructure development in the region around the pipeline: the *Fundación para la Conservación del Bosque Chiquitano* (FCBC – Foundation for the Conservation of the Chiquitano Forest). Four environmental organizations - *Fundación Amigos de la Naturaleza* (FAN-Bolivia), *Fundación Amigos del Museo de Historia Natural Noel Kempff*, Missouri Botanical Garden, Wildlife Conservation Society (WCS), and two energy companies, Shell and Enron, chartered the FCBC. Together they made a commitment to make financial contributions over a period of 15 years in order to design a "Conservation and Sustainable Development Plan" and to guarantee its long-term implementation. Thus, the companies accepted the role of regional development catalysts that acquires a certain responsibility to facilitate a more sustainable model of development. The FCBC has prepared a comprehensive ecoregional Conservation and Sustainable Development Plan and will finance projects to be executed by local stakeholders (Ibisch et al., 2002b). Thus, for the first time in Bolivia a private conservation institution actively promotes and supports regional land-use planning and biodiversity conservation.

The Chiquitano-FCBC episode is also important in the Bolivian history of biodiversity conservation because it was accompanied by the most intense and polemic public debate about oil companies involved in conservation and conservation institutions getting involved with oil companies. The conservationists who were willing to work with the oil companies were blamed for 'selling their soul' and intact forests, while they tried to make the best out of a pipeline project that would have been completed anyway.

Private investment in biodiversity conservation in Bolivia has taken place through the joint implementation of the Framework Convention on Climatic Change. The NGO *Fundación Amigos de la Naturaleza*, together with The Nature Conservancy and the Bolivian government, established a model project for the mitigation of greenhouse impacts through deforestation prevention (the above mentioned Noel Kempff Climate Action Project). After the creation of the new regulations of the Kyoto Protocol, forest conservation projects are not eligible in the first phase of the Clean Development Mechanism. However, the issue of the development of environmental biodiversity services in order to fund its conservation is hotly debated. The NGO PROMETA is exploring and exploiting the topic in the context of protected areas that conserve drinking water for urban centers.

The development of new funding sources is an important current challenge as the traditional views for conservation institutions has changed significantly. In recent years the overall funding has not increased while the number of conservation institutions has. The number of international conservation (donor) NGOs in Bolivia has increased in the last decade. In 1999, for example, WWF-Bolivia became a program office, establishing a larger staff and in 2002 The Nature Conservancy opened its first office in Bolivia.

The enormous efforts in biodiversity conservation sketched above are not well known internationally. Obviously, there is a need for Bolivian conservation actors to link more intensively with international institutions and initiatives.

At the national level, Bolivian society appreciates the national parks as wilderness areas without being well-informed about their full range of values. Biodiversity conservation has gained importance in the local press, especially when mentioned that Bolivia is among the most important biodiversity countries of the world. Biodiversity, however, is not a priority for the general public despite the fact that there is some severe environmental degradation in the arid Andes. Poor rural peasants are well aware of the decreasing supplies of water, soil, pasture, and firewood resources. They are, however, generally kept in a vicious circle of poverty, causing environmental degradation which then leads to more poverty (Ibisch, 1996, 2002). In 2001, the first 'war for water' took place in Cochabamba, an inter-Andean city surrounded by deforested slopes. In 2002, in the deforested and eroded area of La Paz a mud slide caused the death of dozens of people in the center of the city. Flooding and landslides in the rainy season frequently cut off the country's roads.

3. BIODIVERSITY CONSERVATION CHALLENGES

In many tropical countries conservationists are losing the battle against deforestation and extinction (e.g., Whitten et al., 2001). In Bolivia, a country with a very low population density and vast areas with practically virgin forests, there is still much hope for biodiversity. However, after recovering from extreme political instability that impeded investment and economic development, Bolivia is discovering new resources like gas while being depicted as a strategic bi-oceanic transport country, attracting development capital that is rapidly changing the country's realities without necessarily reducing poverty.

Conservation action in Bolivia should be successful once: (a) the best use is made of existing information on threats and values to prioritize most urgent actions and conservation investments, (b) biodiversity conservation becomes part of an integral policy and land-use planning, and, (c) more funding generated by diversifying sources and partners.

3.1 Best use of existing information on threats and values

Although Bolivia is the least biologically understood country of South America, biodiversity research has increased in the last two decades. Yet, there is an urgent need for scientific methods to be developed in order to obtain reliable estimates and predictions of biodiversity dimensions and distribution. There is a need for analysis of spatial socioeconomic patterns such as population density, road access (see, Fig. 2, Ibisch et al., 2002a; Ibisch et al., 2003b), and topographic and climatic diversity to indicate species richness (Ibisch et al., 2001). Biodiversity inventory with traditional methods must continue in order to test and improve the results of the prediction models. Another key challenge is the prediction of climatic change's impacts on biodiversity, especially in combination with possible socioeconomic and land-use changes, because local, regional and global climate changes may multiply their effects on habitat quality for many taxa.

Also, many donors prefer to support the most politically correct and fashionable actions when it comes to certain conservation tools and approaches. In this context protected areas seem to be somewhat old-fashioned and unappreciated. Sustainable use projects are favored, although often is has not been shown that they contribute effectively to conservation. A good example of recent conservation investments with questionable benefits for biodiversity is the large amount of money spent on discussing access to genetic resources without achieving their use.

Conservation projects often target local peasants without regarding more important but less accessible development sectors such as forestry, oil industry, mechanized cash-crop agriculture, and road construction. Often, the goal of projects is to increase the general acceptance of conservation activities instead of assuring the maintenance of unique biodiversity elements and ecological processes and functions. Therefore, it is essential to involve economic sectors like forestry at least in the conservation of ecosystem function. For example, it should be vital to maintain large forest areas in the humid lowlands to protect their climatic role.

3.2 Make biodiversity conservation a part of an integral policy and land-use plan

Probably the most important priority for the conservation sector is to ensure that the government maintains biodiversity conservation as a national priority. The ratification of the CBD and the elaboration of the National Biodiversity Strategy imply this recognition, but it must be acted upon. On the one hand, it is necessary to make environmental legislation more consistent and compatible with competing or controversial laws for other sectors; an integral biodiversity law should be enacted. On the other hand, 'biodiversity-friendly' development requires adequate national, regional, municipal and private property land-use and planning. In this context, the kind of development and resource extraction activities that can be allowed in protected areas and still intact ecosystems must be legally defined. Mitigation and conservation measures must be an integral part of development projects. Land-use planning also requires clear-cut land-tenure information. Municipalities and private land-owners must be empowered and supported in establishing conservation areas that may fill critical gaps in the national protected area system.

3.3 More funding from diversifying sources and partners

Today the traditional charitable funding sources are not sufficient to cover conservation needs. The two main financial challenges are to internalize the environmental costs of development activities and to develop environmental services that pay off. Bolivia does not assign developers their environmental duty in an adequate and sufficient way. Special attention must be given to oil and gas operations and to road construction projects.

Bolivia has the opportunity to be recognized internationally for both its biodiversity and conservation efforts, but that requires government commitment, political stability and good governance. The need for action is now. In 2003, the country immersed itself into a severe social and political crisis that could have led to anarchic chaos or loss of democracy. Easily, this could have meant a break of the systematic governmental and non-governmental conservation efforts. For some accidental and lucky reasons this did not happen. Until now conservation business is working as usually; e.g., a national gap analysis is carried out by the government and NGO. But social movements grow stronger, the territorial organization of the country might be re-organized, and conservation will require more and more good arguments and innovative strategies.

ACKNOWLEDGMENTS

The author, from 1997 until 2003, was an Integrated Expert with the Bolivian conservation NGO, F.A.N., is supported by the German government (CIM/GTZ), and thanks all persons and institutions that made possible the diverse insights and experiences related to Bolivian biodiversity conservation during his work in and for Bolivia, since 1991. Helena Robison revised the English manuscript. An anonymous reviewer provided valuable suggestions.

REFERENCES

Barthlott, W. and Winiger, M., eds., 1998, *Biodiversity - a Challenge for Development Research and Policy*, Springer-Verlag, Berlin, Germany.

Baudoin, M., 1995, Conservation of biological diversity in Bolivia, in: *National Parks without People? The South American Experience*, Amend S. and T., eds., IUCN, Quito, Ecuador, pp. 95-102.

Beck, S.G., 1998, Floristic inventory of Bolivia – an indispensable contribution to sustainable development?, in: *Biodiversity - a Challenge for Development Research and Policy*, Barthlott, W. and Winiger, M., eds., Springer-Verlag, Berlin, Germany, pp. 243-268.

Bruckner, A., Navarro, G. and Fereira, W.J., 2000, Evolución del paisaje y alternativas de ordenamiento sostenible en la región del Chapare, Bolivia, *Revista Boliviana de Ecología y Conservación Ambiental* **7**:47-65.

Camacho, O., Cordero, W., Martínez, I. and Rojas, D. 2001, *Tasa de Deforestación del Departamento de Santa Cruz, Bolivia 1993-2000*, BOLFOR, Superintendencia Forestal, Santa Cruz, Bolivia.

CUMAT (Centro de Investigaciones de la Capacidad de Uso Mayor de la Tierra), 1992, *Desbosque de la Amazonia Boliviana*, La Paz, Bolivia.

3. Biodiversity Conservation in Bolivia

Demeure, J., 1999, Agricultura, in: *Bolivia en el Siglo XX. La Formación de la Bolivia Contemporánea*, Campero, F., ed., Harvard Club de Bolivia, La Paz, Bolivia, pp. 269-290.

DNCB (Dirección Nacional para la Conservación de la Biodiversidad), 1997, *Working for the Conservation of Bolivian Biodiversity*, La Paz, Bolivia.

Ellenberg, H., 1958, Wald oder Steppe? Die natürliche Pflanzendecke Perus, *Umschau* **1958**:645-648, 679-681.

Ellenberg, H., 1981, *Desarrollar sin Destruir. Respuestas de un Ecólogo a 15 Preguntas de Agrónomos y Planificadores Bolivianos*, Instituto de Ecología, La Paz, Bolivia.

Fjeldså, J. and Rahbek, C., 1998, Priorities for conservation in Bolivia, illustrated by a continent-wide analysis of bird distributions, in: *Biodiversity - a Challenge for Development Research and Policy*, Barthlott, W. and Winiger, M., eds., Springer-Verlag, Berlin, Germany, pp. 313-328.

Green, M.J. and Paine, J., 1999, State of the world's protected areas at the end of the 20th century, in: *Partnerships for Protection. New Strategies for Planning and Management for Protected Area*, Stolton, S. and Dudley, N., eds., Earthscan Publications, London, UK, pp. 18-28.

Hanagarth, W. and Arce, J.P., 1986, La situación de los parques nacionales y reservas de vida silvestre en el Departamento de La Paz, en el marco de una planificación regional, *Ecología en Bolivia* **9**:1-67.

Hanagarth, W. and Szwagrzak, A., 1998, Geoecology and biodiversity – problems and perspectives for the management of natural resources of Bolivia's forest and savanna ecosystems, in: *Biodiversity - a Challenge for Development Research and Policy*, Barthlott, W. and Winiger, M., eds., Springer-Verlag, Berlin, Germany, pp. 289-312.

Herzog, S., 2003, Aves, in: *Biodiversidad: Riqueza de Bolivia. Estado de Conocimiento y Conservación*, Ibisch, P.L. and Mérida, G., eds., Editorial FAN, Santa Cruz, Bolivia, pp. 141-145.

Hindery, D.L., 1997, *A History of Tropical Deforestation in the Bolivian Amazon.* Thesis, Faculty of Geography, University of California, Los Angeles, USA; http://thefuturemall.com/academics/theses/tblcntnt.html.

Hutterer, R., 1998, Diversity of mammals in Bolivia, in: *Biodiversity - a Challenge for Development Research and Policy*, Barthlott, W. and Winiger, M., eds., Springer-Verlag, Berlin, Germany, pp. 279-288.

Ibisch, P.L., 1996, "Reparieren" von degradierten Agrar-Ökosystemen in den Anden zwischen Theorie und Praxis - Beispiel Provinz Arque, Bolivien, *Kritische Ökologie* **14**(2):9-15.

Ibisch, P.L., 1998, Bolivia is a megadiversity country and a developing country in: *Biodiversity - a Challenge for Development Research and Policy*, Barthlott, W. and Winiger, M., eds., Springer-Verlag, Berlin, Germany, pp. 213-241.

Ibisch, P.L., 2002, Evaluation of a rural development project in southwest Cochabamba, Bolivia, and its agroforestry activities involving *Polylepis besseri* and other native species – a decade of lessons learned, *Ecotropica* **8**:205-218.

Ibisch, P.L., 2003a, Historia de la conservación de la biodiversidad en Bolivia, in: *Biodiversidad: Riqueza de Bolivia. Estado de Conocimiento y Conservación*, Ibisch, P.L. and Mérida, G., eds., Editorial FAN, Santa Cruz, Bolivia, pp. 348-357.

Ibisch, P.L., 2003b, Apuntes acerca de vacíos de protección en Bolivia, in: *Biodiversidad: Riqueza de Bolivia. Estado de Conocimiento y Conservación*, Ibisch, P.L. and Mérida, G., eds., Editorial FAN, Santa Cruz, Bolivia, pp. 391-417.

Ibisch, P.L. and Beck, S.G. 2003, Espermatófitas, in: *Biodiversidad: Riqueza de Bolivia. Estado de Conocimiento y Conservación*, Ibisch, P.L. and Mérida, G., eds., Editorial FAN, Santa Cruz, Bolivia, 103-112.

Ibisch, P.L., Beck, S.G., Gerkmann, B. and Carretero, A., 2003, Ecoregiones y ecosistemas, in: *Biodiversidad: Riqueza de Bolivia. Estado de Conocimiento y Conservación*, Ibisch,

P.L. and Mérida, G., eds., Editorial FAN, Santa Cruz, Bolivia, pp. 47-88. (Ibisch et al., 2003a)

Ibisch, P.L., Chive, J.C., Espinoza, S.D. and Araujo, N.V., 2003, Hacia un mapa del estado de conservación de los ecosistemas de Bolivia, in: *Biodiversidad: Riqueza de Bolivia. Estado de Conocimiento y Conservación*, Ibisch, P.L. and Mérida, G., eds., Editorial FAN, Santa Cruz, Bolivia, pp. 264-271. (Ibisch et al., 2003b)

Ibisch, P.L., Columba, K. and Reichle, S., eds., 2002, *Plan de Conservación y Desarrollo Sostenible para el Bosque Seco Chiquitano, Cerrado y Pantanal Boliviano*, Editorial FAN, Santa Cruz, Bolivia. (Ibisch et al., 2002b)

Ibisch, P.L., Müller, R. and Nowicki, C., 2001, El bio-corredor Amboró-Madidi – primeros insumos botánicos para un Plan de Conservación, *Revista de la Sociedad Boliviana de Botánica* 3(1/2):64-103.

Ibisch, P. L., Nowicki, C., Gonzáles, R., Oberfrank, T., Specht, C., Araujo, N. and Minkowski, K., 2000, Identification of conservation priorities in the Bolivian Amazon - A new biological-socioeconomic methodology using GIS, in: *Knowledge Partnership. Challenges and Perspectives for Research and Education at the Turn of the Millenium. 14.-15.10.1999. Session Biodiversity, Nature Conservation and Development*, Proceedings "Deutscher Tropentag 1999", Humboldt-University and ATSAF, Berlin, Germany (CD-ROM).

Ibisch, P.L., Nowicki, C., Müller, R. and Araujo, N., 2002, Methods for the assessment of habitat and species conservation status in data-poor countries - case study of the Pleurothallidinae (Orchidaceae) of the Andean rain forests of Bolivia, in: *Proceedings of the First International congress "Conservation of Biodiversity in the Andes and the Amazon Basin", 24.-28.09.2001, Cusco, Peru,* Bussmann, R.W. and S. Lange, eds., pp. 225-246. (Ibisch et al., 2002a).

Jones, J.C., 1995. Enviromental destruction, ethnic discrimination, and international aid in Bolivia. In: Painter, M. and W.H. Durham (eds.): The social causes of environmental destruction in Latin America. The University of Michigan Press, Ann Arbor, USA: 162-216.

Justiniano, H., 1998, Friends of Nature Foundation (FAN Bolivia) – the involvement of NGOs in conservation in Bolivia, in: *Biodiversity - a Challenge for Development Research and Policy,* Barthlott, W. and Winiger, M., eds., Springer-Verlag, Berlin, Germany, pp. 389-398.

Kessler, M., 1995, *Polylepis-Wälder Boliviens: Taxa, Ökologie, Verbreitung und Geschichte*, Dissertationes Botanicae, Vol. 246. J. Cramer in der Gebr. Borntraeger Verlagsbuchhandlung, Berlin, Stuttgart, Germany.

Kessler, M., 1998, Land use, economy and the conservation of biodiversity of high-Andean forests in Bolivia, in: *Biodiversity - a Challenge for Development Research and Policy,* Barthlott, W. and Winiger, M., eds., Springer-Verlag, Berlin, Germany, pp. 339-352.

Kessler, M. and Driesch, P. 1994, Causas e historia de la destrucción de bosques altoandinos en Bolivia, *Ecología en Bolivia* 21:1-18.

Killeen, T.J. and Schulenberg, T.S., eds., 1998, *A Biological Assessment of the Parque Nacional Noel Kempff Mercado*, RAP Working Papers Number 10, Conservation International, Washington D.C., USA.

Köhler, J., Lötters, S. and Reichle, S., 1998, Amphibian species diversity in Bolivia, in: *Biodiversity - a Challenge for Development Research and Policy,* Barthlott, W. and Winiger, M., eds., Springer-Verlag, Berlin, Germany, pp. 329-338.

Marconi, M., 1991, *Catálogo de Legislación Ambiental en Bolivia*, Centro de Datos para la Conservación (CDC-Bolivia), La Paz, Bolivia.

Marconi, M., 1992a, ed., *Conservación de la Diversidad Biológica en Bolivia*, CDC, La Paz, Bolivia.

Marconi, M., 1992b, Marco jurídico e institucional de la conservación, in: Marconi, M. (eds.): *Conservación de la Diversidad Biológica en Bolivia,* CDC, La Paz, Bolivia, pp. 391-438.

MDSP (Ministerio de Desarrollo Sostenible y Planificación), 2001, *El Proceso de Participación en el Diseño y Formulación de la Estrategia Nacional de Conservación y Uso Sostenible de la Biodiversidad de Bolivia*, La Paz, Bolivia.
MDSP, 2002, *Estrategia Nacional de Biodiversidad*, La Paz, Bolivia.
Miller, K., 1980, *Planificación de Parques Nacionales para el Ecodesarrollo en Latinoamérica*, FEPMA (Fundación para la Ecología y la Protección del Medio Ambiente, Spain).
Monheim, F., 1965, *Junge Indianerkolonisation in den Tiefländern Ostboliviens*, Georg-Westermann-Verlag, Braunschweig, Germany.
Montes de Oca, I., 1989, *Geografía y Recursos Naturales de Bolivia*. 1. edition, La Paz, Bolivia.
Moraes, M., 1998, Richness and utilization of palms in Bolivia – some essential criteria for their management, in: *Biodiversity - a Challenge for Development Research and Policy*, Barthlott, W. and Winiger, M., eds., Springer-Verlag, Berlin, Germany, pp. 269-278.
Morales de, C., 1982, Presentación de Instituto de Ecología, *Ecología en Bolivia* **1**:5-7.
Pacheco, P., 1997, *La Deforestación y la Degradación de los Bosques en las Tierras Bajas*, La Paz, Bolivia.
Reichle, S., 2003, Anfibios, in: *Biodiversidad: Riqueza de Bolivia. Estado de Conocimiento y Conservación*, Ibisch, P.L. and Mérida, G., eds., Editorial FAN, Santa Cruz, Bolivia, pp. 133-137.
Robison, D., Salas, H., Linzer, K., Saucedo, R. and Balcazar, K., eds., 2002, *Plan de Manejo de la Reserva Municipal del Valle de Tucavaca*, F.A.N. and Municipio de Roboré, Editorial F.A.N., Santa Cruz, Bolivia.
Salazar, J. and Emmons, L. 2003, Mamíferos, in: *Biodiversidad: Riqueza de Bolivia. Estado de Conocimiento y Conservación*, Ibisch, P.L. and Mérida, G., eds., Editorial FAN, Santa Cruz, Bolivia, pp. 146-148.
Salinas, E., 1995, Beni Biosphere Reserve and Biological Station: education and development, in: *National Parks without People? The South American Experience*, Amend S. and T., eds., IUCN, Quito, Ecuador: pp. 103-116.
Sánchez de Lozada, A., 1998, Conservation of biodiversity: a national task, in: *Biodiversity - a Challenge for Development Research and Policy*, Barthlott, W. and Winiger, M., eds., Springer-Verlag, Berlin, Germany, pp. 369-388.
Seibert, P., 1993, Vegetation und Mensch in Südamerika aus historischer Sicht, *Phytocoenologia* **23**:457-498.
Steininger, M.K., Tucker, C.J., Townsend, J.G., Killeen, T.J., Desch, A., Bell, V. and Ersts, P., 2001, Tropical deforestation in the Bolivian Amazon, *Environmental Conservation* **28**:127-134. (Steininger et al., 2001b).
Steininger, M.K., Tucker, C.J., Ersts, P., Killeen, T.J., Villegas, Z. and Hecht, S.B., 2001, Clearance and fragmentation of tropical deciduous forest in the Tierras Bajas, Santa Cruz, Bolivia, *Conservation Biology* **15**:856-866. (Steininger et al., 2001a).
Stolz, R., Beck, S., Espig, G., Hanagarth, W. and Roth, H.H., 1986, *Möglichkeiten zur Nutzung der Tropenwaldressourcen in Nord- und Ostbolivien unter Einbeziehung ökologischer Aspekte*, Forschungsauftrag des Bundesministeriums für wirtschaftliche Zusammenarbeit, Bonn, Germany.
Suárez M., O., 1985, *Parques Nacionales y Afines de Bolivia*, La Paz, Bolivia.
Villwock, W., 1998, Using Lake Titicaca's biological resources – problems and alternatives, in: *Biodiversity - a Challenge for Development Research and Policy*, Barthlott, W. and Winiger, M., eds., Springer-Verlag, Berlin, Germany, pp. 353-368.
Whitten, T., Holmes, D. and MacKinnon, K., 2001, Conservation Biology: a displacement behaviour for academia? *Conservation Biology* **15**:1-3.
Wilkes, D.R. and Lowdermilk, W.C., 1938, Soil conservation in ancient Peru, *Soil Conservation* **4**(4):91-94.

PART 2: NATIONAL POLICIES, LOCAL COMMUNITIES, AND RURAL DEVELOPMENT

Chapter 4

PEASANT, ENVIRONMENT AND MAIZE "MODERNIZATION"
in Zacapoaxtla, Mexico, 1974-1982

Bert Kreitlow[1]
[1]*Department of History, Carroll College, 100 North East Avenue, Waukesha, WI, 53186*

Abstract: This chapter analyzes farm modernizing efforts placed in the 1970s by the Mexican national government in the Zacapoaxtla, Mexico region of Totonac and Nahua subsistence farmers. The case explores relations between environmental and political change, and the role that peasants play in that change. The postrevolutionary Mexican government began deliberately targeting peasants for such projects in the 1960s not only to increase food output in the country, but also to revive the regime's populist self-representation and to squelch rising dissent. The Zacapoaxtla project began in 1974 with the priority placed on "modernizing" maize. By "modernizing" is meant the introduction of scientific technology such as hybrid seed, synthetic pesticides, and synthetic fertilizer that was the standard toolbox of the Green Revolution modernization programs. However, after the first three years of the program, the top priority placed on maize was abandoned by the technicians implementing the program in favor of the goal to organize peasants into marketing and purchasing cooperatives. Northern Sierra peasants rejected most of these proposed farm technologies, partly because the methods were environmentally and economically inappropriate. In this sense, the Zacapoaxtla environment conditioned the state project and therefore participated in political change

Key words: Green Revolution; Postrevolutionary Mexico; maize

1. INTRODUCTION

The succession of men who claimed to rule Mexico in the aftermath of the 1910-1920 revolution confronted an urgent challenge. They not only needed to mold a nation, but also to feed it. Peasants and their farm fields would prove central to both tasks. That farm modernization played a significant role in the "revolutionary" Mexican regime of 1929 to 2000 is exemplified by the fact the nation hosted in the World War II era, with U.S. assistance, the gestation of the "Green Revolution." This well-known and controversial series of scientific agricultural research and promotion programs first targeted large-scale, commercial farms in Mexico. On these large commercial farms, U.S. technicians introduced and adjusted a package of technology that featured hybrid seed and synthetic chemical inputs.

However, by the 1960s and 1970s the Mexican national government dramatically changed direction in its agricultural policy and began to aim Green Revolution "innovation" not at large commercial operations, but at the humble peasant *milpa*, or corn field. Two contemporaneous crises provoked this dramatic change. One was a crisis of legitimacy for the ruling Institutional Revolutionary Party (PRI). Party leaders hoped to repair the PRI's tattered populist image after the government brutally repressed student and peasant militancy, of which the most infamous case was the massacre at the Tlateloco plaza of Mexico City in October, 1968. The second crisis was one of hunger. Despite the "revolutionary" boosts in Mexico's grain production over the first two decades of the farm modernizing programs, a burgeoning population in the 1960s was again outstripping those harvests (Hewitt de Alcántara, 1976; Fox, 1992, Ochoa, 2000).

This chapter analyzes the changing relations between Northern Sierra peasants, their farming environment, and agricultural technicians during the Plan Zacapoaxtla project of 1974 to 1982, one of the first Green Revolution-style projects that targeted subsistence-farming, indigenous peasants rather than large-scale commercial farmers. The very first example of this new focus on peasants took place in the Puebla valley beginning in 1967 and was called Plan Puebla. There the small farmers were more assimilated and less isolated than the farmers who participated in the subsequent Plan Zacapoaxtla project. Although the Plan Zacapoaxtla project officially extended past 1982, Mexico's severe financial crisis that began that year effectively ended the funding and staffing of the project. Its setting was seven municipalities in the mountainous northern region, sometimes called the Northern Sierra, of Puebla state. Here the city of Zacapoaxtla serves as a political and economic hub.[1] The primary objective stated at the project's outset was to increase harvests of maize and other subsistence crops, with maize being the dominant food source. "The principal objective [of Plan Zacapoaxtla] is that of promoting the increase in the income of the peasants that practice a rain-fed agriculture, through improving the quality and quantity of the harvest of Plan Zacapoaxtla area's most important crops,

---

[1] The seven municipalities, similar to counties, were Zacapoaxtla, Cuetzalan, Nauzontla, Xochitlán, Xochiapulco, Huitzilan and Zoquiapán.

which are maize, beans, potatoes and fruits" (Equipo Técnico, 1976). Due to the agricultural focus of the project, those project technicians sent to work with peasant farming villages found that their work engaged the soil, climate, seeds, and weeds of the farmers' fields as well as the peasants' social and cultural settings.

The project was also propelled by specific political aims. The Zacapoaxtla area was chosen as the site for this project in part due to land invasions and other peasant mobilization there in the early 1970s (Hewitt de Alcántara, 1976; Edelman, 1980). Both the officially expressed aim of the project and the reports by technicians stated the intention, beyond improving their farming, of more fully integrating peasants into national life (Equipo Técnico, 1975; Verdesoto Alvarez, 1977). Therefore, Plan Zacapoaxtla exemplifies the significant role that farm modernization programs played in the postrevolutionary Mexican state. Given that role, to rule the country became in part an environmental project as well as a political one.

After the first three years of the Plan Zacapoaxtla project in the Northern Sierra peasant communities, the technicians abandoned maize as the project's top priority. The technicians thereafter focused on organizing peasants into marketing cooperatives. The central question of this chapter concerns why maize farming was dropped as a top priority. Rather than undergoing a change that was imposed upon them and their biological environment, the Northern Sierra peasant communities and their environment took part in a more complex process. At the same time, the project would bring to bear changes on the Zacapoaxtla region's peasant communities, their relations with local elites, and their environment. It is important to resist any tendency to idealize the peasants and their environmental setting, since farming methods and overpopulation had depleted soil fertility and stripped forest cover before 1974 in many parts of the Northern Sierra, and those problems were to worsen after Plan Zacapoaxtla ended. Peasants and their mountain setting influenced the state project, and were likewise influenced by it.

2. TECHNICIANS' EARLY ENCOUNTERS

The seven original members of the technical team, all male graduate students with an average age of 24, arrived in the main city of Zacapoaxtla in the spring of 1974. They were supervised by the Colegio de Postgraduados, the graduate school of the National Agricultural College to which the national government agriculture ministry had given responsibility to run the program. As state agents, these project technicians stood in an ambivalent position between the national state and the peasants. On the one hand, the technicians shared long-standing motives with the national state that were central to the Green Revolution projects, such as an interest in more closely linking peasants to the central government and the desire to increase maize harvests. (Verdesoto Alvarez, 1977; Equipo Técnico, 1976, 1978). The technicians intended to attempt to introduce the Green Revolution package of hybrid seed and synthetic inputs. On the other hand, the aftereffects of the Tlateloco massacre on Mexican young people in general, and three contemporary intellectual currents specifically, seemed to instill in the technicians some skepticism toward the ruling regime. These currents were pro-peasant populism, inchoate environmentalism, and anti-U.S. dependency theory. One food policy analyst argued that both the populist swing of the national government and youth activism helped shape rural "modernization" programs after 1968.

> There were two results [of the massacre]. First, there was a political decision at the highest levels to take up the issue of popular participation in a democratic way, since the link between the base and the state had been dislocated or broken. That was one of the reasons the state tried to recover its social base by broadening democratization in certain policies and regions. Second, the people who went out to work in the countryside began to work at the grass-roots level to build independent, autonomous social movements . . . I don't think either that the organizing and democratizing happened spontaneously or that it came about as a result of the government's political posture; rather, the [reformist] position from above converged with a movement from below (Fox, 1992).

Such influences led the young technicians to be skeptical toward Green Revolution technology due to its origins in U.S. farming models. Such attitudes were expressed in the introductions to the theses of two technicians who researched peasant farming methods. These technicians also promoted higher esteem for peasant farming methods and for the peasants' own objectives, which contrasted with earlier agronomist attitudes toward peasant farming during the Green Revolution's infancy (Mora Aguilar, 1980, Chagra Guerrero, 1980). The attitudes of Plan Zacapoaxtla technicians mattered because in the coming period of interaction with peasants, the program's organizational scheme granted these technicians considerable flexibility and autonomy.

Another aspect of the technicians' attitudes was their animosity toward the Northern Sierra's nonindigenous elite and the political and economic control they exerted in the area. One Plan Zacapoaxtla technician, Sergio Mora Aguilar, was among those who criticized the local nonindigenous elites who were merchants based in the towns that sold goods to peasants at exploitative rates Mora Aguilar (Mora Aguilar, 1980). Another technician,

Alvaro Aguilar Rayon, noted with derision the ostentation and exploitation of local elites in his thesis:

> In the areas [of the Plan Zacapoaxtla area] of greater marginalization the contrast between merchants and farmers is accentuated: the first arrive to show off their riches, expressing [the socio-economic contrast] with new luxury autos, with celebrating uproarious parties, with the construction of large and expensive homes that mock the thousands of farmers who live in permanent misery residing in very modest huts. These merchants strongly oppose the development of their areas, seeing the roads as a threat to their interests, since farm products could escape by means of those roads or they could allow residents to "open their eyes" (Aguilar Rayon).

One coordinator among the technicians served as leader, while three others were in charge of farm research, two were responsible for disseminating information to farmers and one evaluated the program. The team also hired 25 assistants from the local area, largely for the purpose of helping technicians communicate with villagers in locally spoken Nahuatl and Totonac languages. The work began tenuously, and consisted of learning about the peasant communities and vice versa. The farm researchers first toured the area, observed their farming practices, yields, and pest problems. The maize fields had already sprouted tall stalks that were within weeks of maturity. The researchers were therefore limited that first year to experimenting with the response in one community's nearly mature fields to late applications of nitrogen fertilizer.

Those in charge of spreading information to the peasants also began with a tour. "During the spring and summer of 1974, the principal activities of the information staff were to inform farmers about Plan Zacapoaxtla and to obtain a general understanding of each one of the communities within the area through direct contact with local authorities, for the purpose of penetrating the communities and thereby working with those communities' farmers" (Equipo Técnico, 1976). The early efforts were energetic. By the end of the project's first year, the information specialists held 441 meetings in the rural communities, more than half the time with barely a dozen people in the audience. To draw whatever crowd they could, the staff used filmstrips at 55 of the meetings. The films were from the agricultural college. These were educational films with titles such as "Do You Want To Increase Your Maize Harvest?"

3. INTERCROPPING AND PRODUCTIVITY

The technicians came to the Northern Sierra with a goal that seemed clear at the outset: increase maize harvests. However, in the first weeks of the program, as the investigators walked the trails through the mountain landscape and saw how peasants farmed, the existing ecological setting began to undermine their original strategies. These strategies rested on the basic agronomic notion of productivity. Productivity, or yield, is easy to quantify according to agricultural sciences: how much of the desired product is harvested from a given area of land. Technicians learned early in their research, based on the few farmers who would talk to them in the first months, that productivity in the sense of maize yields was low. Farmers said they harvested barely over 1,000 kilograms per hectare (approximately 16 bushels per acre at 56 pounds/bushel), which is a yield only one-eighth as productive as a typical crop from Mexico's large, commercial maize farms.

As technicians estimated how much farmers produced, they noted the way farmers raised their maize crop as well. What technician Victor Hugo Chagra Guerrero observed in the peasants' fields led him and others to rethink what they understood "productivity" to mean. Northern Sierra farmers had evolved a system that was similar to many other tropical subsistence farmers, but which was contrary to the monoculture, one-crop-for-each-field, approach that defined U.S. commercial farming (Scott, 1998). The Northern Sierra farmers planted many crops on the same plot of ground and made wider use of each plant than merely its kernels or beans. In a majority of the fields, maize was planted with bean crops. A larger "pole" bean was commonly planted in the same hole in which farmers dropped by hand their four or five maize seeds.[2] The bean plant vine climbed the maize stalks. Farmers also planted a bush-style of bean that produced a small black bean in the soil between rows of maize.[3] Less common but still frequently encountered were the farmers who also planted squash on the margins of their field. When the total quantity of the varied human foods harvested from a given area was measured, it obviously exceeded the measure of productivity based solely on the quantity of maize kernels from that same area. In sum, although the yield of each plant in an intercropped field was less than if it were a plant in a monocrop field, a more complex notion of productivity was needed to account for the total quantity of the foods harvested from the intercropped field (Chavez Valencia, 1996; Papendick, et al., 1976).

Technicians noted as well that more products of value were taken from the typical peasant plot than just food. Some farmers used the leaves of the

---

[2] *phaseolus canavalia ensiformis*, known in Spanish as *haba*.
[3] Known in Spanish as *negro pitaleño*.

maize plant, for example, as forage for livestock (although the majority of peasants owned little or no livestock). Farmers often used the stalks as a building material--most often for fences. The cobs were burned as a fire fuel and the silks and tassels could be used in home remedies (*Secretaría de Educación Pública*, 1982). After peasants picked beans picked from their plants, the remaining leaves and stems could also be used as forage. Considering all the uses that an intercropped field served, the relatively low maize yields no longer alarmed technicians such as Chagra Guerrero.

The technicians soon noticed other advantages to the intercropping methods, once they began assessing the peasant methods from a wider perspective than a mere concern with grain yield. For example, the intercropped fields led to longer-term fertility because less of the plot was bare soil, and thus the soil, in most cases on a considerable slope, was less susceptible to erosion.[4] The presence of other crops also reduced pest problems and increased the efficiency of peasant labor, since the work necessary to maintain the bean crop arose at different times than the obligations for maize. Chagra Guerrero said that within the first year the technicians became favorably impressed with the system already in place, despite the soil fertility problems. Peasants "had a technology that was above what we were offering," Chagra Guerrero said (Chagra Guerrero, 1998).

The technicians adapted their research to reflect their interest and appreciation of the intercropped methods of raising maize and beans, as well as the rotation of maize and potatoes that was common in some areas.[5] For example, in the first summer technicians researched the response of intercropped maize plots to added fertilizer and compared those results with the response of maize planted alone. Another advantage of intercropping was already known by the agronomists, namely that bean plant roots fix nitrogen in the soil. Thus, the technicians were not surprised to determine in their experiments that maize planted with beans required less fertilizer. Chagra Guerrero concluded from his early studies of the Northern Sierra peasants' productive intercropped fields that these farming systems did not need to be significantly changed, but merely helped along with fertilizer.

---

[4] An exception to this general practice is found in the hotter, lower lands near Cuetzalan. Farmers in the Jonotla area at an elevation of about 900 meters reportedly do and did plant their maize fields leaving their soil very exposed, with a lack of other crops and with distances of one meter between maize plants. Agronomist Pedro Irigoyen speculates that the farmers believe the practice under the hot, tropical sun burns away weeds that would thrive in shadier fields (Irigoyen, 1998).

[5] Potatoes were common in an area centered around the municipality of Nauzontla that was midway in elevation between the higher Zacapoaxtla-Xochiapulco *tierra fría* and the lower Cuetzalan *tierra caliente*.

4. HYBRIDS VS. OPEN-POLLINATED SEED

Hybrid seed is developed in carefully controlled breeding sequences that isolate and emphasize desirable genetic traits such as high yield or disease resistance. These traits are combined and further enhanced in a final batch of genetically uniform seed by crossing the male pollen of the plants carrying a certain trait with the second batch of plants carrying other traits. In the U.S., the introduction of hybrid seed had more than doubled yields in the first half of the century. Unlike the traditional, non-hybrid seed, also called "open pollinated," the kernels harvested from hybrid seed are essentially sterile. This means that farmers balance the advantages of hybrid seed with the need to purchase it rather than save seed from a harvested crop.

In the first months of the project, the agronomists among the technicians began to research the potential of hybrid maize seed for increasing the peasants' harvests. The enthusiasm for hybrid seed as a tool for boosting peasant maize production was high in the early 1960s among the directors of worldwide farm modernization programs. For example, E. J. Wellhausen, the director of the Mexico-based International Center for the Improvement of Maize and Wheat (CIMMYT), told seed breeders at a conference in 1966:

> If in some additional way this variation could be concentrated in one single race [of maize] I think that we would be in a better position to develop some true super varieties in the future and probably achieve a revolution in the production of maize even greater than what occurred in the past . . . [despite possible pitfalls] you will be surprised by how great will be the dividends of a program of this nature (Wellhausen, 1966).

Nevertheless, Plan Zacapoaxtla technicians had low expectations for hybrids in the Northern Sierra environment even before they planted their first test plots. By the 1970s, the promise for improving peasant maize had lost much of its luster in the course of the previous Green Revolution project called Plan Puebla (Felstenhausen and Díaz-Cisneros, 1985; CIMMYT, 1967; Fox, 1992).

The need to purchase hybrid seed every year was one considerable disadvantage among poor peasant farmers. In addition, hybrid seed is genetically uniform and highly sensitive to changes in soil quality, climate, light, etc. Hybrid seed performance is usually vastly superior to open-pollinated seed, when the hybrid seed grows in the strict conditions for which it was bred. But that yield quickly falls below the open-pollinated seed's yield when environmental conditions stray outside the hybrids' optimum. In the Plan Puebla area, hybrids did poorly because rainfall and soil types varied too much, even over short distances. In the Northern Sierra, environmental conditions were even less uniform.

The Plan Zacapoaxtla technicians nonetheless started a series of experiments in breeding hybrid seed. They incorporated promising local seeds in their breeding so that the resulting seed would hopefully be adapted to at least some of the region's very diverse conditions. However, developing hybrids, especially the "double-cross" versions that were sought during Plan Zacapoaxtla, necessitated several growing seasons.[6] As expected, the results were disappointing in field trials compared to the yields of local, traditional seeds.

There were other, less obvious, advantages of local "open-pollinated" varieties in addition to the fact that they yielded more and provided seed for the next crop without cost. Farmers, in selecting their seeds, had helped to adapt the plants to their specific local conditions. The ecological system in which farmers participated was complex, and not easily "improved" out of a single intent to increase a seed's productivity. In the extremely wet conditions in the municipality of Cuetzalan, average rainfall exceeds 4,000 millimeters, or over 160 inches, per year. Just before the wettest months of July and August, farmers in Cuetzalan, as throughout the Northern Sierra, bend the stalks downward below the ear, so that the ear then points its end toward the ground. The ear is then left in this position on the stalk for two more months while the kernels dry. Technicians failed to appreciate that the husks of local varieties were longer and clung more tightly to the tip of the ear. In the technicians' field trials of nonnative seed, or seed that mixed both outside and native genetic material, the husks around the ear did not completely and tightly cover the very tips of the ears. As a result, the thick moisture of early August penetrated and rotted the ears before they could be harvested (Aguilar Ayon, 1998). Birds could also more easily pillage the kernels.

The futility of using outside seed was obvious to local farmers from the outset. Their world was mountainous and varied. Their ecological sense assumed diversity, even over short distances. Farmer José Miguel de los Santos of San Miguel Tzinacapan, who does not speak Spanish well, understood immediately why the seed that technicians brought with them from far away would struggle in their soil, producing plants much inferior to those from the seed de los Santos kept from his field's previous harvest. *"Cada semilla, allí,"* De Los Santos said, "every seed, from right there (De Los Santos, 1998). Growers in the village of Ixehuatl, just down the slope from the ridge top where Xochiapulco perches, were interviewed one morning, standing in the square of their town. The valley bearing maize

---

[6] A double cross is the product of crossing two so-called "single cross" hybrids that are the product of inbred (homozygous) lines.

fields dropped away behind them to the Apulco River. One farmer gestured around them and explained, as if it were obvious, that a farmer cannot even bring seed from the next village down the road and expect it to do well there in Ixehuatl. Seed from outside "won't take" (*no pega*), they agreed (Cantero, et al., 1998).

To no one's surprise, then, project leaders abandoned the hybrid breeding research in Zacapoaxtla after two years. Plant breeders, for example, had not expected any spectacular results due to the disappointing outcome of the previous Plan Puebla's hybrid research work and their sense – confirmed upon arrival – that the Northern Sierra presented even more extreme ecological diversity than the variations that had doomed the prospect for hybrids in the Plan Puebla area. What they had not expected, but which further diminished the feeble enthusiasm for hybrids, were the superior adaptations of local open-pollinated varieties to conditions the scientists had not even considered, such as the stifling humidity and pillaging birds. Peasants, for their part, were likewise not surprised because they had always put great effort into selecting and caring for their maize seed, and knew as surely as they knew their local mountains that alien seed *"no pega"* (Valadez-Ramirez, et al., 1998). The result was that, by 1976, precisely as the overall emphasis of the technicians' efforts was steering away from a preoccupation with maize production, the staff reported as a hypothesis to guide the coming year that "the seed selection practiced by farmers is the most effective" (Equipo Técnico, 1976b).

5. PESTICIDES VS. TURKEYS AND HOES

The use of synthetic pesticides was the second type of modern technology that most past Green Revolution projects introduced among its targeted farmers. As with hybrid seed, chemical pesticides were used by few Northern Sierra peasants when the technicians arrived in 1974. In a 1976 survey, for example, only 16 percent of farmers in the Plan Zacapoaxtla area said they had used herbicides at least once. Usage tended to be higher in the southern hot country around Cuetzalan because of greater weed pressure in coffee groves there (Equipo Técnico, 1976c). As in the case of hybrid seeds, Plan Zacapoaxtla did not increase pesticide usage, since technicians encountered a local ecological system in the Northern Sierra where such sprays were not appropriate. Furthermore, the farmers' poverty rendered the sprays inaccessible. Since the size of farmer plots was small, weeds could be successfully controlled by hand with their broad hoes (*azadones*), or, in fewer cases, by burro with a plow. Peasants also developed pest-controlling innovations with techniques and objects that were already available and free of cost. The farmer Carlos Díaz Reyes said that when the common pest known as the blindworm was detected in a field, he and his neighbors in Tatzecuala would bring out to the field a plow or hoe along with the turkeys and dogs from the home yard. As the tools turned up the soil, the turkeys and dogs would come behind and devour the exposed worms. Díaz said the system worked very well in averting damage to the crops, and, besides, the worm-killing chemicals were very expensive (Díaz Reyes, 1998). When attacks on the seeds of newly planted fields were expected from birds or rodents, farmers would also use rocks or stakes to cover the seed or to construct a fence. Unlike larger, commercial farmers, peasants did their own work in the fields and thus had no wage costs against which to balance the cost of weed killing sprays and spraying equipment. Even among those peasants who were able to pay cash to help control weeds, it was cheaper to hire other workers to hoe the fields than to buy the herbicides. Therefore, plows and hoes were a more effective, and certainly less expensive, way to control periodic pest problems than synthetic pesticides.

Technicians were impressed with the efficacy of the traditional methods. As mentioned above, intercropping methods had impressed the technicians in part because a variety of plants in one space reduced both the likelihood and the consequences of pest losses. Not surprisingly, the technicians' original intentions to introduce chemical pesticides were turned aside by local ecological and economic conditions. Thus, technicians encountered a farming system that in many ways did not require, and would not benefit from, the standard means of "modernizing" maize farming.

6. SYNTHETIC FERTILIZERS

Although other Green Revolution technology did not meet a hospitable reception in the Northern Sierra, peasants did eagerly accept selected components of the technicians' offer of modernity. Specifically, peasants adapted synthetic fertilizers to compensate for depleted soil fertility. Farmers in the higher elevations recalled in 1998 that the only significant change in the way they raised maize occurred in the 1970s when synthetic fertilizer use became widespread and consistent (Cantero, et al., 1998; Díaz Reyes, 1998). Project researchers found that in many communities the use of fertilizers rose from 30 percent before Plan Zacapoaxtla began in 1974 to nearly 100 percent usage by 1980 (Equipo Técnico, 1977; Chagra Guerrero, 1980; Centro de Investigación del Desarrollo Rural, 1976). Anthropologist Pierre Beaucage likewise observed the widespread change toward synthetic fertilizer use taking place in the early 1980s. The disappearance of the longstanding custom to share labor among community members, known in

Spanish as *mano vuelta*, was accompanying the appearance of synthetic fertilizers (Beaucage, n.d.). Unlike pesticides, fertilizers were also affordable, costing 130 to 140 pesos for 100 kilograms around 1980, which all landowners could afford (Centro de Investigación del Desarrollo Rural, 1976, Archivo Municipal Xochiapulco, presidencia, 1973, n.d.). The increase in fertilizer use in turn modified local ecological systems.

The two most troubling changes to the Northern Sierra physical environment during the twentieth century had been deforestation and lost soil fertility (Nava and Islas, 1998). The most important reason peasants embraced a change to synthetic fertilizer use was the crisis in soil fertility. A Plan Zacapoaxtla technician noted in 1976 that it was in these areas where "year after year, the productive capacity of the land is being reduced, which as a result harms national production" (Equipo Técnico, 1976b). Observers had been noting problems of erosion for decades. Many fields were planted on slopes measuring from 30 to 60 percent. This exposed the land to erosion, particularly when heavy rain of several inches per day arrived in July and August. The most eroded areas were in Zacapoaxtla, Xochiapulco and Cuetzalan municipalities. Furthermore, in the lower, hot country, the spread of cattle grazing in the early 1970s added pressure for land and led farmers to clear fields into hillsides that had been timbered. These sloped fields quickly eroded once the trees were removed. However, a specific study of the community of Jilotepec in Xochiapulco found that the farmers' methods had prevented "strong erosion problems" (Equipo Técnico, 1976b).

Even where erosion had not been significant, soil fertility was often depleted. In other words, even when the soil stayed put, its fertility declined. Maize, like other crops such as tobacco or cotton, is a hungry consumer of nutrients, especially nitrogen, and depletes soils continually. Peasants were motivated to work with technicians to at least acquire fertilizers because they experienced growing pressure to produce more food from their fields. At the same time the harvests were dwindling, a growing peasant population added to the pressures. Local population growth exceeded 16.5 percent in the 1970s, exacerbating problems derived from a population density that ranked highest among rural areas of the state (Censo General de Población y Vivienda, 1960 & 1970). Technician Chagra Guerrero believed that erosion problems were not severe in the Northern Sierra as a whole, but that fertility had been depleted to a critically low state. "The soils were exhausted," Chagra Guerrero recalled (Chagra Guerrero, 1998).

For many farmers, the dwindling fertility was palpable. They noted smaller plants and fewer baskets of maize that were harvested from the same area of land where their ancestors' crops seemed to be much more bountiful (Cantero 1998, Díaz Reyes 1998, de los Santos, 1998). An elderly peasant in the Cuetzalan village of San Miguel Tzinacapan recalled that the use of

4. Peasant, Environment and Maize "Modernization"

fertilizers beginning in the 1970s was a necessary, but merely stop-gap, measure:

> Before . . . [the maize plant] really gave good ears, up to two or three ears per plant and it put out leaves at the tip of the ears. It did not produce *molkat* [small ears] . . . That's the way it was. We have always cultivated maize here, but before there was more. Before the ground produced a good harvest without fertilizer or anything. But now that we apply fertilizer, the ground is used to it. If we don't apply, [the ground] doesn't want to put out anything (Taller de Tradición Oral de la Sociedad Agropecuaria del CEPEC, 1994).

A young San Miguel Tzinacapan farmer, Fernando Flores Carillo, noted a similar decrease in productivity on the plot his family owns outside of town that has been used to raise maize for the past thirty years. When first cultivated by his grandfather, the plot produced two harvests per year. Ten years later the plot only produced one harvest, even with the then recently adopted use of fertilizers. Flores Carillo believed the lost productivity was due mainly to erosion (Flores Carillo, 1998).

Many peasants may have felt more comfortable with the idea of obtaining fertilizers rather than using other technology such as hybrid seed or pesticides, because surveys showed Northern Sierra already had some limited experience with fertilizer use. In one such survey taken in the project's first year in 1974, 86 percent of the farmers questioned knew about fertilizers and 60 percent had used them at least once. That compares with the 20 percent who had planted hybrids at least once and the 12 percent who had tried pesticides at least once (Equipo Técnico, 1978). Just before Plan Zacapoaxtla began, the municipal president of Xochiapulco reported that peasants there used either synthetic fertilizers or "natural inputs" (i.e. manure) and "in that way obtain higher yields" (Archivo Municipal Xochiapulco, presidencia 1973, n.p.). In response to the detectable decline in harvests, many farmers had for decades been using manure from their poultry or sheep (if they had them) as a fertilizer. Although prior to Plan Zacapoaxtla the results of fertilizer use were disappointing, and the prices offered by the local merchants were prohibitive, there nonetheless existed more prior knowledge and more comfort toward the idea of fertilizers than toward nonlocal seed or expensive pesticides.

Among the nineteen peasants who participated in the first round of fertilizer testing in 1974, the results of added fertilizer were significant (Equipo Técnico, 1976). The yields were 78 percent better than the average yield measured in the region. In other words, the test participants had an average yield that was 800 kilograms per hectare better than the region's overall average of 1,010 kilograms per hectare. Unlike hybrids and pesticides, technicians had found one standard Green Revolution technology that did in fact serve the project's top goal of increasing maize harvests. "[Research] shows us an inadequate use of inputs (i.e. fertilizers) have as an effect low productivity. This shows that a technologically adequate use would permit a substantial increase in the yields of the area in a short period of time" (Equipo Técnico, 1976a).

In the second year, Northern Sierra peasants began eagerly accepting the offer by technicians to use synthetic fertilizers. The Plan Zacapoaxtla state-peasant interaction brought this significant change to Northern Sierra agriculture in tandem with a change it brought to Northern Sierra power relations, namely the organization of peasant cooperatives. In the second year, one of the original reasons that technicians organized peasants was to obtain synthetic fertilizers. One organizing effort was to obtain fertilizers without resorting to local, exploitative merchants or to government lending programs that had aroused the suspicion of farmers. The technicians also organized a trip in late 1975 attended by nineteen peasants to a government-subsidized fertilizer plant in Veracruz. The nineteen peasants on the trip talked with authorities and initiated direct purchases by peasant cooperatives (Chagra Guerrero, 1980).

The number of farmers who purchased fertilizer through the technician-led programs would increase dramatically throughout the program's lifespan. Before the technicians set up such programs, there already existed ninety three farmers in the area who participated in a cumbersome offer by the government farm credit bank to fund fertilizer purchases. In their first year the technicians aided farmers in using the credit bank, and the number of participants shot up to 416. Then in 1976, the technicians more aggressively acted on inclinations to organize peasants by forming an organization of farmers formed that obtained fertilizer directly from the government-controlled, or *paraestatal*, fertilizer distributor known as FERTIMEX (*Fertilizantes Mexicanos*). In this second year of Plan Zacapoaxtla the number of participants almost doubled, to 801. The overall use of fertilizers by all farmers in the area increased from 30 percent to 40 percent, in the first three years of the project (Equipo Técnico, 1978). Eventually, participation in Plan Zacapoaxtla fertilizer purchasing programs would reach 1,446 farmers in 1979.

It is important to point out, however, that peasants were not fastidious about using the synthetic fertilizers according to technicians' recommendations. In the same way the farmers selected only the fertilizer component of the Green Revolution technology, peasants selectively accepted and rejected various parts of the fertilizer recommendations. In the Cuetzalan region, for example, farmers would use types of fertilizer designed for coffee on their maize crop in place of maize-designed fertilizer

(Aguilar Ayon, 1998). As in the Plan Puebla project, peasants would use fertilizer, but not at the rates recommended. Thus, the state's effort to "modernize" maize farming did lead to one significant change, but the extent and manner of fertilizer use was ultimately determined by peasants and their choices.

The change in peasant farming in turn further exacerbated ecological problems. Rufina Edith Villa Hernandez, who farms with her husband in the village of San Andrés Tzicuilan in Cuetzalan municipality, attributed many many ecological problems to the effects of synthetic fertilizers that intensified in the 1970s. The use of fertilizers, as well as pesticides, is more intense in the Cuetzalan area where Villa Hernandez resides because coffee cultivation is concentrated there. Also, coffee farmers operate in a cash economy and participate in government coffee program support, both of which encourage greater chemical use. Previously, more diverse types of plants and more vegetable crops existed in the area.

> When the chemicals arrived, when the fertilizers and all that entered, it began to impoverish the soil. The soil was worsening and there stopped being other (sweet potato and cucumber) crops that had been there before because the soil is finished. Crops are no longer diversified and organic fertilizers (manure) are no longer used. What happens is that with the fertilizers it gives more production but it impoverishes the soil faster. It over-exploits the soil. (Villa Hernandez, 1998).

Not everyone agrees with Villa Hernandez. Agronomists, such as Alvaro Alguilar Ayon, who now works for Cuetzalan's largest cooperative, dispute that there are downsides to the fertilizer use that Plan Zacapoaxtla encouraged. Leaving aside the issue of the extent of ecological damage from synthetic fertilizers, it is nonetheless clear that Plan Zacapoaxtla affected soil quality and land use, at the same time that the environmental conditions, and in particular their extreme variability, affected Plan Zacapoaxtla by rendering hybrid seed unworkable as a viable innovation.

7. THE PROGRAM CHANGES

Technicians formalized in a 1977 annual report the change in program priorities that had been taking place gradually since the first year. An emphasis on organizing marketing cooperative among peasants supplanted the original top priority placed on increasing maize harvests. Their experiences in the first year in the Northern Sierra had planted the seeds of the change toward this emphasis on peasant organizing, when they began to observe the problems of local elite exploitation at the same time they observed the environmentally adapted nature of many of the peasant farming methods. Local non-indigenous families based in the cities of Zacapoaxtla and Cuetzalan controlled commercial channels, charging peasants exorbitant rates for goods such as sugar and fertilizer, while offering less than prevailing market rates to those relatively few peasants who sold crops such as coffee and fruits. The experience of technicians in organizing the fertilizer-buying cooperative, and the enthusiastic response of peasants to this alternative, further compelled the technicians to expand cooperative organizing. "The experiences obtained in that year (1976) led the technician team to analyze the problem of commercialization and of peasant organization, and this analysis meant, as a result, that the technicians of the National Agricultural School would consider in the following years the peasant organization as the most important factor in the strategy." The wording of this added objective also implied the underlying political dimensions of the program, i.e., the desire for tighter links between peasants and the national state. The new objective was meant "to incorporate the small-holding peasant of the region into the development of the country, basing this upon the increase in production and peasant organization (Chagra Guerrero, 1980).

The emphasis on organizing cooperatives led to heightened tensions in the Northern Sierra, and increased animosity by elites toward the technicians. Elites who were economically threatened by a cooperative formed in Cuetzalan exerted their influence on politically powerful figures and groups in Puebla City. The state-level chief of the federal agriculture ministry and the governor, Alfredo Toxqui Fernández de Lara, heard complaints from Northern Sierra elites that the cooperative was comprised of drug dealers, drunks, and agrarian radicals. Responding to the complaints, Toxqui Fernandez in 1979 asked the Colegio de Postgraduado leaders to pull the Plan Zacapoaxtla technicians out of the region, and the Colegio carried out the request. Fortunately for the technicians, in the following year a new state leader of the agriculture ministry arrived who was enthusiastic about peasant cooperatives and prevailed upon the governor to reinstate the Plan Zacapoaxtla program (Sanchez Hernandez, 1987). Thus, similar to many other intended effects of the Plan Zacapoaxtla program, the original political goal to bring peasant quiescence to the Northern Sierra did not work out as anticipated.

Other scholars who have researched the Plan Zacapoaxtla program have so far neglected the role of maize farming methods and their environmental context in explaining the change in Plan Zacapoaxtla priorities. One peasant cooperative of the Northern Sierra that originated during Plan Zacapoaxtla, and a women's cooperative that split from the original group after the end of the Plan Zacapoaxtla program, have attracted considerable study (Sánchez Hernandez, 1987; Fox, 1992). In tracing the origin of the cooperatives, studies emphasize the economic context of the region and point in particular to the exploitative control by local elites of commercial channels. The exploitation by local elites of peasants is highlighted as a factor that pulled technicians and peasants toward the goal of organizing peasant cooperatives.

These explanations overlook that the original intention to modernize maize farming encountered many obstacles, and these encounters pushed technicians and peasants from farming as a basis of collaboration. Maize farming methods were already largely appropriate, or at least more successful, than most innovations offered by technicians. The widely variable environmental conditions, as well as the peasant farming methods specific to their local conditions, largely explain the reluctance of peasants to adopt most of the Green Revolution package of innovations. However, many Northern Sierra peasants overcame resistance in principle to collaborating with the Plan Zacapoaxtla technicians. Environmental problems of high population density and low soil fertility had created crises of desperation that drove peasants to possible solutions, including ideas offered by the sympathetic technicians. Thus, while maize farming innovations were largely abandoned, technicians and farmers instead found peasant organizing to be a more successful basis for collaboration instead.

The technician Chagra Guerrero conceded as much when he compared the dramatic change in the direction of Plan Zacapoaxtla with the sustained focus on maize farming in the preceding Plan Puebla maize modernizing project. "The announced strategies at the beginning of the plan (Zacapoaxtla) were the same as those of Plan Puebla, but the *ecological* and socio-economic conditions in the area (of Zacapoaxtla) were not the same as those in Puebla, and therefore the strategy over time has experienced changes" [emphasis mine]" (Chagra Guerrero, 1980). Farmers, who adapted their methods and seed to their local conditions, negotiated between their environmental setting and any viable opportunities offered by the technicians. In this sense, then, the Northern Sierra environment participated in modifying the state project.

8. CONCLUSION

By tracing how maize farming changed and was changed by the Plan Zacapoaxtla initiative, we learn about the influence local conditions exert on the nature of the postrevolutionary Mexican state. Farm modernizing was one way Mexican national leaders attempted to gain and control peasant support for their regime. A reason Plan Zacapoaxtla was located in the

Northern Sierra was in response to recent peasant mobilization and even violence in the region. The young technicians who were in charge of implementing Plan Zacapoaxtla, although in some ways sympathetic to peasants and ambivalent toward national leaders, did express a desire, beyond the goal of maize modernizing, to see peasants better integrated within national life.

In the course of interacting with Northern Sierra peasants and other local conditions, technicians presided over changes in the farm-modernizing project. While peasants were reluctant to adopt hybrid seed and synthetic pesticides because they were too expensive and not appropriate in their farming systems, technicians gained an appreciation for the methods and rationales used by farmers. The change in the main priority of the project did not mean that all innovations offered by the technicians and modern technology were rejected, since chemical fertilizers did in fact become widely used in the course of the technicians stay in the Northern Sierra. The widespread adoption of synthetic fertilizers, which exacerbate the sustainability problem still facing the region's agriculture, is among a number of changes that the project touched off in the Northern Sierra. The peasants' marketing cooperatives that confronted the oppression by local elites is another example.

Overall, the case of the Northern Sierra of Puebla during the Plan Zacapoaxtla program suggests that the type of rule that actually resulted from farm-modernizing programs was an ongoing, negotiated result. This was not a top-down process of invasion, but instead a process in which not only peasants participated, but their environmental setting as well.

ACKNOWLEDGMENTS

I wish to single out for particular gratitude the staff at the Cholula Campus of the Colegio de Postgraduados and the villagers of Tatzecuala and San Miguel Tzinacapan among the many who helped me research this article in the state of Puebla, Mexico. I also thank Aldemaro Romero, Sarah West and two anonymous reviewers for their comments.

REFERENCES

Aguilar Ayon, Alvaro, 1998, interview with author, Cuetzalan, Mexico.
Beaucage, Pierre, n.d., "Costos de producción y comercialización de cinco cultivos," Unpublished study for the regional agricultural cooperative Tosepan Titataniske.
Cantero, Artemio, 1998, interview with author, Ixehuaco, Mexico.

4. Peasant, Environment and Maize "Modernization"

Centro Internacional de Mejoramiento de Maíz y Trigo (CIMMYT), 1967, *Project for the Promotion of Increased Production of Maize in Mexico*, CIMMYT, Mexico City, Mexico.

Centro de Investigación del Desarrollo Rural, 1976, *Estudios de Caso: Localidades de Jilotepec, Municipio de Zacapoaxtla y de Cuauzimaloyan, Mun. de Xochiapulco de la Region de Zacapoaxtla, Edo. de Puebla,* Tomo II, Colegio de Postgraduados Chapingo, Mexico City, Mexico.

Díaz Reyes, Carlos, 1998, interview with author, Tatzecuala, Mexico.

Chagra Guerrero, V. H., 1980, Organización campesina y decisiones de los productores para la adquisición de fertilizantes: Estudio de caso en dos comunidades del Plan Zacapoaxtla, master's thesis, El Colegio de Postgraduados de Ciencias Agrícolas, Chapingo, Mexico.

Chagra Guerrero, V.H., 1998, interview with author, Xalapa, Veracruz.

Chavez Valencia, Gabriela, 1996, *Respuesta a la fertilización en la asociación Maíz-Jengibre en Cacatecuautan Xiloxochico, Cuetzalan, Puebla*, agronomic engineer thesis, Universidad Autónomo Chapingo.

De los Santos, José Miquel, 1998, interview with author, San Miguel Tzinacapan, Mexico.

Edelman, Marc, 1980, "Agricultural modernization in smallholding areas of Mexico: A case study in the Sierra Norte de Puebla." *Latin American Perspectives* **7(4)**: 29-49.

Equipo Técnico, 1975, *Plan Zacapoaxtla, informe anual I, 1974-1975,* unpublished report. El Colegio de Postgraduados, Cholula, Mexico.

Equipo Técnico, 1976a, *Plan Zacapoaxtla, informe anual II, 1975-76.* unpublished report. El Colegio de Postgraduados, Chapingo, Mexico.

Equipo Técnico, 1976b, *Programa de actividades para 1977*, unpublished report, Colegio de Postgraduados, Chapingo, Mexico.

Equipo Técnico, 1976c, *Encuesta Base, Marco de Referencia, Sexta Parte, Tecnología Tradicional Zacapoaxtla*, Colegio de Postgraduados, Cholula, Mexico.

Felstenhausen, Herman and Heliodoro Díaz-Cisneros, 1985, The strategy of rural development: The puebla initiative, *Human Organization* 44(4): 285-291

Flores Carillo, Fernando, 1998, interview with author, San Miguel Tzinacapan, Mexico.

Fox, Jonathon, 1992, *The Politics of Food in Mexico: State Power and Social Mobilization,* Ithaca, N.Y.: Cornell Univ. Press, Ithaca, N.Y.

Hewitt de Alcántara, Cynthia, 1976, *Modernizing Mexican Agriculture: Socioeconomic Aspects of Technological Change, 1940-1970,* United Nations Research Institute for Social Development, Geneva.

Irigoyen, Pedro, 1998, interview with Author, Cuetzalan, Mexico.

Kreitlow, Bert, 2002, *State and Peasant: Maize and Modernizing in Zacapoaxtla, Mexico, 1930-1982*, Ph.D. dissertation, University of Iowa.

Mora Aguilar, Sergio, 1980, La organización campesina en el desarrollo rural: Una experiencia de la Sierra Norte de Puebla, *Sociología del desarrollo rural* 2(18): .

Ochoa, Enrique C., 2000, *Feeding Mexico: The Political Uses of Food since 1910*, Scholarly Resources, Wilmington, Delaware.

Papendick, R.I. et al., 1976, *Multiple Cropping: Proceedings of a Symposium*, American Society of Agronomy, Madison, Wisconsin.

Valadez-Ramirez, Mario, Antonio Maxías-López and Aguiles Carballo-Carballo, 1998, Traditional Technologies Used by Farmers to Obtain, Select and Maintain Local Maize Seeds in Mexico, Paper presented at the symposium "Indigenous Knowledge Systems on Biodiversity Management and Utilization," Cebu City, The Philippines.

Rosa Nava, Alma and Marina Islas, "Procesos demográficas y deforestación en la sierra norte del estado de Puebla", *Cotidiano* 88: 57-62

Sánchez Hernandez, Miguel, 1987, *Local Organization in Rural Development Programs: The Case of the Puebla Project*, Ph.D. dissertation, University of Wisconsin-Madison.

Scott, James, 1998, *Seeing Like A State*, Yale University Press, New Haven, Conn.

Secretaría de Educación Pública, 1982, *Nuestro Maíz: treinta monografías populares*, 2 volumes. Mexico City, Mexico.

Taller de Tradición Oral de la Sociedad Agropecuaria del CEPEC Taller de Tradición Oral de la Sociedad Agropecuaria del CEPEC, 1994, *Tejuan tikintentkakiayaj. Les oíamos contar a nuestors abuelos: Etnohistoria de San Miguel Tzinacapan*, Instituto Nacional de Antropología e Historia, Mexico City, Mexico.

Verdesoto Alvarez, Galo, 1977, *La posición estructural de la región de Plan Zacapoaxtla, Puebla, México,* master's thesis, Colegio de Postgraduados de Ciencias Agrícolas.

Villa Hernandez, Rufina Edith, 1998, interview with author, Cuetzalan, Mexico.

Wellhausen, E.J., 1966, *Germoplasmo Exótico para el mejoramiento del maíz in los Estados Unidos*, folleto de investigación no. 4, CIMMYT, Chapingo, Mexico.

Chapter 5

PLANTING TREES, BUILDING DEMOCRACY: SUSTAINABLE COMMUNITY FORESTRY IN MEXICO

Ross E. Mitchell[1]
[1]*Sustainable Ecosystems, Integrated Resource Management, Alberta Research Council, 250 Karl Clark Road, Edmonton, Alberta, Canada T6N 1E4*

Abstract:	This chapter examines the intersection of forestry management, forest trade, and local democracy in Mexican communities. It traces the historical development of environmental policy and the Mexican forest industry that eventually led to community control of forest resources. Implications of the North American Free Trade Agreement (NAFTA) and the forest certification process on trade and local people are also discussed. This chapter examines various case study examples to explore the hypothesis that local control and democracy are necessary for environmental sustainability, especially in forest-based communities of Mexico. It concludes that communal forestry management offers new hopes for environmental and democratic sustainability.
Key Words:	certification; common property; environment; forest; governance; NAFTA; Oaxaca; policy; trade

1. INTRODUCTION: FORESTS AS PATRIMONY AND POLICY

Forests have long provided for humanity. As civilization advanced over the ages, the economic and political importance of forests came to be prioritized. Many realized that the real 'gold in them hills' could be found in the trees themselves. Not just local woodcutters, but entrepreneurs and politicians also entered the lucrative lumber business, drastically changing the scale of operations. Forest use and regulation gradually became intertwined with politics and trade, and many countries enacted national forestry legislation for both use and protection.[1]

Complex factors are considered in making pulp and lumber or enclosing vast forested expanses as protected areas. In the process, the vital role that forest ecological systems play as global life support mechanisms may be ignored. Global warming and air pollution are partially attributed to releases of carbon particulates into the atmosphere when forests burn. Both commercial logging operations and agricultural encroachment on forestlands have had huge impacts worldwide. Reckless cutting down of trees can cause severe soil erosion, watershed damage, and loss of wildlife, among other problems.[2]

Above all, and germane to this chapter, marginalized rural people with little influence in environmental and forestry policymaking are those most affected by rapid ecological changes. Local people are often treated as an afterthought, or worse, as a hindrance to forestry development. Government agencies and industrially focused trade policies can hasten these potentially negative consequences. Funding institutions may link financial aid packages to large-scale forestry projects and polices, often to the detriment of rural communities and the forests they depend upon. Moreover, forest communities are often excluded from discussions over trade, logging concessions, and the rights to use various areas of the forest.

Times are changing, though. In countries like Mexico, new trade and forest management arrangements that involve local people affect environmental and social systems with surprising results. Rigid and environmentally damaging practices are giving way to a more community-oriented focus, with high potential to benefit both forests and people.

2. WHY MEXICO?

In this chapter, I explore the interstices of forest-related policy, trade, and democracy in Mexican communities. Building on environmental and forestry related literature, I treat democracy and environment as crosscutting themes in forest-dependent communities. Democracy is a particularly nebulous but crucial concept that has not been rigorously applied to link policy and management in the Mexican community forest sector. My primary concern is whether rural or indigenous forms of democracy, as

---

[1] Among Latin American countries, Mexico and Chile were rather progressive in their policy approach to 'wise forest use' during the early twentieth century, perhaps akin to the Conservation (or utilitarian) movement associated with Gifford Pinchot, first U.S. Chief Forester. Costa Rica's Forestry Law is also praised for its advancement of forest conservation.

[2] Many Latin American countries may actually better conserve their forested areas by carefully controlled logging, but the reverse is true where forestry management is lacking.

evidenced by Mexican communal forest property systems, can be linked to environmental sustainability. In the face of mounting pressures for privatization of land and international trade liberalization, can collective decision-making and communal property systems benefit both forests and local people?

At least five reasons stand out for selecting Mexico as a case study. First, the country has paradoxically experienced a history of both devastating and encouraging environmental consequences. These range from radically altered landscapes to extensive community ownership of forested lands. Second, Mexico's long-standing authoritarian rule is testimony to the lack of democratic privileges extended to its resource-dependent communities. Such circumstances seem to explain why democracy has rarely been linked to forestry trade and management in Mexico. Still, Mexico's political history should not prevent us from exploring the challenge of incorporating local people in forestry decision-making. Third, Mexico's forests are home, and a source of supplementary income, to some 17 million mostly indigenous and poor people (Merino-Pérez, 1997). It stands to reason that they be involved in forestry-based decision making. Fourth, Mexico has placed high importance on community-based forestry since at least the mid-1980s. Finally, Mexico's entry into the North American Free Trade Agreement (NAFTA) in 1994 placed pressure on its once protected forest industry and forest-dependent communities.

3. DEMOCRATIC FORESTRY, ENVIRONMENTAL SUSTAINABILITY

Although it may seem unlikely given the above considerations, some hope for democratic forestry is on the horizon. Unique forms and expressions of local democracy have come to the forefront of the Mexican political landscape in recent years. Indigenous democratic regimes are mixing with sustainable forestry management, sometimes with excellent results. Yet these few positive examples are fragile at best in a country long characterized by authoritarianism, corruption, and opportunistic politics.

Local control and democracy are necessary for environmental sustainability in the Mexican forestry sector, unique for its high degree of community ownership. This premise is contrary to Garret Hardin's 'tragedy of the commons' argument, which states that environmental degradation occurs when common property is managed in a decentralized fashion (Hardin, 1968). From this perspective, monopoly (i.e., private or corporate) ownership of common resources would solve the problem of environmental degradation. In contrast, local (i.e., communal) ownership would only exacerbate environmental degradation due to the 'rational' use of individuals to maximize perceived personal benefits. This has been perceived to be the case particularly in the developing country context where local coordination is often difficult. Policies promoting privatization of the rural commons in Mexico were justified with just such 'tragedy' rationales, blaming forest degradation on collective tenancy (World Bank, 1995).

In contrast, I demonstrate that it may *be* communal ownership and management under the right circumstances that increase environmental wellbeing. I argue that such circumstances must include democratic participation in natural resource management. Likewise, appropriate socio-political mechanisms must be available for people to exercise both individual and group rights. Environmental justice is part of this egalitarian package. With the opening of democracy in Mexico, environmentally related popular movements are becoming more commonplace. Politicians and trade boosters should not fear local demands for greater management control and fair market practices, but instead work with them to mutual benefit. Nor should environmentalists and foresters worry that communities will, given the chance, degrade or destroy communal forest property. If democracy really is environment's best friend (Lowe Morna, 1993), then it stands to reason that forest ecosystems should benefit from increased public involvement. Local contentment with democratic forestry processes and practices should be incentive enough to carry out fair and environmentally friendly logging activities. Local involvement would prevent costly local forms of resistance (that is, costly both in economic and cultural terms for all sides), not to mention deferment of industrial pollution or forest degradation to future generations. In this light, any action or policy involving Mexico's forests should be a shared, democratic initiative.

Sustainable social progress in Mexico must rest on a mutual foundation of trust, citizenship, inclusiveness, respect, and participation. These values are not only attributable to contemporary democracies, but also to indigenous communities. If such qualities are lacking, local people may turn to unsustainable, even illegal, activities for livelihoods, or may react in unpredictable and possibly violent ways. In turn, they may be met with threats, coercion, or state-sanctioned violence. Conversely, renewed local governments would not only support dialogue and conciliation, but also move toward the sustainable achievement of justice, democracy, sane economic growth, and environmental wellbeing.

Hence I maintain that local control and democracy is necessary for environmental sustainability, especially in forest-based communities of Mexico. To begin, I take a brief look at the current state of the environment in Mexico and the historical development of its forestry policies, trade, and land use. Institutions and policies created as a result of NAFTA are considered in relation to Mexican environmental issues and forestry. Community-based forestry examples from Oaxaca and Chihuahua are examined with democracy (including justice and participation), forest trade,

5. Planting Trees, Building Democracy 99

and socio-environmental wellbeing as key themes. Lastly, I conclude with some observations on the relevance of community-centered forestry politics and practices.

4. MEXICO, LAND OF GREEN AND BROWN

Surprisingly very little has been written on environmental change in Mexico. This seems odd given that the destruction of its environment is at the root of many seemingly unrelated problems (Bray, 1991; Simonian, 1995; Arizpe et al., 1996; Klooster, 1997; Simon, 1997; O'Brien, 1998). Mexico experienced an impressive economic expansion from the 1940s until the debt crisis of the 1980s and peso crisis of the mid-1990s. Not only has poverty increased, the state of the environment has substantially deteriorated in Mexico during the twentieth century. Critical environmental problems today include severe air and water pollution in urban areas, declining water availability, badly eroded farmland, and deforestation.

Mexico's environmental troubles may be more easily observed in large urban areas compared to less populated sites. Mexico City, in particular, is nothing short of an environmental disaster. Its lack of basic services and infrastructure has had a severe impact on urban living conditions and population health. The sprawling city of an estimated 20 million inhabitants has among the world's worst air quality, so bad that cars are forced to drive on alternate days. Mexico City was built only by vastly transforming the formerly marshy and lake-filled landscape with disastrous results. Today, enormous quantities of water must be pumped uphill to meet the city's consumption needs.

Negative environmental impacts on rural ecosystems include agricultural and ranching encroachment onto forested lands, degraded forests due to improper forestry practices, and soil erosion. This is especially troubling when we consider that Mexico is ranked fourth in the world in forest species diversity (CEC, 2001), with the world's largest number of oak species (Jaffee, 1997). Closed, mainly temperate forests cover approximately 26% (49.7 million hectares) of Mexico's total land area (191 million hectares) (Torres-Rojo and Flores-Xolocotzi, 2001). With an annual deforestation rate of 510,000 hectares (just over 1%) according to a conservative United Nations estimate, Mexico is ranked fifth in the world in terms of annual forest loss (Roper and Roberts, 1999).[3] Some estimates indicate that well

---

[3] Recent estimates from a study of satellite images show that annual forest cover loss from 1993 to 2000 averaged 1.13 million hectares (about 2.3%), and cost Mexico around 12% of its Gross National Product, or roughly US$38 billion annually (People & the Planet, 2002).

over half of Mexico's forests have been lost (Guerrero et al., 2000a).[4] Areas experiencing severe deforestation include the Lacandona Rain Forest of Chiapas, the Chimalapas region of Oaxaca, and the Meseta Purepecha region of Michoacán (Jaffe, 1997).[5]

5. FOREST POLICY TWISTS AND TURNS

Since at least the mid-nineteenth century, Mexican public policies have overwhelmingly favored large-scale agricultural crop and livestock production over forest conservation. The Liberal government of Benito Juárez (1858-72) enacted the first national forestry law in Mexico in 1861, requiring woodcutters to plant ten mahogany and cedar trees for every tree they chopped down (Simonian, 1995:54). Nonetheless, this statute only applied to public lands and was blatantly ignored by woodcutters and mining companies. Further Liberal reforms in the 1860s facilitated the concentration of land in the hands of a privileged few, which intensified speculative land exploitation (Barry, 1995).

After the Revolution (1910-17), and recognizing that land distribution would contribute to regime stability, Article 27 was incorporated into the reformed 1917 Constitution. In this groundbreaking precedent, land in Mexico was to be 'socially' defined.[6] Over time, most large landholdings were parceled out to impoverished peasants under communal land holdings called *ejidos*; these could be inherited but not rented, sold, or mortgaged outside of the ejido.[7] Ejidos grew in both area and symbolic importance from the mid-1930s until the 1980s.[8] Today, about 49% of the total land area in Mexico is comprised of ejidos, pertaining to 3.1 million ejidatarios (registered ejido members) and members of agrarian communities, or 70% of Mexican farmers (Barry, 1995:119).

While the ejido system appeared to be a democratic institution (e.g., consensual decision-making and elected authorities), many were also hotbeds of corruption and *caciquismo* (bossism). Also, ejido lands were

---

[4] About 24 million hectares of tropical humid forests and 25 million hectares of temperate forests remain.

[5] About half of the Meseta's native forests have been lost since 1963.

[6] Mexico's two types of 'social' land tenure include the *ejidos*, in which land was allocated to a group of people who jointly share the land rights, and *indigenous communities* (or agrarian communities), in which the state recognizes a community's ancestral rights to land that they had occupied before colonialism.

[7] In 1992, restrictions on selling or renting ejido lands were lifted to permit individual ejido owners to join the private sector, although most have not done so until now.

[8] In practice much of the most fertile and accessible land remained in the hands of rich and powerful interests, although limits were placed on the maximum amount of land that could be held as private property.

composed of individual parcels that were too small for subsistence, and ejidatarios lacked incentives to manage their few acres well (DeWalt and Rees, 1994). As a consequence, overutilization led to ecological degradation on their farming plots (Yetman and Búrquez, 1998).

Forest exploitation was also conducted under less than democratic conditions in the past, although not because of the ejidos or indigenous communities. The Mexican government virtually controlled forest production throughout the 1940s and 1950s by alternating concession-granting periods with bans on forest-product extraction. The first concessions were granted to privately owned industries, then later to state-run enterprises. A 1947 forestry law established that communities could only sell wood products to Forest Exploitation Industrial Units (UIEFs). Companies operating in these units were required to plant ten trees for every cubic meter removed (Simonian, 1995:123). The government bans were placed under the rationale that forest resources were dwindling and environmental degradation was increasing. By 1958, some 21 states representing 32% of the country's forests were affected by the bans. Although there may have been some justification, these restrictions severely hurt community and ejido members who depended on forests for fuelwood and building materials.

Although the restrictions made it seem that the government was genuinely concerned about conserving forested areas and promoting wise forest use, the reality was much different. The undemocratic practice of awarding long-term concessions (generally 25 years) was environmentally and socially destructive. Communities generally received below-market prices for timber. Furthermore, the parastatals responsible for providing education, training, and social services to communities largely reneged on their commitment. The logging bans and concessions denied communities of any opportunities to learn how to utilize their own forest resources (Bray, 1991). Facing poverty and marginalization, many communities illegally extracted forest products and intensified deforestation.

Not only were the social costs high, the type of forestry carried out was not environmentally sustainable. It was driven mainly by access to the resource and equipment capacity, not conservation (WRI, 1996). Forestry was based on 'scientific' practices for logging - the so-called *Método Mexicano* (Mexican Silviculture Technique). This technique was based on 'high grading,' or selectively removing the biggest and healthiest trees, which severely degraded forest quality. In the Sierra Juárez in the southern state of Oaxaca, planting was often not done since it was believed that (poor quality) pine seedlings and saplings under the closed canopy would grow once a few large diameter trees were removed. Unfortunately, this technique did not work well with shade intolerant pine trees. Not all the companies did can be labeled as harmful, however. Forest investment in silviculture, technology, and research was high in the days of the concessions (pers.

comm. Torres-Rojo, J.M., September 16, 2002). Still, more than investment was needed given the ongoing forest degradation and social neglect of forest-based communities.

The first stirrings of community protest against the forest concessions emerged in the 1960s. Movements for local control and management of forests in Mexico began in the southern part of the country. In 1968, 14 communities in the Sierra Juárez of Oaxaca led a five-year boycott of a parastatal paper factory, protesting against mistreatment of the communities and their forests (Bray, 1991). In 1979, 26 indigenous communities in Oaxaca created an organization to 'defend together our natural resources, principally our forests, to develop our people and defend our organization from the political and educational apparatus of the state' (SEDUE, 1986).

A collective 'ecological consciousness' emerged among certain forest-based communities. Some recently interviewed Oaxacan sources state that certain ejidos and indigenous communities came to realize the importance of their natural, local patrimony- in so doing, perhaps achieving a conservation land ethic (Mitchell, 2005). On the other hand, this emergent consciousness may have masked the desire to favor economic interests and a perceived inherent right to local autonomy. In particular, other sources thought it highly unlikely that indigenous communities demanded forest control because of a deep-seated feeling for conserving nature. Instead, poor economic circumstances led Mexican indigenous communities to take advantage of the 'ecological' argument made by certain conservationist organizations. Whatever the case, protests in the 1970s and early 1980s eventually bore fruit for communal forest control.

6. NEW FORESTRY LAWS AND INSTITUTIONS

In the mid-1970s, a new policy emerged from a division within the Secretary of Agriculture and Hydraulic Resources (SARH). The Head Office for Forestry Development (*Dirección General de Desarrollo Forestal*) began a determined and temporarily well-funded effort to train communities to manage their own forest resources and industries. In 1986, the federal government passed a forestry law that ended private concessions and parastatals, and thus returned control of most of Mexico's forests to indigenous communities and ejidos. It also introduced more systematic environmental regulation for the forestry sector. As before, communities and ejidos would have to develop forest management plans by enlisting the aid of qualified foresters and submit permits for transport, processing, and sales of wood. But for the first time, seeds were sown for democratic forest management.

For all the good that was beginning to emerge, community-focused political attention would not last long. Mexico's forest legislation shifted from an emphasis on both industrial use and grassroots development to a market-friendly, neoliberal perspective during the 1988-94 administration of Carlos Salinas. New agrarian reforms fit both the conditions placed upon Mexico by the NAFTA treaty and the general policy of free-market economic restructuring (Silva, 1997). In 1992, amendments to Article 27 terminated the government's historic commitment to provide land to petitioning campesinos and opened the door to privatization of the country's social sector (Barry, 1995). Neoliberal democracy, not social democracy, was the primary focus.

The 1992 Agrarian Law reforms had tighter forest management requirements than the 1986 reforms. The reforms required forest management plans to be written by qualified foresters, physical and biological characteristics of the forest ecosystem described, and measures to conserve and protect natural habitat specified (World Bank, 1995). Applicants seeking permission to harvest timber were required to either hold title to the land or hold a legal right to harvest its timber. The 1992 law, however, also deregulated controls on logging and the transportation of forest goods. This activity was previously controlled by *guías forestales* (forestry documentation) that served as both an official permit and a way to calculate the volume of wood being extracted. Now only a hammer mark on the logs was required. This new approach made monitoring of annual wood production virtually impossible, and instead increased illegal logging in some regions (pers. comm. Torres-Rojo, J.M., September 16, 2002). Some Mexican foresters described the new law as making it 'easier to be good, but also easier to be bad' (Jaffee, 1997:10).

The Mexican government later reformed its General Law of Ecological Equilibrium and Environmental Protection in 1994 and the Forestry Law in 1997. The 1994 reforms combined forest management and general environmental responsibilities into a new, centralized Ministry of Environment, Natural Resources and Fisheries (SEMARNAP).[9] Under the rubric of 'sustainable development,' SEMARNAP was charged with defining the principles for ecological policy and ecological management; preservation, restoration, and improvement of the environment; protection of natural areas, wild and aquatic flora and fauna; and prevention and control of air, water and land pollution. The 1997 Forestry Law reforms focused on solving the problems of *tala clandestina* (illegal cutting), unregulated commercial forest plantations, and technical forestry services. Some

---

[9] The fisheries component was dropped in 2000 and the office became the Ministry of Environment and Natural Resources (SEMARNAT). Before SEMARNAP, the Attorney General's Office for Environmental Protection (PROFEPA) was created in 1992 to enforce environmental regulations and respond to citizen complaints.

regulations eliminated in 1992 were reestablished to again require documentation and control of timber harvesting, transport, storage, and processing.

Other recent forest sector developments in Mexico include more financial incentives for commercial forest plantations, an influx of private (and occasionally, foreign) investment in timber production, international financing of community forest development projects, and growing worldwide consumer preferences for socially and environmentally responsible forest products (e.g., certified wood). These recent priorities and pushes do not necessarily favor industrial forestry management. Nor do they automatically lead to a more democratic, 'environmentally-friendly' type of forestry. It does appear that while national forest policy in Mexico has taken on a market-oriented shape and direction, many rural communities and ejidos continue to define how to manage their forest resources. This is a negotiated definition, to be sure. Communal forestry management must involve relevant civil actors, government agencies, and private industry. Mexican communities and ejidos need assistance to sustainably manage their forests for social, economic, and environmental benefits - for today and tomorrow, and for all to enjoy.

With all of these complex legal and political changes since the Revolution, communities have finally begun to realize increased benefits from managing their forests. But what kind of benefits are these? Are they economic, social, or environmental? For instance, some forestry-based communities in the state of Durango were the first to start managing their forests in the 1960s, and enjoyed certain economic benefits from roundwood sales before the 1986 Law. Yet as much as the Head Office for Forestry Development tried to help, only a few 'pilot communities' in Durango received support from the government or timber buyers (pers. comm. Torres-Rojo, J.M., September 16, 2002).

It is important to make a distinction between short-term pecuniary benefits to long-term gains in social capacity and ecological health on communal lands. Those motivated by the short-term pecuniary approach tend to degrade or destroy their forest resources, often irreparably, in true Hardinian fashion. An example of this is the *oro verde* (green gold) or avocado plantations of Michoacán. With the 1997 United States Department of Agriculture (USDA) ruling that allowed Mexican avocados into the continental U.S., not only were large forested areas converted to plantations; deforestation also took place due to the huge demand for one-time use, pine packing crates (Jaffee, 1997).

Conversely, communities that have a long tradition of healthy cooperative relationships (internally and externally), strong communal decision-making practices, and sufficient forested lands are less likely to engage in such destructive practices. These groups tend to *collectively* think of multiple benefits, and with deference to future generations. The forests are not just for today but must also serve tomorrow's needs. Several individuals from Oaxacan forest-based communities have described to me the importance of their forests as persistent providers of clean water, as necessary to prevent soil erosion, and as 'givers of life.' Other non-fiber uses that provide supplementary incomes to many Mexican forest-based communities include the collection of ornamental plants and mushrooms, pine resin tapping, and provision of ecotourism services. Furthermore, the decision-making mechanisms communities employ to manage and enjoy these immediate and future benefits are generally participatory ones, especially in many indigenous communities of Oaxaca. Here, community assemblies are the main venue where majority voting (albeit, normally by male members) is conducted on crucial agrarian and forestry issues.

These benefits cannot be achieved without concerted effort and support, both internally and externally, and many challenges remain. Most communities across Mexico lack adequate knowledge, financial resources, and, in most cases, even sustained interest for managing their forests. On the other hand, well-organized communities with a historically transparent governance structure and a strong conservation ethic now have the opportunity to demonstrate their commitment to responsible forestry. This means that actions must be taken not just on a profit-oriented basis, but also on a socially and environmental sustainable path.

7. RURAL DEMOCRACY: A SAPLING TO NURTURE

Can local expressions of democratization help shape and guide forestry policies and practices in Mexico's current pro-trade context? There are certainly no guarantees. On the surface, the fact that Mexican peasants control the majority of the country's forested resources seems ideal for democratic decision-making and shared benefits. Part of the problem lies in the rampant corruption and *clientelism* (client-patron relationships) that predominate forestry practices in many ejidos and communities (although much less likely in the latter due to strong cooperative and communal ties). Unfair trade arrangements are another factor that may favor cheaper wood imports over sustainably produced, domestic forest products. These issues need to be addressed if real progress is to be made. Likewise, many rural citizens in forested regions and their supporters will have to accept that less national support probably exists for ejidos and community-based forestry. Perhaps because of increased global awareness of environmental problems nowadays, many urban Mexicans feel less comfortable with traditional logging practices, no matter how sustainable they may be in reality. There seems to be little faith in community-based forestry.

In a country where authoritarianism and violence have long prevailed, the words of John McGowan offer some insight. He says that democracy 'is a sapling that must be nurtured in a fairly hostile environment' (McGowan, 1998:79). Democracy is also about trust, negotiation, and compromise. Shared decision-making may still occur under less than ideal conditions such as fair process or economic parity. Under such circumstances, it may be trust that is most important in community-based forestry, residents and non-residents alike. That is, trust in local authority, in decision-making, in improved forest techniques, and in the 'rules of the game.' Communal acceptance of rules is already a cultural reality in many indigenous communities, along with a social obligation for both men and women to participate in local governance. Communal forms of governance offer at least the potential for democratic decision-making. In this vein, we next consider the relevance of transnational trade to Mexican environmental and forestry policy.

8. NAFTA AND THE ENVIRONMENT: TRADING TREES

Commercial trade of forest products may have taken precedence over small-scale, sustainable forestry in Mexico. For instance, the 1992 amendments to Article 27 of the Mexican Constitution may have paved the way for privatization of communal, socially defined land, even though this has yet to occur. Under the right conditions, however, pro-trade interests can overlap with local interests for mutual benefit - and favor both democratic and environmental sustainability. To consider this premise, some background is useful on North America's most important transnational trade agreement - NAFTA.

Through its supplemental environmental cooperation agreement,[10] NAFTA marked the first time that environmental concerns were addressed in a comprehensive trade agreement. Mexico is the only developing country to enter a free trade agreement that incorporates an environmental clause (Schatan, 2000). Yet appearances can be deceiving. A significant weakness is the exclusion of natural resource policy from the definition of environmental law in the North American Agreement on Environmental Cooperation (NAAEC) side accord (Condon, 1995). Instead, the main environmental issue is pollution control such as transport or waste disposal of hazardous materials. Environmental impacts caused by natural resource-

---

[10] The North American Agreement on Environmental Cooperation between the Government of Canada, the Government of the United Mexican States and the Government of the United States of America (NAAEC).

based industry activities (e.g., forestry, fishing, and mining) are mostly 'off limits' as points of contention.

Are there democratic outlets within NAFTA's framework for environmental concerns? Some feel that NAFTA continues to generate unjust and ecologically destructive economic practices, leading to actively coordinated grassroots mobilization (Dreiling, 1997). Yet precisely because of its environmental challenges, Mexico may be the main beneficiary of cooperation with the other NAFTA partners, and may find solutions to its ecological problems (Husted and Logsdon, 1997). But if national government agencies or private firms remain indifferent, citizens may attempt to resolve environmental problems either through formalized (but democratic) procedures, or by popular means. As unwieldy and drawn-out as it may be, one way for concerned citizens to get involved is through the Montreal-based Commission for Environmental Cooperation (CEC). Developed as part of NAFTA, the CEC can respond to any submissions it receives through the preparation of a 'factual record.'[11] Any resident of a NAFTA signatory may file a submission that their country has failed to effectively enforce its environmental laws. Forestry activities are not considered, technically speaking, but a submission could instead deal with a failure to effectively enforce environmental laws by highlighting logging's effects on water quality or endangered species.

If democratic process is at least partially involved in the CEC's work, NAFTA has other problematic issues that may, in effect, increase environmental hazards and risks with severely punitive consequences. For example, corporations have the right to sue any of the concerned governments under the investment provisions of NAFTA's Chapter 11. These provisions were designed to ensure a corporation's investment would not be 'expropriated.' Unfortunately, companies use the provisions to prevent the expropriation of their anticipated profits by the imposition of 'unfair' environmental standards. This allows corporations to legally attack domestic environmental laws, even if such laws are designed to protect citizens and the environment alike. Hence Chapter 11 poses a substantial threat to Mexico's ability to adequately regulate forestry or forestry product operations of companies from the U.S. and Canada. Closed-door settlement negotiations in Washington boardrooms that can affect the lives of millions are neither democratic nor desirable, and their environmental benefits are dubious.

These considerations aside, how does NAFTA relate to forestry? As its boosters so often reminded the negotiators (many continue this line of

---

[11] Most submissions are eventually rejected, however. See http://www.cec.org for details of all submissions received.

reasoning today), NAFTA was not about environment but trade.[12] Forestry trade has been a contentious issue between Canada and the United States for many years, but Mexican forestry has received much less attention. The U.S. (and lately, Chile) is among the principal suppliers of forest products to Mexico, especially for coniferous lumber, plywood, and temperate hardwoods. The balance of trade in forest products between the U.S. and Mexico has traditionally tipped in favor of the U.S. due to Mexico's dependence on U.S. pulp and paper exports (Lyke, 1998). Due to the drastic peso devaluation as a result of the 1994-95 economic crisis, Mexican forest exports temporarily gained a larger market share relative to U.S. products. Today, large stockpiles of domestic forest products compete with cheap lumber imports. Efficiency has become a major concern for Mexican sawmill owners.

Can free trade benefit local, forest-based communities and their environment? At first glance it would appear unlikely. The social institutions underlying community forestry face high pressure to promote privatization, and position elite as key decision makers for market choices. With inadequate training, poor financial capacity, and often-geographical isolation, forest-dependent communities of Mexico are faced with major hurdles. But there are alternatives to this export-led model of *laissez faire*-inspired policy. If new trade opportunities can be linked to a growing worldwide awareness for indigenous, sustainably produced products, both fiber and non-fiber, and if supportive institutional and financing arrangements were put in place, more communities might gain entry into niche markets and succeed. Such possibilities also involve the socially and environmental responsible forestry certification process, still poised to take off as a viable, domestic consumer market.

Under the right organizational and cultural conditions, pro-trade interests may actually be compatible with sustainable forest harvesting and encourage democratic consolidation in communities. Mexican forest-based ejidos and communities have long been engaged in collective decision-making, and many have local forestry knowledge. Several communities are not only capable of modern, sustainable forestry management; some are positioning themselves to enter domestic and international markets 'when their own capacity to produce the quality and quantity required is reached, or when regional buyers begin to actively seek certified timber' (Robinson, 2000a).

---

[12] This also includes forest products, although sectors such as manufacturing and agriculture have much greater volumes traded among the NAFTA countries.

5. Planting Trees, Building Democracy 109

9. THE COMMUNITY FORESTRY EXPERIENCE

Trade in forest products with its NAFTA partners may be relatively small in proportion to other sectors, but forestry is far from insignificant to Mexicans. Of the total country's population of 92 million, 23 million are considered rural, and of these approximately 10 million people live in forest areas (CONAPO, 1996). With individual community land holdings ranging from 100 to 100,000 hectares, Mexico's community-managed forests appear to be at a scale and level of maturity unmatched anywhere else in the world. Partially as a result of aggressive community organizing in the 1980s, Mexican community forestry has seen high growth; numbers, however, still remain relatively low overall. Recent estimates indicate that about 8,000 communities have forests, but no more than 12% are legally engaged in forest commercialization (pers. comm. Torres-Rojo, J.M., September 16, 2002). Moreover, only a small percentage of Mexico's community forest enterprises are commercially competitive.

Independent, third party certification of well-managed forests emerged in Mexico in the 1990s. The certification scheme currently predominant in Oaxaca is the Forestry Stewardship Council (FSC), based in Oaxaca until late 2002.[13] The FSC believes that endorsed certification can expand market opportunities for well-managed community forest enterprises. Mexico is now a world leader in community forest certification. As of August 2004, its total certified forest area was 608,117 hectares (1.2% of total closed forest), and almost 95% of this area was communally managed by either communities or ejidos (31 out of 36 certificates).[14]

The earliest forest certification experiences in the Mexico temperate forest zone were the community forests of *La Union de Comunidades Forestales Zapoteco-Chinanteca* (UZACHI) in Oaxaca state, founded in 1989. As a union of four indigenous communities,[15] UZACHI formed to protect its community forest resources in the Sierra Juárez Mountains north of Oaxaca. This enterprise had its origins in the early 1980s grassroots movement. UZACHI convinced the government to drop its long-term concession to the state-owned pulp and paper company and transfer forest stewardship responsibility for 21,000 hectares of upland pine-oak forest to local communities. They continue to operate and are one of the best examples of community forest enterprises in Latin America.

Still, many of the presumed benefits from forest certification, such as obtaining higher prices for 'green stamped' forest products, have yet to be

---

[13] Its new headquarters are in Bonn, Germany, but FSC Oaxaca remains as a regional office.

[14] See website http://www.fscoax.org for more information.

[15] These communities are San Mateo Capulalpan de Méndez, Santiago Comaltepec, Santiago Xiacui, and La Trinidad Ixtlán.

realized. This is due to a variety of reasons such as outdated milling equipment, low production volumes, domestic reluctance to pay more for certified wood, and inadequate marketing. Yet the gains made until now go far beyond increased log sales and higher prices in niche markets, as follows:

> '...certification has proven to be extremely useful as an internationally accredited seal of approval for the advances made in community forestry nationally. This has helped to reduce intimidation from skeptical politicians and environmentalists who previously had barely distinguished between community based timber extraction and illegal logging' (Robinson, 2000a).

In effect, certification counters the common perception that forestry is associated with environmental degradation. Recognition for responsible community forest management may ultimately be more important than any additional financial benefits for local people.

The real question is if democracy and environmental health can co-exist in a global climate so geared toward quick profits, or with government bodies that encourage colonization of native tropical forests (e.g., Chimalapas in eastern Oaxaca). Lack of true democracy in Mexico is not just at the political level; it extends, and even arises to a certain extent, at the grassroots level. Protests and rebellions are no longer confined to socio-economic or cultural issues. Momentum also appears to be growing in environmentally-related mobilizations among many Mexican citizens (Hindley, 1999). Intentionally planned to coincide with the 1994 implementation of NAFTA, for instance, the Zapatista rebellion in Chiapas brought global attention to poverty and injustice within Mexico. The dispute also involved environmental issues (e.g., propriety rights for agricultural and forestry purposes) and democratic principles (e.g., liberty, citizenship, and other freedoms).

Before singling out a positive example of community forestry, it is worth looking at the more typical scenario throughout Mexico: uncontrolled tree cutting for profit combined with intense pressure to convert forests into pasture or cropland.

10. DEFORESTATION: A LEGACY OF CORRUPTION AND DESPERATENESS

Deforestation in Mexico is often due to factors such as corruption and rampant poverty rather than poor forestry management. Still, in some states such as Chihuahua, owners of lumber, paper, and pulp companies generally profit from forest exploitation. In contrast, Chihuahuan ejidos and indigenous communities have received little benefit from their forest resources (Guerrero et al., 2000). Ejidos may formally control the forest's timber, but the ejidatario typically receives only a small annual dividend from wood sales. As shown later in this chapter, this situation differs greatly in southern Mexico where indigenous communities predominate. Unfortunately, politicians often ignore that the forests are important to their inhabitants in diverse ways, not just as a source of commercial timber or as a contentious point of rural conflict. Mexican forests, for instance, also provide construction materials for local dwellings, are the source of many edible plants and medicinal herbs, and can sustain livelihoods.

In the last few years, indigenous leaders, ejido residents, non-governmental organizations, and others of the Sierra Tarahumara in Chihuahua have filed hundreds of citizen complaints about illegal cutting and other unsustainable forestry practices (Guerrero et al., 2000a).[16] Chihuahuan forestry and environmental laws are inadequate and enforcement is deficient. Intensive pressures to harvest the forest, corrupt socio-political control structures in forestry ejidos, inadequate resources and personnel, and a lack of political will have all contributed to these problems. The regional SEMARNAT office has been asked to conduct full and public audits of whether forestry operations in the Sierra Tarahumara comply with their forestry management plans; to conduct and make public land use and ecological studies to identify which areas of the Sierra should be off-limits for further harvesting; and to identify areas that should be protected to sustain the Sierra's biodiversity and indigenous communities.

What does this melding of politics, forestry, and popular mobilization tell us? Above all, ecological resistance emerges among those cultures that have expressed, or may express, strong participatory action at the local level. This phenomenon has been borne out in several instances in Mexico over the past decade, and has often taken a reactionary tone against perceived injustice. Environment as a key variable for social justice is becoming commonplace.

Guerrero, Michoacán, Sinaloa, Chiapas, and Oaxaca, among other states, also experience deforestation for diverse causes. Three critical regions of Oaxaca where uncontrolled logging occurs include Chimalapas, the Sierra Sur, and the southern coast. For example, forest coverage on the Oaxaca coast has been reduced by 50% since 1958, because of the sustained demand for tropical hardwoods, inadequate planting and conservation, and the scarcity of income-generating alternatives (Barkin, 2000). Traditional communal management practices once restricted forest access, but opportunities for employment in tourism create a heavy flow of migrants from the central highlands and other regions. Unfortunately, peasants are forced to eke out an existence, which dismembers their communities and devastates their environment. Massive tourism developments such as at Bahías de Huatulco must also share some responsibility, since speculation and investment accelerated the process of social and spatial polarization.

---

[16] The rest of this example in Chihuahua is taken mainly from this source.

Instead of respecting and working closely with indigenous groups in the region, a clientelist political structure and outside investors shunt locals aside to make room for commercial development.

11. DEMOCRACY IN THE SIERRA JUÁREZ

Not all is bad news for Mexico's forests and forest-based communities. In contrast to its coastal forestry problems as discussed above, Oaxaca in southern Mexico also stands out with many positive examples of successful community forestry projects.

With almost three and a half million inhabitants, about 37% of Oaxaca's population five years of age and over speaks an indigenous language.[17] Over one-third (41%) of the economically active population is involved in agriculture, fishing, and forestry activities. Indigenous politics and cooperative practices are readily seen among Oaxaca's culturally diverse communities.

Before the Spanish Conquest, the mountains of Oaxaca were covered with forests. Today, they are Mexico's most devastated landscape; for example, 70% of the once-arable land in the Mixteco region has been ruined by erosion (Simon, 1997:36). The southern Mexican dry forests are among the top 10 most threatened forests in the world with only 2.1% listed as protected or managed (ENS, 2001). One of the principle causes of deforestation in Oaxaca is agricultural and ranching expansion, with an annual forest loss of about 40,000 hectares (González, 2001). Yet an incredible 97% of Oaxaca's remaining forests are distributed among 280 communities and ejidos (Zabin, 1992).

The community forests in the mountains of the Sierra Juárez state of central Oaxaca were one of the earliest certifications in the Mexico temperate forest zone. In Ixtlán de Juárez, an indigenous community of 2,500 people, about 384 commoners share the rights to 19,500 hectares of land that includes over 60% of productive pine-oak forest (Robinson, 2000b). The community has worked hard to increase job opportunities.[18] Ixtlán has many professionally trained forestry workers, and is known for its organizational strength and internal capacity.

Around 1996, the World Wildlife Fund (WWF)-Oaxaca program offered to help finance Ixtlán's certification under FSC standards, but the community initially declined. They later decided that certification would be worth pursuing. In 2000, Ixtlán leaders agreed to revisit an outstanding territorial dispute with a neighboring community so that a forest certification

---

[17] Total of 1,120,312 indigenous speakers (2000 census).
[18] There are now over 200 permanent jobs in forestry.

evaluation could be undertaken. A professional forester explained that certification would help them to find an international niche market for their products and indicated that little demand existed for certified wood in Mexico. The community made clear that they first wanted to consolidate a fledgling door-production workshop that targeted local and national markets, but eventually hoped to enter the international market. Their overall goal was to improve their livelihoods through sustainable forest management. Besides generating employment opportunities, forestry is now the largest income generator for the community. Proceeds from forestry sales have allowed for investments in service provision (schools, roads, etc.), social security payments to the elderly and the sick, and annual profit shares to both workers and commoners. The forest business has also become a source of regional employment opportunities. Women have been recruited to work in the palette factory and kiln operation.

The community enjoys more than just socio-economic benefits. Significantly, their forests are regaining both health and quality after many years of degradation during the concession years. Approved logging plans respect wild mushroom areas and cutover areas are kept small. Annual planting with fast-growing pine trees raised in community-run nurseries improves reforestation success, even where *arboles padres* (seed trees) are left for natural regeneration. Severe sanctions are applied to those who abuse forest privileges by, for example, cutting trees without permission. The maximum annual allowable wood volume may not even be met if the community concludes that enough timber has been harvested to meet their needs.

Several other flourishing examples of community forestry exist in Oaxaca (e.g., Santa Catarina Ixtepeji, Textitlán, and San Pedro El Alto). But what makes these examples so successful, and how is 'success' measured? Perhaps most importantly, are they environmentally sustainable? Due to the uniqueness of each community, such questions would have to be answered on a case-by-case basis. Yet there are some commonalities that can be mentioned. At least four reasons stand out for why some of these communities have been so successful both democratically and environmentally speaking, and which appear to refute Hardin's 'tragedy of the commons' argument.

First, strong traditions of communal management continue to predominate in successful communities with well-preserved forests. Several Oaxacan rural communities have a shared tradition of strong cooperative relationships, collective land ownership and management, abiding support for local forms of communal organization, and well-engrained cultural patterns that reinforce long-held decision-making mechanisms. Second, many communities have a well developed conservation ethic that prevented them from wiping out their forests once turned over to them. Many are acutely aware of their responsibility to the forest, abiding by the principle that the forest is for all to use, including visitors and future generations. Third, they combine modern scientific forestry techniques with new strategies for increased sales of forest products. Several communities take full advantage of technological advancements to improve administrative efficiency, and use, for example, the Global Positioning System (GPS) for mapping purposes. They hope that forest certification will guarantee them a fair share of domestic and international markets. Non-governmental organizations, such as FSC and the WWF, support several of these communities, not only in their efforts to certify their forests, but perhaps to find niche markets for their products. Fourth, the communities diversify their forestry activities. Many communities rely on more than roundwood or sawn lumber sales. They supplement earned (and most importantly, shared) income from primary and secondary wood production with wild mushroom exports to Japan, bottled spring water for regional consumption, and even ornamental plants for specialized markets in Oaxaca. These non-fiber, low-impact activities provide opportunities for all community residents to get involved and earn extra income.

Finally, returning to Hardin's contested 'tragedy of the commons' model, we can rightly ask if this were true (namely, that communal ownership inevitably leads toward environmental degradation), then why have several community-based forestry operations been so successful in Mexico? By 'successful,' this has to mean more than just economic efficiency or amount of wood produced and exported. Success in social and environmental terms can also be measured by whether it provides meaningful employment, generates supplementary incomes, builds community pride and trust, contributes to democratic decision-making, increases environmental awareness, and improves forest health. Success according to these measures has been achieved in special cases such as UZACHI, Ixtlán, and Ixtepeji in the Sierra Juárez in Oaxaca. Admittedly, they are not typical even in Oaxaca, but they do represent a viable alternative at least for other communities.

When given adequate access to resources, supportive policies and programs, and transferable technical assistance, rural poor in developing nations will engage in direct action to protect and improve the environment. As David Barkin suggests, rural producers should be encouraged in their efforts to become once again a vibrant and viable social and productive force, and ultimately promote sustainability (Barkin, 2000). The key to forest use and protection may well rest with an empowered citizenry, regardless of nationality, ethnicity, class, or gender. Local communities are generally well placed to determine what is ultimately best for them and their environment. If democracy can be encouraged in resource-dependent communities, and when supported by favorable trade arrangements, then nature has a better chance to thrive compared to its alternative: forest harvesting on a 'for profit' basis, often with little genuine desire to sustain either forests or livelihoods.

12. CONCLUSION: LOCAL, DEMOCRATIC FORESTRY: A PATH TO SUSTAINABILITY?

This chapter demonstrates why policies, agreements, and other measures involving forestry must incorporate local concerns and democratic decision-making. This is not only true of Mexico or the rest of Latin America but in any country with substantial forests. Many rural residents depend on healthy forests for livelihood, or at the very least as a complement to family incomes, as well as for their own health. Both situations are particularly relevant to impoverished regions. In these areas, revenues and income are worthy of protective efforts, but cultural survival may also be at stake.

Improper or illegal forestry practices should be addressed with swifter and more effective enforcement of existing forestry and environmental laws. This is imperative not only at the federal level, but through state and municipal levels as well. Likewise, rural people that depend on communal-based forest tenure systems have to be treated on an inclusive, democratic basis, and provided with opportunities for training to manage their forests better. Fair trade in locally produced, sustainable forest products should be encouraged and facilitated for greater local self-sufficiency. These situations have not been typical, unfortunately. Given the hegemonic control of global markets and trade policies, many governments have exempted forestry from popular debate. Voting privileges and power sharing are often not extended to the people whose very lives depend on healthy forest ecosystems. Community-oriented forestry stands out as an alternative model, in which voting and participation in governance are central features.

We should take care not to romanticize communities. Community-based forestry decisions are not always the most ecologically sound ones, whether inclusive approaches are taken or not. But market-oriented policies without local support or lacking sound management practices can lead to diminished natural resources, increased political instability, and worsened poverty. When rural people are ignored, they are often left with little alternative but to exploit the land and its resources to the maximum. In this context, Hardin may be right. But we cannot pin the blame entirely on marginalized forest-based people. Corruption, *caciquismo*, and clientelist politics have taken their toll in Mexico. Unfair trade policies and practices have also forced forest-dependent communities to pay any associated social and environmental costs.

Local expressions of democratization are not only shaping environmental policies and forestry practices in Mexico's current pro-trade context, but may in fact preclude environmental sustainability. Responsible forestry that strives to serve a greater number of citizens, yet ultimately protects and nourishes the environment for future generations, is only possible through democracy. Just how democracy is to be defined and implemented is another question. The important thing is that democratic examples already exist in some of Mexico's fragile forested regions, and are boldly spreading roots for a greener, shared tomorrow.

ACKNOWLEDGMENTS

This chapter was based on my dissertation research *Ecological Democracy and Forest-Dependent Communities in Oaxaca, Mexico* from 2000-2004 at the University of Alberta. Financial support was generously provided by the University of Alberta, the Social Sciences and Humanities Research Council (SSHRC), the International Development Research Centre (IDRC), and the Organization of American States (OAS). David Bray, Debra Davidson, Dawn Robinson, and Juan Manuel Torres-Rojo merit special thanks. I also extend my gratitude to Sarah West, Aldemaro Romero, and the two anonymous reviewers for their many valuable suggestions. I accept full responsibility for any errors or emissions.

REFERENCES

Arizpe, L., Paz, F., and Velázquez, M., 1996, *Culture and Global Change: Social Perceptions of Deforestation in the Lacandona Rain Forest in Mexico*, Ann Arbor: University of Michigan Press.
Barkin, D., 2000, Social tourism in rural communities: an instrument for promoting sustainable resource management, Paper presented at the 2000 meeting of the Latin American Studies Association, Hyatt Regency Miami, FL.
Barry, T., 1995, *Zapata's Revenge: Free Trade and the Farm Crisis in Mexico*, Boston: South End Press.
Bray, D.B., 1991, The struggle for the forest: conservation and development in the Sierra Juárez, *Grassroots Development* 15(3):13-25.
CEC, 2001, *The North American Mosaic: A State of the Environment Report*, Montreal, Canada: Secretariat Commission for Environmental Commission.
CONAPO, 1996, *Programa Nacional de Población 1995-2000*, cited in Gerardo Segura, *Mexico's Forest Sector and Policies: A General Perspective*, México, D.F.: Instituto de Ecología, Universidad Nacional Autónoma de México.
Condon, B.J., 1995, The impact of the NAFTA, the NAAEC, and constitutional law on environmental policy in Canada and Mexico, in: *NAFTA in Transition*, S.J. Randall and H.W. Konrad, eds., Calgary, Canada: University of Calgary Press, pp. 281-294.
DeWalt, B., and Rees, M., 1994, *The End of Agrarian Reform in Mexico: Past Lessons, Future Prospects*, San Diego, CA: Center for U.S.-Mexican Studies, University of California, San Diego.
Dreiling, M., 1997, Remapping North American environmentalism: contending visions and divergent practices in the fight over NAFTA, *CNS* 8(4):65-98.
ENS (Environment News Service), 2001, WWF lists top 10 most vulnerable forests (15 November 2001); http://ens.lycos.com/ens/feb2001/2001L-02-22-11.html.
González, A.R., 2001, Los bosques de Oaxaca: una visión de fin de siglo, *Calidoscopio Identidades* Num. 5:37-44.

5. Planting Trees, Building Democracy

Guerrero, M.T., Reed, C., Vegter, B., and Kourous, G., 2000a, The timber industry in northern Mexico: social, economic, and environmental impacts, *Borderlines* **8**(2):1-4, 15-16.

Guerrero, M.T., de Villa, F., Kelly, M., Reed, C., and Vegter, B., 2000b, The forestry industry in the state of Chihuahua: economic, ecological and social impacts post-NAFTA, Paper presented at North American Symposium on Understanding the Linkages between Trade and Environment, 11-12 October 2000, Washington: North American Commission for Environmental Cooperation.

Hardin, G., 1968, The tragedy of the commons, *Sc.,* **162**:1243-1248.

Hindley, J., 1999, Indigenous mobilization, development, and democratization in Guerrero: the Nahua people vs. the Tetelcingo Dam, in: *Subnational Politics and Democratization in Mexico,* W.A. Cornelius, T.A. Eisenstadt, and J. Hindley, eds., San Diego, CA: Center for U.S.-Mexican Studies, University of California, San Diego, pp. 207-238.

Husted, B.W., and Logsdon, J.M., 1997, The impact of NAFTA on Mexico's environmental policy, *Growth & Change* **28**(1):24-48.

Jaffee, D., 1997, Confronting globalization in the community forests of Michoacán, Mexico: free trade, neoliberal reforms, and resource degradation, Paper presented at Latin American Studies Association (LASA) Congress, April 1997, Guadalajara, Mexico.

Klooster, D., 1997, Conflict in the commons: commercial forestry and conservation in Mexican indigenous communities, Unpublished Ph.D. Diss., University of California - Los Angeles.

Lowe Morna, C., 1993, Africa: introduction, in: *The Power to Change: Women in the Third World Redefine Their Environment,* Women's Feature Service, London and New Jersey: Zed Books, pp. 25-51.

Lyke, J., 1998, The impact of the North American Free Trade Agreement on U.S. forest products trade with Canada and Mexico: an assessment, *For. Prod. J.* **48**(1):23-28.

McGowan, J., 1998, *Hannah Arendt: An Introduction,* Minneapolis, MN: University of Minnesota Press.

Merino-Pérez, L., 1997, Organización social de la producción forestal comunitaria, in: *Semillas para el Cambio en el Campo: Medio Ambiente, Mercados y Organización Campesina,* L. Paré, D.B. Bray, J. Burstein, and S. Martínez Vázquez, eds., México, D.F.: UNAM, Instituto de Investigaciones Sociales.

Mitchell, R.E., 2005, Ecological democracy and forest-dependent communities of Oaxaca, Mexico. Unpublished Ph.D. Diss., University of Alberta, Edmonton, Canada.

O'Brien, K.L., 1998, *Sacrificing the Forest: Environmental and Social Struggles in Chiapas,* Boulder, CO: Westview Press.

People & the Planet, 2002, Mexico's forests falling fast (5 April 2002); http://www.peopleandplanet.net/doc.php?id=1407.

Robinson, D., 2000a, Certification in communally managed forests - perspectives from Mexico, *Forests, Trees and People,* Newsletter No. **43**:28-31.

Robinson, D., 2000b, The actual and potential impacts of forest certification and fair trade on poverty and injustice: the case of Mexico. Unpublished draft report prepared for the Community and Resource Development Unit, Ford Foundation, New York.

Roper, J. and Roberts, R.W., 1999, Forestry issues: deforestation, tropical forests in Decline, CIDA Forestry Advisors Network (5 April 2002); http://www.rcfa-cfan.org/english/issues.12.html.

Schatan, C., 2000, Lessons from the Mexican environmental experience: first results from NAFTA, in: *The Environment and International Trade Negotiations: Developing Country Stakes,* D. Tussie, ed., New York: St. Martin's Press in association with International Development Research Centre, Ottawa, Canada, pp. 167-185.

SEDUE (Secretaría de Desarrollo Urbano y Ecología), 1986, *Ciudad y Medio Ambiente: Calidad de Vida y Percepción Ambiental,* México, D.F.: Cetamex, Centro de Estudios de Tecnologías Apropriados para México, pp. 89, cited in Simonian, 1995:208.

Silva, E., 1997, The politics of sustainable development: native forest policy in Chile, Venezuela, Costa Rica and Mexico, *J. L. Am. Stud.* **29**:457-493.

Simon, J., 1997, *Endangered Mexico: An Environment on the Edge,* San Francisco, CA: Sierra Club Books.

Simonian, L., 1995, *Defending the Land of the Jaguar: A History of Conservation in Mexico,* Austin, TX: University of Texas Press.

Torres-Rojo, J.M., and Flores-Xolocotzi, R., 2001, Deforestation and land use change in Mexico, *J. Sus. For.,* **12**(1-2):171-191.

World Bank, 1995, *Mexico Resource Conservation and Forest Sector Review* (Report No. 13114-ME): Natural Resources and Rural Poverty Operation Division, Country Department II, Latin America and the Caribbean Regional Office, World Bank.

WRI (World Resources Institute), 1996, Lessons from community development based on forest resources in Mexico and Brazil (3 October 2004); http://pubs.wri.org/pubs_content_text.cfm?ContentID=1259.

Yetman, D., and Búrquez, A., 1998, Twenty-Seven: a case study in ejido privatization in Mexico, *J. Anthro. Res.* **54**:73-95.

Zabin, C., 1992, *El Mercado de la Madera en Oaxaca,* Oaxaca, México: Fundación Interamericana y SALDEBAS, A.C.

PART 3: GETTING THE PRICES RIGHT: MECHANISMS FOR PROTECTING PUBLIC GOODS

Chapter 6

MARKET-BASED POLICIES FOR POLLUTION CONTROL IN LATIN AMERICA

Sarah E. West[1] and Ann Wolverton[2]
[1]*Macalester College, Department of Economics, 1600 Grand Avenue, St. Paul, MN 55105;*
[2]*National Center for Environmental Economics, U.S. Environmental Protection Agency, 1200 Pennsylvania Avenue NW, Washington, D.C. 20460*

Abstract: Rapid urbanization and increased industrialization have led to high pollution levels throughout Latin America. Economists tout policies based on market-based economic incentives as the most cost-effective methods for addressing a wide variety of environmental problems. This chapter examines market-based incentives and their applicability to Latin America. We first review the market-based incentives traditionally used to address pollution and compare these instruments to command-and-control policies. We then discuss two sets of factors that affect how feasible and efficient pollution control policy will be in Latin America: practical considerations, and the violation of standard modeling assumptions. Finally, we compare Latin American experiences with market-based incentives to those in the U.S. and Europe and conclude with several policy recommendations.

Key words: market-based incentives; Latin America; pollution control; environmental policy

1. INTRODUCTION

Rapid urbanization and increased industrialization have led to high levels of pollution throughout Latin America. As economic development continues, household incomes will increase and domestic firms will increasingly participate and compete in the global economy. As these changes occur, Latin Americans and their governments will be predisposed to dedicate more funds to the alleviation of pollution. Since most Latin American countries have little in the way of institutional pollution control infrastructure, they still have the opportunity to determine exactly how to allocate funds and approach their pollution problems. They have the opportunity to decide whether to base their pollution abatement strategies on standard-oriented, command-and-control approaches, on market-based incentives, or on some combination of both.

Policies based on market-based economic incentives have long been touted by economists as the most cost-effective method for addressing a variety of environmental problems. Successful application of such policies, in contrast to implementation of command-and-control regulations, is of much more recent vintage and has largely been accomplished in developed countries. Research on these experiences confirms that in the United States and Europe, incentives can often attain pollution reduction goals at much lower costs.

This chapter examines market-based incentives and their applicability to the Latin American context. We first review market-based incentives traditionally used to address pollution and compare these instruments to command-and-control policies. Such incentives include emissions taxes, environmental subsidies, tax and subsidy combinations, tradable pollution permits, and hybrid instruments.

We then discuss two sets of factors that affect how feasible and efficient pollution control policy will be in Latin America. We focus on practical considerations such as monitoring and enforcement, distributional issues, political feasibility, institutional considerations, administrative costs, and compliance costs. We also examine what the violation of standard modeling assumptions implies for the success of pollution control policy in Latin America. In particular, we focus on non-competitive market structures, imperfect information or uncertainty, the effects of regulation on global competitiveness, and the compatibility of environmental goals with the goals of growth and development.

The third section examines countries' experiences with market-based instruments. We begin with a discussion of the use of several types of incentives in the United States and Europe. We then extend this discussion to examine Latin America's experience with these same incentive-based policies. While market-based instruments have been used more extensively in the United States and Europe, interest in the use of such instruments in Latin America is growing and several experiences are highlighted.

Finally, based on our comparison of Latin American experiences with those in the U.S. and Europe, we conclude with several policy recommendations.

2. COMMAND-AND-CONTROL REGULATIONS AND MARKET-BASED INSTRUMENTS

When firms or consumers decide how much to produce or consume, they weigh the costs of their activity against its benefits. Without proper incentives, however, producers and consumers will not include the costs that they impose on the environment and others in their decision of how much and what to produce or consume. This failure to include these "external" costs in decisions results in a market failure: polluters engage in an inefficiently high amount of polluting activities. When market failure occurs, usually a case can be made for government intervention.[1] Governments have typically used both command-and-control (CAC) regulation and market-based incentives to resolve environmental market failures. We discuss each of these below as well as the possibility of combining aspects of CAC and market-based incentives in a hybrid policy instrument.

2.1 Command-and-Control Regulations

Prior to 1990 virtually every environmental regulation in the U.S. and elsewhere took the form of command-and-control regulations, and they are still commonly used. These regulations "command" that emissions be "controlled" to meet a given minimum quality or maximum emissions standard. As such, they tend to be either technology or performance-based.

Technology standards mandate the control technology or production process that polluters must use to meet the emissions standard set by the government. One problem with this type of CAC regulation is that it applies a one-size-fits-all policy to firms that may differ widely in size and cost structure. Thus, while pollution is abated to the desired level, it is accomplished at a higher cost to firms and consumers than might have occurred if firms were allowed to determine the most cost-effective means for meeting the standard. Alternatively, if more flexible policies were used, *higher* environmental quality could be achieved at the same cost. Technology CAC policies do not encourage firms to find new and innovative abatement strategies nor do they provide incentive to firms to abate beyond the set level.

Performance-based regulations are more flexible CAC policies; they mandate that polluters reach an emissions standard but allow them to choose

---

[1] Only in a narrow set of circumstances is government intervention unwarranted. In 1960 Ronald Coase in "The Problem of Social Cost", *The Journal of Law and Economics,* listed three conditions that must be met for an externality problem to be resolved without government intervention. There must be well-established property rights, a willingness of affected parties to bargain, and a small number of parties affected.

the method by which to meet the standard. Still, once polluters have reached the level specified by the standard, they face little incentive to reduce pollution any further.

2.2 Market-Based Instruments

Market-based policies are regulations that "encourage behavior through market signals rather than through explicit directives regarding pollution control levels or methods." (Stavins, 2000) Many market-based instruments function as follows: A polluting firm or consumer faces a potential penalty in the form of a tax or permit price per unit of emissions. The polluter can choose to pay for existing emissions via the tax or permit or reduce emissions to avoid paying the penalty. Other market-based policies subsidize pollution abatement or combine taxes and subsidies.

Market-based policies give polluters more flexibility than most command-and-control policies. First, the method for reducing pollution is not specified, giving polluters with heterogeneous costs the flexibility to use the least costly abatement method. Polluters that face the same regulation may reduce pollution by recycling, installing new equipment, switching fuels, using labor-intensive methods, or reducing production or consumption. Second, when abatement is relatively costly, polluters can opt not to abate and to instead pay for their higher emissions. Polluters with relatively high costs continue to pollute at a higher level but pay more in taxes or on permits. Those with relatively low costs reduce their pollution when it is cheaper than paying the tax or permit price and pay for residual emissions that are more costly to abate.

Since market-based incentives force polluters to pay taxes, buy permits, or forgo subsidies when they pollute, they provide an always-present incentive to abate. Such incentives therefore also promote innovation in pollution control technologies (Jaffe and Stavins, 1995; Laffont and Tirole, 1996; Parry, 1998).

In the remainder of this section, we examine five commonly prescribed market-based incentives: emissions taxes, environmental subsidies, tax and subsidy combinations, tradable permits, and hybrid policy instruments. In each section we describe the principal advantages and disadvantages of using each incentive.

2.2.1 The Emissions Tax

An emissions tax is exacted per unit of pollution emitted and forces a firm or consumer to internalize the external cost of its emissions. The tax is set so that, for each unit of pollution emitted, a polluter pays the value of the marginal (additional) external damage caused by that unit of pollution. These external damages may include the costs of worsened human health, reduced visibility, lower property values, or loss of crop yields or biodiversity. To avoid the emissions tax a polluter finds the cheapest way to reduce pollution. For any residual pollution, the polluter pays the tax. In addition, the government earns revenue that it can use to reduce other pollution or other taxes.[2]

Despite the apparent usefulness of such a tax, true "Pigovian" emissions taxes – those set equal or close to marginal external damages – are relatively rare.[3] Often it is impossible to tax emissions directly because they are difficult to measure; it is difficult to define and monetarily value the marginal external damages of a unit of pollution; and the taxes are difficult to enforce since they are often exacted on goods that are not directly bought and sold. In addition, attempts to measure and tax emissions may lead to illegal dumping (Fullerton, 1996).

2.2.2 Environmental Subsidies

A subsidy per unit of pollution abatement establishes incentives for emission reductions identical to a tax per unit of pollution. For instance, if use of a cleaner fuel or purchase of control technology is subsidized at the appropriate level it induces firms to switch from a dirtier fuel or to install control technology until the same level of abatement is reached as under the emissions tax. Unlike an emissions tax, however, the subsidy distorts long-run economic incentives of firms. Since a subsidy adds to firms' revenue streams, firms have an incentive to enter the industry or to appear "dirtier" to qualify for the subsidy. The result could conceivably be that, while each individual firm decreases its pollution, the overall level of pollution actually increases (Baumol and Oates, 1988). Because of these long-run effects, subsidies are usually recommended for use over limited periods of time. Once subsidies have been given, however, they are often quite difficult to remove. Another potential disadvantage is that instead of collecting revenue, as with a tax, the government pays firms or consumers and funds it through another revenue-raising device.

---

[2] In particular, environmental tax revenue can be used to reduce labor taxes and thereby reduce the distortion in the labor market caused by these taxes (see Goulder, Parry and Burtraw, 1997).
[3] These taxes are called "Pigovian" taxes after the economist, Arthur Pigou, who first formalized them (see Pigou, 1932).

2.2.3 Tax and Subsidy Combinations

A tax and subsidy also can be combined to achieve an efficient level of pollution. The tax is applied to output under the presumption that all production processes pollute or that all consumption goods become waste. A subsidy is then provided to the extent that a firm provides proof of the use of a cleaner form of production or consumers demonstrate proper disposal. In the case of the firm, the tax increases the cost of output and induces the firm to reduce its use of both clean and dirty inputs. The subsidy provides the firm with an incentive to switch into cleaner forms of fuel or install more control technology.

The main advantage of a tax-subsidy combination is that both parts apply to a market transaction. Instead of attempting to monitor emissions and control illegal dumping, which may be difficult or infeasible, policy makers can immediately observe the taxed and subsidized items. Also, polluters have an incentive to reveal information on abatement activity to qualify for the subsidy instead of hiding information to cover up illegal activity. Disadvantages include potentially high implementation and administrative costs and the political temptation to set the tax or subsidy too low to induce proper behavior.

2.2.4 Permits

While an emissions tax sets the price of pollution, a permit system sets the quantity of allowable emissions. Permits are distributed or auctioned, generally to firms, and represent the right to pollute some set amount of pollution. Firms then buy and sell permits to each other as needed. The market-clearing price is established through this buying and selling and, if the government chooses the optimal level of pollution, will be the same amount as an emissions tax. As in the case of a tax, a firm can reduce pollution to avoid the cost of purchasing permits. For any residual pollution, the firm purchases the needed number of permits. If they are auctioned to firms, then the government collects the proceeds of the auction. If permits are allocated to firms, then firms are given a one-time subsidy equal to the value of the permits.

Permits have several advantages over emissions taxes. First, they do not require a policy maker to measure the marginal external damages of a unit of pollution. Instead, the policy maker decides what the acceptable level of pollution is. Second, unlike an emissions tax or environmental subsidy, permits allow policy makers to determine with certainty the level of pollution that will result. This is particularly important in cases where pollution over a certain amount potentially causes a lot of damage (Weitzman, 1974).

6. Market-Based Policies

Permits also have a number of disadvantages. First, since the government sets the quantity of pollution, there is some degree of uncertainty regarding price. If abatement measurement and permit purchases are much more expensive than expected, they can have large and costly effects on particular industries.[4] Second, permits can be interpreted as giving firms a license to pollute, particularly when permits are allocated to firms for free. Third, transaction costs to complete a trade may be high. Significant search costs or strategic behavior could seriously inhibit the permit trading that would normally ensure the least-cost outcome.

Permit systems tend to work best when there is a sufficiently large market in which firms can trade, they are designed as simple systems that apply to either inputs to production or emissions, and when government acts to reduce regulatory uncertainty and barriers to trade (Stavins, 1995).

2.3 Hybrid Instruments

In addition to the four market-based instruments discussed above, two hybrid policies– those that combine aspects of CAC and market-based policies– are often found in the literature and in practice. These hybrid policies are: combining standards and pricing approaches, and the use of information as regulation.

2.3.1 Combining Standards and Emissions Taxes

Pollution standards set specific emissions limits, and thereby reduce the chance of excessive health and environmental damages. Such standards, however, may impose large costs on polluters. Emissions taxes, on the other hand, restrict costs by allowing polluters to pay a tax rather than undertake excessively expensive abatement. Such taxes, however, do not set a limit on emissions, and leave open the possibility that pollution may be excessively high.

Might there be a policy that sets a limit on both costs and pollution? Some researchers suggest a "pressure-valve" approach to regulation that combines standards with emissions fees (Roberts and Spence, 1976; Spence and Weitzman, 1978). Such a policy combination imposes the same emissions standard on all polluters then subjects all polluters to a unit tax for emissions in excess of the standard. This policy combination has several attractive features. First, if the standard is set properly, proper protection of

---

[4] For instance, the Regional Clean Air Management program (RECLAIM) in Los Angeles, California experienced an astronomical increase in the price of SO_2 permits in 2000 due to a marked increase in the demand for electricity. Prices for SO_2 permits increased from $4,284 per ton in 1999 to $39,000 per ton in 2000. (For more information, see the South Coast Air Management District website at www.aqmd.gov or Goldenberg, 1993.)

health and the environment will be assured. Second, high abatement cost polluters can defray costs by paying the emissions fee instead of cleaning up.

2.3.2 Information as Regulation

An instrument that has been used increasingly in industrialized countries as a method for regulation is the requirement that firms provide the government and public with information on pollution and abatement activities. These reporting requirements attempt to minimize inefficiencies in regulation associated with asymmetric information, where the firm typically has more and better information on what and how much it pollutes than the government or public (Tietenberg and Wheeler, 2001). When expensive emissions monitoring is required to collect such information, switching the burden of proof for monitoring and reporting from the government to the firm can increase the effectiveness of government regulation while substantially lowering the cost. If accompanied by spot checks to ensure that monitoring equipment functions properly and that firms report results accurately, this can be an effective form of regulation.

Use of information as regulation also creates a role for the community. A community with information on a nearby firm's pollution activities can exert pressure on it to internalize at least a portion of the costs of pollution, even when formal regulations are weak or nonexistent (Pargal and Wheeler, 1996). This type of "private enforcement" effectively increases a firm's expected penalty of polluting, and the firm will react as if it were being inspected and fined by a government agency (Naysnerski and Tietenberg, 1992). As in the case of a pure market-based instrument, plants still have the flexibility to respond to community pressure for emission reductions by abating in the cheapest way possible.

Still, it is important to keep in mind two caveats when discussing the use of information as regulation. First, the use of information as regulation is not costless: U.S. firms spend about $346 million/year to monitor and report releases. Any required investments in pollution control are in addition to this amount (O'Connor, 1996). Second, the amount of pressure a community exerts on a plant is related to socioeconomic status. Poorer, less educated populations tend to exert far less pressure than communities with richer, well-educated populations (World Bank, 2000).

3. THE CONDITIONS IN LATIN AMERICA: IMPLICATIONS FOR POLLUTION CONTROL POLICY

Two sets of factors affect how feasible and efficient the use of CAC or market-based incentives will be in Latin America. The first section below discusses a number of practical considerations that should be integrated into any evaluation of policy options. The second section examines the implications of numerous violations of the basic modeling assumptions in the Latin American context.

3.1 Practical Considerations

Any discussion of pollution control policy options must account for a number of practical considerations, particularly in a developing country context. For example, implementation of a pollution control policy may be complicated by monitoring and enforcement issues, distribution of costs and benefits, political feasibility, institutional constraints, administrative costs, or compliance costs.

3.1.1 Monitoring and Enforcement

Effective monitoring and enforcement is a basic requisite for pollution control policy. However, governments first must be able to measure emissions. In developing countries, facility-level emissions data are rarely available. The implementation of policies that require detailed emissions data for monitoring and enforcement are usually prohibitively expensive and therefore likely to be infeasible (O'Connor, 1996).

Monitoring and enforcement is complicated by evasion through activities such as illegal dumping of waste, falsification of emission records, and generation of counterfeit proof of proper disposal. Evidence suggests that countries often set the number of audits and penalty levels too low to induce proper compliance. In Latin America, these difficulties are compounded by political pressure by large firms and the difficulty of monitoring a large informal sector that operates outside the legal tax structure.

A more targeted monitoring and enforcement process may allow governments with few funds and little manpower to increase the compliance of firms substantially. For instance, studies in Rio de Janeiro show that by targeting only 50 plants, 60 percent of the state's industrial pollution is effectively monitored. While the smallest firms tend to be more pollution intensive per unit of output, they are dwarfed as contributors to Rio de Janeiro's pollution by the scale of the largest plants (World Bank, 2000).

Since these small plants are more numerous and more difficult to monitor, particularly those in the informal sector, policies that focus on compliance of large plants are likely to be more effective.

In addition, the number of audits and the penalty levels should account for monitoring device quality and expected returns to the firm or consumer from cheating. Market-based instruments that assess fees on goods bought and sold in markets or that provide incentives to provide proof of compliance also reduce the opportunities for cheating and require less monitoring and enforcement.

3.1.2 Distributional Issues

Pollution control policies that induce polluters to internalize the costs of their actions and increase efficiency are not necessarily equitable. Empirical evidence indicates that environmental regulations typically favor the rich in their redistributive effects, and as such exacerbate existing income inequality (Baumol and Oates, 1988). The costs of environmental regulation tend to be regressive, borne disproportionately by the poor, and come in the form of job losses in the short run and higher consumer prices in the long run.

The distribution of benefits from environmental policies depends on which pollutants are targeted. For instance, policies that preserve pristine wilderness, create national parks, or clean water for recreational use usually benefit the rich comparatively more than the poor, since both the willingness-to-pay and use of these types of environmental goods increase with income levels.[5] Policies that target urban air or water pollutants tend to benefit households at the lower end of the income spectrum since such households cannot afford to move out of industrial areas or buy more expensive, privately acquired water sources.

Many countries in Latin America have highly unequal income distributions and large portions of their populations living in poverty. It is therefore particularly important to design pollution control policies that do not disproportionately affect the poor and exacerbate income inequality. Policy makers should target pollutants that have the largest effect on poor household's lives. Policy instruments that are progressive in their distribution of costs or, when they are not, are coupled with a redistributive subsidy will help to offset costs borne by the poor (Huber et al., 1998).

---

[5] Although as air quality in these neighborhoods improves, the poor may be pushed out of the neighborhood as rising property values attract more affluent families.

3.1.3 Political Feasibility

Politicians often prefer command-and-control policies to market-based instruments since they are relatively straightforward, easy to negotiate with industry, and successfully cut pollution to a pre-determined level (Baumol and Oates, 1988). The political feasibility of market-based instruments is also complicated by the fact that they often raise prices on basic consumer necessities such as food or heating oil, or impose taxes on politically powerful firms.

Given the stringent budgetary constraints under which developing countries operate, policy instruments that rely on the existing institutional framework and infrastructure, that are self-financing, or that raise revenue for governments are likely to be preferred to those requiring substantial outlays. Earmarking revenue collected for a particular program may also increase support since it increases government accountability (Huber et al., 1998).

3.1.4 Institutional Considerations

While most Latin American countries have environmental protection agencies, they are usually of relatively recent vintage, under-funded, and understaffed in environmental protection expertise (World Bank, 1996). Low salaries may lead to negligence or corruption on the part of enforcement agents and weaken the ability of the government to control pollution effectively. Given these constraints, policy instruments that do not require extensive government oversight are more feasible in a developing country context.

Market-based incentives may be more feasible because they are less likely to be constrained by a long history of reliance on CAC regulation. Environmental policy makers in developing countries may have greater freedom to experiment with market-based incentives. Within the set of available market-based incentives, instruments that tax or subsidize goods that have established markets may be more successful. For example, policy makers could take advantage of existing value-added tax systems to increase tax rates on consumption goods such as gasoline.

Effective implementation of many environmental policies may require conditions not present in some Latin American countries. For instance, many market-based instruments require that property rights are clearly assigned and enforced, taxes are collected and administered effectively, a well-functioning and stable financial system is in place, and contracts are fairly negotiated and enforced (O'Connor, 1996).

3.1.5 Administrative Costs

Market-based instruments are used with increasing frequency in Latin America under the assumption that they do not add substantial administrative costs (Huber et al., 1998). The administrative costs of a pollution control policy depend on the level of government involvement in emissions measurement, tax collection, and monitoring and enforcement. If the "burden of proof" lies with the government, then administrative costs tend to be high. For instance, an emissions tax requires the government to collect data on emissions of polluters, set the tax based on this information, collect revenue, and minimize tax evasion. Enforcement of a standard requires proper measurement of emissions as well as effectively established fines and regular monitoring. If the burden of proof is shifted to the consumer or firm, the administrative costs will be lower. For example, a deposit-refund system places the data and reporting requirements on the individual consumer or firm to provide proof of proper disposal.

3.1.6 Compliance Costs

Relative to the United States and Europe, Latin American countries are in the early stages of environmental protection. For example, catalytic converters were required on all new automobiles in Mexico starting in the early 1990s. Since Latin American countries have not exhausted the lower-cost methods of pollution abatement, it will generally cost them less to reduce an additional unit of pollution than it costs Europe or the United States.

Industries in Latin America have less technologically-advanced production processes than U.S. industries, and so may be able to purchase cheaper abatement technologies considered outdated in the United States. Abatement costs, however, are certainly higher relative to overall production costs than in developed country industries. In Latin America, capital, including tools, machines, and abatement technologies such as scrubbers and monitors, is more expensive relative to labor costs.

In the current regulatory environment in Latin America, other compliance costs may also be excessively high. Given the bureaucratic structures common in these countries, it is not unreasonable to expect a large amount of paperwork and other compliance requirements. The costs of maneuvering the bureaucratic mazes in Latin America can stifle economic activity (DeSoto, 2000).

Given that abatement and compliance costs are likely to be high relative to overall production costs, the most cost-effective policies are ones that permit the maximum level of flexibility in abatement choice and require the fewest bureaucratic acrobatics.

3.2 Challenging the Underlying Assumptions

Economic theories of CAC regulations and market-based instruments rest on a number of assumptions that may not hold in the Latin American context. For example, implementation of pollution control policy may be complicated by noncompetitive market structures, lack of perfect information or uncertainty, effects of environmental regulations on global competitiveness, or compatibility of environmental goals with the pressing goals of growth and development. We consider each of these potentially complicating factors below.

3.2.1 Non-Competitive Market Structure

The effectiveness and efficiency of pollution-control policies depend critically on the nature of the markets in which they are implemented. Two market structure distortions that affect the efficiency of policy making in Latin America are monopoly power and the informal sector.[6] Since domestic markets in Latin America are small relative to those in the United States and Europe, it is more likely that one or a few firms in an industry serve the entire market. Since monopolies already restrict output relative to what would be produced in competitive markets, care must be taken to ensure that pollution control policies do not further restrict output to the point where the costs of the policy exceed the benefits. Policies that decrease total pollution through reductions in pollution per unit, rather than through reductions in output, may be a solution.

On the other end of the spectrum sits the informal sector, composed of many very small firms. This sector produces legal goods but avoids government regulation and taxation. The value of Latin American countries' informal sector outputs range from 25 to 60 percent of Gross Domestic Product (GDP) (Schneider and Enste, 2000). This sector also includes a number of industries that are quite pollution-intensive, which means that the potential environmental effects of informal sector production may be large (Blackman, 2000). The size and importance of Latin American informal sectors presents three main problems for environmental regulators. First, since polluters in the informal sector are very practiced at avoiding government fees, it is nearly impossible to impose direct regulations or taxes on informal sector polluters. Second, taxes or regulations that are

---

[6] As mentioned previously, market distortions are also created by pre-existing taxes on labor (see Eskeland and Jimenez, 1992).

successfully implemented in the formal sector may push more polluters into the informal sector. Third, firms in this sector often support families at the low end of the income distribution, meaning that even small increases in the costs to the firm may mean large impacts on the poor.

Indirect taxes on polluting inputs that are produced in formal markets but used in informal sector production may be one way of inducing pollution reduction among firms that can easily evade direct regulation or taxation.

3.2.2 Imperfect Information and Uncertainty

Environmental policy makers in both developed and developing countries may face a significant amount of uncertainty regarding the costs and benefits of pollution reduction. Since regulatory structures in developing countries are generally not as well-developed as those in the United States and Europe, and since limited resources may preclude extensive quantification of costs and benefits, policy makers in Latin America will likely face an even greater level of uncertainty about the relative costs and benefits of pollution reduction (O'Connor, 1996).

Weitzman (1974) provides criteria for choosing between "price" and "quantity" based policies in the face of uncertainty. If policy makers are uncertain about the costs of abatement and fear that polluters may be saddled with high costs as a result of regulation, they can limit these costs by using a price instrument such as an emissions tax or a tax on polluting goods. If, on the other hand, policy makers are more uncertain of the benefits of controlling pollution and fear that excessively high environmental damages may occur, they can limit these damages by using a quantity instrument such as tradable pollution permits. Alternatively, policy makers may opt for the certainty of a standard. Concern for the cost-effectiveness of the policy should lead them to consider a performance-based standard or a hybrid policy that combines a market-based instrument with a pollution limit (Eskeland and Devarajan, 1996; Eskeland and Jimenez, 1992; Roberts and Spence, 1976).

3.2.3 Global Competitiveness

Pollution control policy that involves significant compliance costs for firms may alter the international structure of relative costs and affect patterns of specialization and world trade. In particular, economic theory predicts that in a country that forces firms to internalize an environmental externality, production costs increase and domestic firms' comparative advantage in the production of pollution-intensive goods decreases. Domestic firms then have difficulty competing internationally, particularly with firms that do not face these costs (O'Connor, 1996). In the case where a country has little or no

6. Market-Based Policies 135

effect on world price, passage of environmental regulations that increase firm production costs may result in reduced revenue and increased unemployment in the short run (Baumol and Oates, 1988). In the long run, increased environmental regulation may means decreased competitiveness and growth.

Less-developed countries that have neglected pollution control to focus on more pressing development concerns are often said to develop a comparative advantage in pollution-intensive industries. As trade barriers decline these countries become "pollution havens" for industrialized countries' dirtiest industries and may set low environmental goals to attract firms to relocate there. Only limited evidence of this effect has been found (Birdsall and Wheeler, 1993; Levinson, 1996; Xing and Kolstad, 2002). Usually differences in labor and capital costs and concerns over political and economic stability outweigh differences in environmental costs.

3.2.4 Dynamic Considerations

Since the successful implementation of environmental policies requires significant funding, environmental sustainability requires faster economic growth. Air and water pollution in cities can be reduced only through increased public investment, while high rates of soil erosion and deforestation are "unlikely to be reversed without a rapid improvement in living conditions in rural areas" (Ros et al., 1996). Without economic growth, pollution in cities and degradation in rural areas will worsen.

But are Latin America's current growth and development policies conducive to environmental protection? And what kinds of pollution control policies might better complement current economic growth policies? Given the focus that most Latin American governments currently place on balancing federal budgets and paying down debt, their Ministries of the Environment will be hard pressed to obtain large funding increases in the short run. To the extent that fiscal responsibility keeps interest rates low and promotes economic growth, federal tax revenues will increase in the long run, and some of these new revenues may be channeled into environmental protection. Pollution control policies that generate revenue as they reduce pollution seem to be most complementary to the goal of fiscal responsibility.

Huber et al. (1998) find "that public firms are least accountable and have little internal incentive to meet even their own environmental standards and guidelines." Since government-owned firms are able to insulate themselves from regulatory oversight, they tend to be more polluting than privately-owned firms (Blackman and Harrington, 2000). Privatization therefore would appear to be very conducive to pollution control initiatives. While firms may prefer the certainty of command-and-control regulations, market-based incentives may allow these newly profit-oriented firms to stay competitive while they reduce pollution.

4. EXPERIENCE WITH MARKET-BASED INCENTIVES IN THE U.S., EUROPE, AND LATIN AMERICA

Market-based instruments have only recently been considered alongside command-and-control regulations as options for reducing pollution. However, as policy makers become more familiar with these instruments, witness their successful implementation, and notice some advantages over CAC regulation, they become more open to the use of market-based instruments. This section briefly discusses some examples of the use of market-based incentives for solving market failures related to pollution in the United States and Europe, and throughout Latin America.[7]

4.1 Experiences in the United States and Europe

Until recently, governments did not consider market-based instruments as an option to solve pollution problems. In the U.S., the Environmental Protection Agency (EPA) first allowed plants to "trade" emissions in 1975 with a simple offset program: new or existing plants that made major modifications were allowed to locate in non-attainment areas if they made more than equivalent reductions in emissions at other existing plants (Tietenberg, 1985). In 1988 two U.S. senators sponsored a report that identified innovative policy approaches to environmental policy including the use of economic incentives. And in 1990 amendments to the Clean Air Act were passed with a tradable permit system outlined to control acid rain (Environmental Law Institute, 1997).

Europe's first experiences with market-based instruments occurred in the 1970s, when many countries used "cost-covering charges," designed to cover the cost of monitoring or controlling an environmental asset (European Environment Agency, 1996). Environmental taxes were introduced in the 1980s and their use has accelerated over the last decade. By 2001, for example, eight countries in the European Union had carbon taxes, up from four in 1996 (European Environment Agency, 2000).

---

[7] Standard-and-pricing approaches are discussed in the context of the market-based instrument with which the examples are most closely associated.

4.1.1 Emissions Taxes

The U.S. Internal Revenue Service defines only four U.S. taxes as emissions taxes: these taxes are on petroleum, chemical feedstocks, ozone-depleting chemicals, and motor fuels (Fullerton, 1996). Many of these are not direct taxes on emissions but taxes on another input under the presumption that the input is directly related to the amount of pollution produced. At the state and local level, there are numerous examples of water discharge and user fees. However, most of these are motivated by the need to raise revenue to cover the cost of supplying the service. Most local communities also charge fees for disposal of solid waste either through variable rates (also referred to as pay-as-you-throw) or fixed fees. Unlike the case of fixed fees, which are unrelated to the amount of waste a household disposes, variable rates are based on volume and designed to charge a positive marginal cost for collection and disposal (EPA, 2001).

In Europe, there are many instances of taxes on air and water pollutants and for the purpose of noise abatement, although many of these are not designed to limit emissions but to raise revenue. In such cases, the taxes are often set far too low to illicit serious reductions in pollution and are often combined with command-and-control policies that lessen economic incentives to switch fuels or invest in innovative control technologies (Cropper and Oates, 1992).

Some European countries have had noted success in reducing pollution. For instance, Norway experienced up to a 21 percent reduction in carbon emissions in certain industries as a result of its tax. Sweden taxes the sulfur content of fuel and refunds it to plants when a scrubber is installed. Since the tax is calculated to reflect the marginal cost of abatement, it is noticeably higher than similar taxes in other countries and has had a significant effect: the average sulfur content of fuel declined from 0.65 to 0.40 percent between 1991 and 1993. Sweden also has a tax on NOx that, in an attempt to be revenue-neutral, refunds revenues from the tax to plants in proportion to the amount of energy produced. Total emissions fell by 40 percent in the first two years that plants were subject to the tax (Blackman and Harrington, 2000; Stavins, 2001). In other countries tax exemptions have muted the possible impact on emissions (OECD, 1999).

4.1.2 Environmental Subsidies

Environmental subsidies in the U.S. include grants to local government and businesses for proper waste management and recycling, grants to public bus systems and businesses for use of alternative fuel vehicles, subsidies to purchasers of alternative fuels, and cost-sharing with property owners for land conservation. The largest conservation subsidy program pays farmers up to $50,000 annually to take land out of cultivation and to place it in a conservation "reserve" for ten to fifteen years. The payment is designed to cover all foregone rents from alternative uses of the land as well as half the cost of soil erosion control measures. As of 1992, 36.5 million acres – almost 10 percent of all U.S. land used for agriculture – had been placed in the conservation reserve. The U.S. Department of Agriculture estimates the net social benefits from conserving these acres as between $4 billion and $9 billion (Environmental Law Institute, 1997).

Other industrialized countries have used subsidies to encourage a variety of abatement activities. Canada has a subsidy in place to encourage the reuse and recycling of used tires (OECD, 1999). Germany, Finland, Norway, and Sweden subsidize farmers if they convert from traditional to organic farming. Belgium, the United Kingdom, Finland, Portugal, Spain, and Turkey all have programs in place that are designed to subsidize reforestation.

4.1.3 Tax and Subsidy Combinations

In the U.S., ten states have bottle bills (Container Recycling Institute, 1992). Return rates range from 75 percent to 90 percent (Wahl, 1995). Litter has reportedly decreased by 79 to 83 percent. However, deposit-refund systems can be quite costly to implement and administer. The design of the system is therefore of paramount importance and is the primary determinant of a system's success or failure. The U.S. General Accounting Office (1990) reports that firms face increased costs of between 2.4 and 3.2 cents per container in states other than California, where individual stores are responsible for collecting bottles and refunding money to customers. In California the state government has organized redemption centers within convenience zones of a half-mile radius, so that retailers do not have to be directly involved in the redemption of cans and bottles. As a result, costs in California are much lower than in other states, approximately 0.2 cents per can or bottle (Palmer et al., 1996). Deposit-refund systems in the U.S. have also been applied to used motor oil and lead-acid car batteries.

Deposit-refund systems have been used in Europe to promote proper disposal and recycling of bottles, plastic containers, car hulks, motor oil, and car batteries. Return rates are between 40 and 100 percent. Often refunds are too small to induce the desired amount of returns. For instance, Sweden has attempted to mandate a 75 percent return on car hulks without much success. Because promised refunds are low, other alternatives such as storing and selling parts and then dumping the remaining hulk are preferred to returning it for a refund. Norway's deposit-refund system on car hulks performs quite well due to a much higher refund. The return rate is between 90 and 99 percent.

6. Market-Based Policies

Other countries impose presumptive charges on water pollutants based on their process and equipment. Polluters that reduce their emissions below the presumed level and can prove it by supplying acceptable monitoring data can get a refund based on the size of the reduction. This system has the advantage of shifting the monitoring burden onto the polluter. One such system in the Netherlands resulted in dramatic reductions in water pollution (World Bank, 2000).

4.1.4 Permits

While Europe makes greater use of taxes, the United States has extensive experience with tradable permit systems. The original Emissions Trading Program, introduced in 1975, was based on a system of credits. Later programs allowed for the trade of pollution allowances. Permit systems for leaded gasoline phasedown, water pollutants, ozone-reducing chlorofluorocarbons, and the Regional Clean Air Incentives Market (RECLAIM) in California provide valuable lessons for the development of current and future permit systems.

The Fox River water permit system in Wisconsin was designed to control biological oxygen demand (BOD). Initial studies suggested that the use of a permit system would save over $7 million per year compared to other policy alternatives (O'Neil, 1983). However, only one trade occurred over a six-year period. What went wrong? Only a small number of firms could conceivably trade with each other. Transaction costs of trading were also high. Each trade required a modification or re-issuance of the existing permit. A firm had to justify its need for a permit to establish eligibility. Trades that only reduced operating costs were not allowed. Also, firms faced great uncertainty when attempting to evaluate the future value of a permit (Hahn, 1989).

A more successful tradable permit system for sulfur dioxide emissions was put into place in 1995 in the U.S. to address the problem of acid rain. Emissions of sulfur dioxide were limited to 10 million tons, 50 percent of 1980 levels. Of the plants that participated, most were coal-fired units located east of the Mississippi River. Permits were allocated to units on a historical basis. Units can use the permit, sell the permit to other units, or "bank" the permit for use in subsequent years. Continual emission monitoring (CEM) systems have allowed the government to easily monitor and enforce emission restrictions in accordance with the permits (Stavins, 1998). The second phase of the program, begun in 2000, imposes further limits on the level of sulfur dioxide emissions and brings almost all sulfur dioxide generating units into the system (Stavins, 2001).

Initial evaluations of the first phase of this implementation suggest that it has significantly reduced emissions at a low cost. A significant level of trading has occurred resulting in savings of over $1 billion per year in comparison to command-and-control alternatives. Emissions in 1995 were almost 40 percent below the 10 million-ton limit. One reason for such large reductions in sulfur dioxide emissions below the allowable limit is the banking of permits for future use (Schmalensee et al., 1998). Overall, U.S. experience with tradable permits suggests that a market with low transaction costs and "thick" with buyers and sellers is critical if pollution is to be reduced at lowest cost.

4.1.5 Information as Regulation

One of the most touted information regulatory programs in the United States is the Toxics Releases Inventory (TRI). The TRI requires that manufacturing firms report emissions to land, air, and water on an annual basis at the plant level if emissions of a toxic chemical exceed a threshold. Evidence indicates that the most polluting firms experienced significant declines in stock prices on the day TRI emissions were released to the public (Hamilton, 1995). Firms that experienced the largest drop in their stock prices reacted by reducing their reported emissions most in subsequent years (Konar and Cohen, 1997).

The United Kingdom, the Netherlands, Norway and Sweden all have publicly available pollution emission inventories in place. The European Union and many individual European countries have plans to implement their own pollution reporting systems in the near future (Sand, 2002).

4.2 Experiences with Market-Basket Instruments in Latin America

Market-based instruments have been used far less in Latin America than in either the United States or Europe. However, their use is on the rise. Many Latin American countries have passed laws that grant the right to use emission charges or permit trading systems to address pollution problems. While these countries have encountered complications in implementing market-based incentives, they have greatly increased their understanding of the necessary conditions for success, and continue to improve implementation of incentives in policy making.

4.2.1 Emissions Taxes

The primary experiments with direct pollution taxes in Latin America are water charges. Policies in Brazil, Chile, Colombia, and Mexico, for example, set water charges to reflect use and effluent discharges. These policies have

6. Market-Based Policies

generally failed due to lack of enforcement. Local authorities, though many times constitutionally empowered to assess pollution-based water charges, find that institutional constraints and polluter opposition prevent them from properly monitoring pollution and assessing charges. Colombia managed to implement a moderate exception to this failure. In 1993 it implemented fees related to biological oxygen demand (BOD) and total suspended solids (TSS) levels in water while maintaining a pre-existing minimum effluent standard. A small charge per kilogram of water is assessed that translates to a 5 to 10 percent increase in a household's monthly water bill. These charges are reportedly still too low to cover treatment costs for public utilities, and collection of fees is still one third of the total amount charged, but both BOD and TSS have seen moderate declines since the implementation of the program (Ardila and Guzman, 2002).

Other than these water charge experiments, there has been little experience with direct emission taxes in Latin America. Some conventional taxes designed to generate revenue act as indirect taxes on emissions. For example, Mexico taxes older cars but does not set the tax to reflect pollution differences across car vintages. Generally, however, the Latin American public strongly opposes taxes on polluting goods on the grounds that such taxes are regressive and may hinder local economic development.

4.2.2 Environmental Subsidies

Environmental subsidies in Latin America are more common and often take the form of subsidized credit and tax relief:

> They cover abatement investments or clean technology adoption in the industrial sector in Brazil, Mexico, and Colombia, ... reforestation activities in Chile and Colombia, mercury emission control in Ecuador, cleaner energy uses in ... Ecuador, and Brazil (solar, wind, and gas/hydroelectricity sources, respectively) and chlorofluorocarbon (CFC) phaseout in Colombia, Chile, and Brazil (Huber et al., 1998).

Brazil also redistributes revenue back to states from the state value-added tax in partial accordance with the land-use restrictions and conservation provisions in each state. Such policies, perhaps because they involve a transfer from the public to the private sector and not vice-versa, appear to be more successful than emissions taxes. Subsidies to abatement technology have had limited success in inducing proper behavior due to low enforcement of existing emission standards, and lack of monitoring to ensure proper investment of funds (Huber et al., 1998).

4.2.3 Tax-Subsidy Combinations

Taxes in combination with subsidies are used to address environmental problems in Latin America. As in the United States and Europe, these tax-subsidy instruments most often taken the form of deposit-refund systems. Huber et. al find well-developed deposit-refunds in most major Latin American countries. Most of these are voluntary systems and apply to items such as glass bottles or aluminum cans that have a high resale or reuse value. Informal systems also exist in many countries for paper, plastic, and other recyclable materials such as used motor oil and tires. Mexico is the only Latin American country that has a mandatory deposit-refund system in place, for car batteries (Huber et al., 1998). Costa Rica also has a performance bond requirement that covers the cost of reforestation of developed or logged land (World Bank, 1996).

No formal evaluations of these deposit-refund systems exist, but analysts point to the feasibility of these systems due to low legal, institutional, and political barriers (Huber et al., 1998). Deposit-refund systems in Latin America also are associated with potential employment and income benefits from collecting and returning containers, particularly for individuals outside the formal job market.

4.2.4 Permits

Several countries participate in the Clean Development Mechanism, an international offset program for carbon emissions under the Kyoto Protocol. Firms that find reducing their own carbon emissions relatively expensive can obtain carbon credits through the support of sequestration programs in countries where carbon abatement is relatively inexpensive. Costa Rica has had a system of tradable offsets since 1995. It has arranged the sale of recognized offsets to both U.S. and Norwegian firms and in return has transferred pastureland into managed forest. Belize, Bolivia, Honduras, Nicaragua, and Panama are all developing similar programs (World Bank, 1996). Permit systems to deal with air emissions are also under consideration in both Peru and Mexico.

The only country in Latin America that has its own tradable permit system in place is Chile (Huber et al., 1998). The permit system is for a daily allowable emissions capacity of particulate matter and is coupled with a command-and-control emissions standard. Since monitoring of plant-level emissions does not occur in Chile, these permits are based on current capacity of a plant and the type of fuel used rather than on emissions. These permits are also issued in perpetuity, so that new sources must purchase permits from existing plants already in the system (Montero et al., 2002).

6. Market-Based Policies 143

Since trading started in 1995, most of it has been within firms and not between them (Huber et al., 1998). A number of problems with Chile's tradable permit system stem from institutional deficiencies and poor market design and development. Initial allocations of permits were not well-defined, and property rights were not well-established. Legal uncertainties exist regarding enforcement. High transaction costs and lengthy approval processes for trades, and fear on the part of plants that if they sell permits they will not be able to purchase them back when conditions change have severely limited the number of trades that actually take place (Montero et al., 2002). In spite of these problems, introduction of the permit system has resulted in improved compliance with the emissions standard and better air quality for Santiago (Huber et al., 1998).

4.2.5 Information as Regulation

Information about plants' emissions has not been used by countries in Latin America as a form of environmental regulation.[8] However, Mexico is currently developing an emission reporting system similar to the TRI in the U.S.[9] The World Bank has confirmed that stock markets in developing countries react similarly to those in industrialized countries when information about environmental performance is made public, and in fact that the response is larger than those reported for the TRI (Dasgupta et al., 1997).

5. CONCLUSIONS AND POLICY RECOMMENDATIONS FOR LATIN AMERICA

While there is little direct evidence on how well market-based instruments perform in Latin America, a number of policy recommendations can be made. We base these recommendations on the experiences of the United States and Europe, on the Latin American experiences that do exist, and on how environmental policy making in Latin America differs from that in the United States and Europe.

---

[8] At least one developing country, Indonesia, has had success with the use of information as regulation. Indonesia implemented a public disclosure program of plants' environmental performance and a rating system referred to as PROPER (for more information, see World Bank, 2000; Pargal and Wheeler, 1996; Afsah et al., 2000). There is limited evidence of success with "private enforcement" lawsuits, a possible byproduct of public information on emissions, in Colombia, Chile, and Mexico (see Tietenberg, 1996).
[9] For more information on the development of Mexico's reporting system, see www.cec.org.

1. Because monitoring and enforcement are both difficulty and costly, pollution control policies should attempt to minimize the need for large amounts of direct monitoring. Market-based incentives such as a combined tax and subsidy that shift the burden of proof to the consumer or firm through market transactions are often superior to the use of a tax or subsidy alone.

2. Latin America governments are trying to meet important economic development goals while constrained by a decidedly limited budget. Preference should be given to policies that deliver the most benefits relative to the costs of the policy and are revenue-increasing on net, so that weak environmental protection institutions can become self-sustaining.

3. Instruments that collect environmental tax revenue may allow governments to address distributional concerns, and to increase efficiency in other markets by allowing for the option to recycle revenue (for instance, by reducing the tax burden borne by the poorest people).

4. Funding and the exchange of expertise are needed to help build the institutional capacity of environmental protection agencies throughout Latin America. This will increase effectiveness in policy-making, and allow consideration of a wider number of policy options.

5. Based on the experiences of the United States and Europe, no one policy instrument is suitable for addressing every pollution problem. In fact, political preferences may be a great determinant of the feasible use of certain instruments: for instance, permits are generally preferred in the United States, while environmental taxes are preferred in Europe.

6. In many cases, hybrid instruments that combine aspects of command-and-control standards with market-based instruments allow governments to minimize large uncertainties in the distribution and magnitude of costs and benefits while dealing effectively with practical considerations. In particular, the use of information as regulation and standard and pricing approaches may have promising futures in Latin America.

Continued research into the applicability of market-based instruments to environmental problems in developing countries is needed. Systematic evaluation of Latin America's experiences with these instruments are few. However, discussions to-date indicate that policies that incorporate at least some aspects of incentive-based policies have the potential to allow for more cost-effective pollution control policy than offered by traditional command-and-control.

ACKNOWLEDGMENTS

For their useful comments and suggestions, the authors thank Charles Griffiths, Raymond Robertson, Tom Tietenberg, and an anonymous referee. The views expressed in this paper are those of the authors and do not reflect the official views or policies of the U.S. Environmental Protection Agency or Macalester College.

REFERENCES

Ardila, S., and Guzman, Z., 2002, Implementation of Water Discharges in Colombia: Problems and Achievements, Presentation, Second World Congress of Environmental and Natural Resource Economists. June.

Afsah, S., Blackman, A., and Ratunanda, D., 2000, How Do Public Disclosure Pollution Control Programs Work? Evidence from Indonesia, Discussion Paper 00-44, Resources for the Future, Washington, D.C.

Baumol, W., and W. Oates, 1988, *The Theory of Environmental Policy*, Cambridge University Press, 2nd ed., Cambridge.

Birdsall, N., and Wheeler, D., 1993, Trade Policy and Industrial Pollution in Latin America: Where are the Pollution Havens?, *Journal of Environment and Development* **2**(1):137-149.

Blackman, A., 2000a, Informal Sector Pollution Control: What Policy Options Do We Have?, *World Development* **28**(12):2067-82.

Blackman, A., and Harrington W., 2000b, The Use of Economic Incentives in Developing Countries: Lessons from International Experience with Industrial Air Pollution, *Journal of Environment and Development* **9**(1):5-44.

Coase, R., 1960, The Problem of Social Cost, *Journal of Law and Economics* **3**:1-44.

Container Recycling Institute, 1992, *Beverage Container Deposit System in the United States*, Container Recycling Institute, Washington, DC.

Cropper, M., and Oates, W., 1992, Environmental Economics: A Survey, *Journal of Economic Literature* **30**:675-740.

Dasgupta, S., Laplante, B., and Mamingi, N., 1997, Capital Market Responses to Environmental Performance in Developing Countries, Development Research Working Paper, no. 1909, World Bank, Washington, DC.

DeSoto, H., 2000, *The Mystery of Capital: Why Capitalism Triumphs in the West and Fails Everywhere Else*, Basic Books, New York.

Environmental Law Institute, 1997, *The United States Experience with Economic Incentives in Environmental Pollution Control Policy*, Environmental Law Institute, Washington, DC.

EPA, 2001, *The United States Experience with Economic Incentives for Protecting the Environment*, U.S. EPA, Washington, DC.

Eskeland, G., and Jimenez, E., 1992, Policy Instruments for Pollution Control in Developing Countries, *The World Bank Research Observer* **7**(2):145-169.

Eskeland, G., and Devarajan, S., 1996, *Taxing Bads by Taxing Goods: Pollution Control with Presumptive Charges*, World Bank, Washington, DC.

European Environment Agency, 1996, *Environmental Taxes: Implementation and Environmental Effectiveness*, Environmental Issues Series Report, no. 1, European Environment Agency, Copenhagen.

European Environment Agency, 2000, *Environmental Taxes: Recent Developments in Tools for Integration*, Environmental Issues Series Report, no. 18, European Environment Agency, Copenhagen.

Fullerton, D., 1996, Why Have Separate Environmental Taxes?, in: *Tax Policy and the Economy*, J. Proterba, ed., MIT Press, Cambridge.

Goldenberg, E., 1993, The Design of an Emissions Permit Market for RECLAIM: A Holistic Approach, *UCLA Journal of Environmental Law and Policy* **11**(2):297-328.

Goulder, L., Parry, I., and Burtraw. D., 1997, Revenue Raising vs. Other Approaches to Environmental Protection: The Critical Significance of Pre-Existing Tax Distortions, *RAND Journal of Economics* **28**(4):708-731.

Hahn, R., 1989, Economic Prescriptions for Environmental Problems: How the Patient Followed the Doctor's Orders, *Journal of Economic Perspectives* **3**(2):95-114.

Hamilton, J., 1995, Pollution as News: Media and Stock Market Reactions to the Toxics Release Inventory Data, *Journal of Environmental Economics and Management*, **28**(1):98-113.

Jaffe, A., and Stavins R., 1995, Dynamic Incentives of Environmental Regulations: The Effects of Alternative Policy Instruments on Technology Diffusion, *Journal of Environmental Economics and Management* **29**(3):S43-S63.

Konar, S., and M. Cohen, 1997, Information as Regulation: The Effect of Community Right-to-Know Laws on Toxic Emissions, *Journal of Environmental Economics and Management*, **32**(1):109-124.

Laffont, J., and Tirole, J., 1996, Pollution Permits and Environmental Innovation, *Journal of Public Economics* **62**(1,2):127-140.

Levinson, A., 1996, Environmental Regulations and Industry Location: International and Domestic Evidence, in: *Fair Trade and Harmonization: Prerequisite for Free Trade?*, J. Bagwati, and R. Hudec, eds., MIT Press, Cambridge.

Montero, J., Sanchez, J., and Katz, R., 2002, A Market-Based Environmental Policy Experiment in Chile, *Journal of Law and Economics* **45**(1):267-287.

Naysnerski, N. and Tietenberg, T., 1992, Private Enforcement of Federal Environmental Law, *Land Economics* **68**(1):28-48.

O'Connor, D., 1996, *Applying Economic Instruments in Developing Countries: From Theory to Implementation*, Special Paper, OECD Development Centre, Paris.

OECD, 1999, *Economic Instruments for Pollution Control and Natural Resources Management in OECD Countries: A Survey*, OECD Development Centre, Paris.

O'Neil, W., 1983, The Regulation of Water Pollution Permit Trading under Conditions of Varying Streamflow and Temperature, in: *Buying a Better Environment: Cost-Effective Regulation Through Permit Trading*, E. Joeres, and M. David, eds., University of Wisconsin Press, Madison, pp. 219-231.

Pargal, S., and Wheeler, D., 1996, Informal Regulation of Industrial Pollution in Developing Countries: Evidence in Indonesia, *Journal of Political Economy* **104**(6):1314-1327.

Parry, I., 1998, Pollution Regulation and the Efficiency Gains from Technology Innovation, *Journal of Regulatory Economics* **14**(3):229-254.

Palmer, K., Sigman, H., and Walls, M., 1996, The Cost of Reducing Municipal Solid Waste, Discussion Paper 96-35, Resources for the Future, Washington, DC.

Pigou, A., 1932, *The Economics of Welfare*, MacMillan and Company, 4th ed., London.

Roberts, M., and Spence, A., 1976, Effluent Charges and Licenses Under Uncertainty, *Journal of Public Economics* **5**(3,4):193-208.

Ros, J., Draisma, J., Lustig, N., and Ten Kate, A., 1996, Prospects for Growth and the Environment in Mexico in the 1990s, *World Development* **24**(2):307-324.

Sand P., 2002, The Reality of Precaution: Information Disclosure by Government and Industry, 2nd Transatlantic Dialogue on *The Reality of Precaution: Comparing Approaches to Risk and Regulation*, Airlie House, VA, 15 June.

Schmalensee, R., Joskow, P., Ellerman, A., Montero, J., and Bailey, E., 1998, An Interim Evaluation of Sulfur Dioxide Emissions Trading, *Journal of Economic Perspectives* **12**(3):53-68.

Schneider, F., and Enste, D., 2000, Shadow Economies: Size, Causes, and Consequences, *Journal of Economic Literature* **38**(1):77-114.

Spence, A., and M. Weitzman, 1978, Regulatory Strategies for Pollution Control, in: *Approaches to Controlling Air Pollution*, A. Frielander, ed., MIT Press, Cambridge, MA.

Stavins, R., 1995. Transaction Costs and Tradeable Permits, *Journal of Environmental Economics and Management* **29**(2):133-148.

Stavins, R., 1998, What Can We Learn from the Grand Policy Experiment? Lessons from SO2 Allowance Trading, *Journal of Economic Perspectives* **12**(3):69-88.

Stavins, R., 2000, Market-Based Environmental Policies, in: *Public Policies for Environmental Protection*, P. Portney and R. Stavins, eds., Resources for the Future, Washington, DC, pp. 31-76.

Stavins, R., 2001, Experience with Market-Based Environmental Policy Instruments, Discussion Paper 01-58, Resources for the Future, Washington, DC.

Tietenberg, T., 1985, *Emissions Trading: An Exercise in Reforming Pollution Policy*, Resources for the Future, Washington, DC.

Tietenberg, T., 1996, Private Enforcement of Environmental Regulations in Latin America and the Caribbean: An Effective Instrument for Environmental Management?, Working Paper No. ENV-101, Inter-American Development Bank, Washington, DC.

Tietenberg, T., and Wheeler, D., 2001, Empowering the Community: Information Strategies for Pollution Control, in: *Frontiers of Environmental Economics*, H. Folmer, H. Gabel, S. Gerking, and A. Rose, eds., Edward Elgar, Cheltenham, UK, pp. 85-120.

US GAO, 1990, *Trade-Offs Involved in Beverage Container Deposit Legislation*, US GAO, Washington, DC.

Wahl, M., 1995, Letter to State Representative Lisa Naito, Oregon Department of Environmental Quality, Portland.

Weitzman, M., 1974, Prices versus Quantities, *Review of Economic Studies* **41**(4):477-491.

World Bank, 1996, *Five Years After Rio: Innovations in Environmental Policy*, World Bank, Washington, DC.

World Bank, 2000, *Greening Industry: New Roles for Communities, Markets, and Governments*, Oxford University Press, New York.

Xing, Y., and Kolstad, C., 2002, Do Lax Environmental Regulations Attract Foreign Investment?, *Environmental and Resource Economics*, (forthcoming).

Chapter 7

A DEEPER SOLUTION FOR THE INTERNATIONAL COMMONS
Building an effort control regime for the Eastern Tropical Pacific tuna fishery

Brian Potter[1]
[1]*Department of Political Science, The College of New Jersey, Ewing, NJ 08628-0718*

Abstract: The parable of the tragedy of the commons tells that resources held under open access conditions are prone to over-exploitation. For fisheries, regulations to limit aggregate catch improve resource use yet promote over-investment. Optimal regulation would limit the investments and labor dedicated to harvesting, a task quite difficult in a global forum. Motivated by three reasons, the member-states of the Inter-American Tropical Tuna Commission have attempted such limits for the purse-seine yellowfin fishery, in addition to implementing other conservation measures. First, increases in fishing capacity by existing participants and newcomers have resulted in financial losses for the major fleets. Second, the resolution of the tuna-dolphin controversy encouraged additional harvesting capacity in an already-crowded fishery. Finally, bureaucratic changes in some countries have empowered fishery professionals who balance sustainable resource use with economic development. The strength of these influences, as well as efforts and concessions to create a regime, vary among the states involved.

Key words: fishery management, natural resource policy, common property, open access, over-capitalization, Inter-American Tropical Tuna Commission, Latin America.

1. INTRODUCTION

Successful international regimes to manage common property resources remain rare (Miles, 2002). Most resources shared internationally have fallen victim to Garret Hardins' tragedy of the commons, whereby lack of cooperation to prevent overexploitation of a renewable resource leads to the demise of not only the resource but also the livelihood of those who depend on it (Hardin, 1968). While creating common property institutions remains a formidable challenge within countries, international regimes to address the tragedy of the commons must navigate the added obstacles of sovereignty.

This chapter presents the most successful case to date of international management of a common property resource. Recently, participants in the Eastern Tropical Pacific (ETP) tuna fishery created and approved a binding regime that not only addresses Hardin's tragedy yet goes further to ensure that the result is economically efficient and legitimate among users. The ETP Tuna Regime addresses the heart of the common property problem, restricting the economic inputs directed at a resource. Regimes to limit harvesting effort are more difficult to create yet will yield increased legitimacy and economic efficiency. For global fisheries, the ETP Tuna Regime is the first example of a capacity-restricting agreement.

The success of the regime holds importance not only for the ETP tuna stocks and industry but provides an example for other trans-boundary resources. The fishery is just one of many natural resources that require cooperative management by sovereign states. The behavior of countries in managing ETP tuna offers lessons to future negotiations concerning multilateral common property resources (i.e. other fisheries, biodiversity, carbon sinks, the atmosphere). In managing the global commons, states are not free to autonomously negotiate international commitments. Instead, elected officials balance the needs and power of constituents against foreign policy objectives. This chapter contributes to the investigation of international environmental conflict and cooperation by explaining how and why Latin American countries have strived to create a regime to manage trans-boundary resources.

Many wonder how Latin American countries, most burdened by reoccurring economic crises and challenges to government capacity, were able to create the most effective international fishery regime to date. I argue that the success of fishery development efforts, the resolution of trade disputes, and changes in government structure forced member governments and their industry constituents to make painful concessions to insure the sustainability of the fishery. Since the early 1980s several Latin American countries have aggressively developed harvesting and processing capacity as a means of development. Their efforts have proven all too successful and the increase in fishing capacity has led to a "race for fish" and declining profits. Second, the resolution of the tuna-dolphin trade row between US preservationist groups and Latin American tuna firms offered improved marketing prospects for Latin American fishers. However, removing US import restrictions would allow the return of the US and other fleets. The specter of additional participants in a fishery already over-pressured by Latin American and multinational harvesters provides reason to create limitations on entry.Finally, the intergovernmental climate for negotiations has improved. In some states, jurisdiction over high seas fishery issues has been gradually transferred from foreign ministries to environmental or fisheries bureaus and their mission of protecting resources. The advances of the fishing industries, multilateral trade harmonization and environmental ministries provide reason for the creation of an effort-based regime, although this process varies by country.

After explaining the importance of the ETP tuna fishery for national economic development, this chapter distinguishes between two types of common property regimes. The participant countries opted for the more stringent of these regime types. Next, the specific regulations of the regime and the bargaining positions of the participating countries are described. The following section explains how pressure from the end of the "dolphin-safe" embargo, and changes in profitability and the international leadership structure encouraged international cooperation. The conclusion summarizes the argument and explores the contribution of the ETP tuna regime to other efforts to manage global common property.

2. DEVELOPMENT MODELS AND THE EASTERN TROPICAL PACIFIC TUNA FISHERY

Since the late 1970s, many Latin American countries have invested in the Eastern Tropical Pacific tuna fishery in pursuance of economic development. The tuna fishery stands out as a significant contributor to the exports of nations with abundant resources or large fishing fleets. Furthermore, the impact on local economics and politics is far greater. Without an international agreement to limit and allocate catches, the fishery may fall prey to the tragedy of the commons, whereby tuna stocks become depleted to the point where they are no longer commercially harvestable. The profitability of past development efforts hinges on the success of an exclusive international accord.

The United Nations Convention on the Law of the Sea (UNCLOS) granted coastal countries partial property rights over resources up to 200 miles from shore.[1] The most promising marine resource for many Latin American states was the US-dominated tuna fishery in the Eastern Tropical Pacific. Mexico, Ecuador and Costa Rica respectively found themselves nominal stewards of the three national zones holding the largest stocks of

---

[1] The fifth United Nations Conference on the Law of the Sea (UNCLOS) allowed states to extend their jurisdiction of economic resources from twelve to 200 miles, resulting in a transfer of approximately one-fifth of the world's surface area. These areas of jurisdiction are commonly called Exclusive Economic Zones (EEZs).

tuna, an abundance based on two geographical reasons. First, ocean currents and upwellings offer biologically rich fishing grounds. Second, all three countries enjoy an enhanced coastal zone due to their sovereignty over islands: respectively the Guadalupe and Revilla Gigedos, the Galapagos and the Cocos Islands. The promise of lucrative industry fueled optimism not only on the part of the three tuna-abundant countries but also for neighboring states anxious to develop harvesting, processing and marine services sectors directed at US markets.

While all Latin American countries had extended their jurisdiction by 1979, they would have to wait another decade to expel US vessels from their waters. Arguing that their targeted fishery defied international boundaries, the powerful US tuna lobby influenced the State Department to declare tuna as exempt from national jurisdiction and requiring international management. As tuna stocks migrate both among countries' national zones and into international waters, an international agreement is indeed necessary; the abundance of tuna in each national zone varied by year to year, creating a resource that might be temporarily excludable yet over the long term represented a common fishery. The State Department policy, however, implied that the US would dictate the terms of the organization. This controversial decision announced the era of "tuna wars" where US vessel owners flagrantly violated the UNCLOS rights of Latin American states. Latin American states seized over 220 tuna vessels in the following decade. The US Congress responded first by generously compensating tunamen for any losses incurred and second by boycotting all marine products from countries hostile to the US tuna industry. Frustrated by the loss of market and international business links, Mexico, among others, seemed ready to compromise in 1989.

A well-publicized 1990 videotape showing the death of dozens of dolphins in a tuna net marked the entry of a new class of actors which would frustrate the ambitions of US and Latin American tuna firms alike. Almost all of the Eastern Pacific yellowfin catch is caught by purse-seining (pronounced sane-ing), whereby a large vessel and supporting boats encircle a surface school of tuna with a net, which is then closed on the bottom and hauled on board. Surface schools in the Eastern Pacific are associated with dolphin, meaning that adult tuna almost always closely swim underneath schools of dolphin. Seiners use dolphin sightings to find tuna, harming dolphin stocks as they encircle them and then inevitably catch dolphins in the seine. While most dolphins are released, any harm to their population violates the US Marine Mammal Protection Act. Led by the Earth Island Institute, the US environmental movement succeeded in prohibiting the import of any tuna by countries that encircled dolphin beginning in February 1991. A secondary boycott blocked the sale of tuna products from countries that purchase raw material from primary embargoed nations. The boycott was informally strengthened two years later when the three major US tuna processors agreed to buy only "dolphin-safe" tuna.[2] Unwilling to challenge environmental interests with the same zeal with which they battled Latin American countries, most US harvesters moved to the Western Pacific, where tuna are not associated with dolphin. Latin American firms remained tied to their national zones and faced at least another decade without their most important market.

Throughout these diplomatic troubles, Latin American tuna industries found ample support from their respective governments. Early growth of the tuna industries occurred under the economic program of Import-Substituting Industrialization, a state-led effort to replace dependence on industrial imports with increased national self-sufficiency. Latin American governments provided preferred financing, favorable exchange rates, infrastructure investment and other support for their tuna industries. Some countries such as Mexico invested in state-owned, vertically integrated tuna firms. Although this support drastically diminished with the neo-liberal reforms of the 1980s and 1990s, leaders of the tuna industries maintained close ties with officials in the fisheries, finance and foreign ministries. Such links would prove important during the dolphin-safe controversy and in the negotiations of the tuna regime.

3. PREVENTING COMMON RESOURCE EXPLOITATION THROUGH CATCH AND EFFORT CONTROLS

The tragedy of the commons results from overexploitation of resources held under open access conditions. Fisheries are a renewable resource that, if managed effectively, can produce indefinitely sustainable harvests. However, when catch rates exceed the growth of the stock, the size of the population dwindles, thereby producing decreased harvests for following seasons. If overharvesting continues, resource depletion will worsen and may lead to extinction in extreme and very rare cases. When access is difficult to exclude, restraint by any one user is fruitless, as her sacrifice only enhances the yields of competitors. The focus of the tragedy is on declining abundance of resources held under common access.

The non-regulation of common property poses a tragedy for resource users as well as the fish stock. As the stock population declines, so do

---

[2] Even if the embargo were lifted, few US consumers would have purchased tuna without the dolphin-safe label.

catches and hence revenues.[3] The costs of harvesting rise when fish become harder to find, forcing fishers to spend more time at sea in search of the vanishing resource. The profits enjoyed before over-harvesting decrease (and may turn negative for a brief period of time or longer if subsidized) as harvesting costs increase while revenues decline. Munro and Scott have termed this a Class I common property problem, examples of which are all too evident (Munro and Scott, 1985). One solution to over-harvesting is the abidance to overall catch limits, whereby harvesting for the season stops when catches meet the sustainable yield.

This parable and policy describe only part of the over-harvesting issue. Even if regulators could impose catch limits, the fishery would fall prey to Munro and Scott's Class II common property problem: excess harvesting capacity or fishing effort (the sum of all inputs, labor and investment, used in harvesting). Setting a limit on total catch may prevent resource decline yet will exacerbate a 'race for fish.' Fishers anticipate the annual closing of the fishery and, in turn, compete to harvest their largest share of the resource before the season ends. This competition draws excess capital and labor into the fishery and, over time, the increased inputs necessitate shorter seasons. The market failure of quota-based common property policy results in increased investment to harvest a diminishing resource.

The presence of excess fishing effort diminishes the political legitimacy of the regulatory regime. Fisheries are a fugitive and often remote resource and pose high costs of monitoring and compliance. A common belief in the utility of regulations is necessary to minimize the cost of enforcement efforts. When catch regulations force a halt to harvesting yet fixed costs (i.e. mortgage payments, vessel maintenance, lack of employment to generate alternative wage opportunities) continue, individual temptation to exceed catch limits may prove irresistible. This is especially true when harvesters experience mixed success during the legal season, which provokes

Table 7-1. Members of the Inter-American Tropical Tuna Commission

Country	Years
Costa Rica	1949-79, 1989-now
Colombia	1962-68
Ecuador	1961-68, 1997-now
El Salvador	1997-now
European Community	2000-now
France	1973-now
Guatemala	2000-now
Japan	1970-now
Mexico	1964-78, 1999-now
Nicaragua	1973-now
Panama	1953-now

---

[3] This assumes a stable price, which is not always the case.

7. A Deeper Solution for the International Commons

Table 7-1. (continued)

United States	1950-now
Vanuatu	1990-now
Venezuela	1992-now

resentment against a system that rewards some more than others. As a few harvesters begin to 'free ride' on the restraint of others, the common belief that catch regulations serve the greater good erodes. Support for the catch limits will diminish unless effort is restricted.[4]

Working under the auspices of the Inter-American Tropical Tuna Commission (IATTC), countries participating in the ETP yellowfin fishery have long recognized the need to abide by catch limits. The resource is common property: availability in each country's Exclusive Economic Zone (EEZ) varies by year and roughly 60 percent of catch comes from the high seas of the Commission's Yellowfin Regulatory Area (CYRA) (Inter-American Tropical Tuna Commission, 1999b). Table 7-1 lists countries participating in the IATTC while Table 7-2 shows the Commission's recommended yellowfin quota and total actual catch in the CYRA for the years 1967 to 1998. Catches at times exceed and in others fail to meet recommended limits. Over-harvesting has remained modest, save for the late 1980s when the US clashed with the resource owners over fishing access (and the lack of an El Niño weather effect permitted large catches). Total allowable catch has more than tripled since 1967, reflecting both a better knowledge of the resource and a willingness to travel farther from shore. The international quota system has worked well for decades, yet the historical participants fear that recent and anticipated increases in fishing effort (by established tuna industries and newcomers both), propelled by the opening of protected markets, will threaten the legitimacy and efficacy of catch regulations. Hence, the fisheries ministries and industry representatives of the leading states have sought to negotiate a regime to limit effort.

4. EVOLVING NORMS AND DISTRIBUTION IN THE ETP YELLOWFIN FISHERY

While all actors agree on the need to limit the expansion of effort, the process of creating an agreement has necessarily been slow for two reasons. First, international agreements on the distribution of a fixed share are subject

---

[4] One exception may be the use of individual transferable quotas, whereby the individual vessel owner controls her own investment decisions and the timing of effort over the season.

to satisfaction of all sovereign parties and global development norms. The sovereignty of states requires unanimous agreement, providing opportunistic actors the opportunity to hold out for a larger share. Both UNCLOS and the United Nations Food and Agricultural Organization's (FAO) Code of Conduct for Responsible Fishing allow privileges for poorer countries to

Table 7-2. Recommended catch and actual purse-seine catch for CYRA, 1967-1998

Year	Quota	Actual catch
1967	76.7	80.0
1968	84.4	100.9
1969	108.9	111.4
1970	108.9	127.8
1971	127.0 + (2 x 9.1)	102.2
1972	108.9 + (2 x 9.1)	136.5
1973	117.9 + (2 x 9.1)	160.3
1974	158.8 + (2 x 9.1)	173.2
1975	158.8 + (2 x 9.1)	158.8
1976	158.8 + (2 x 9.1)	190.2
1977	158.8 + (18.1 + 13.6)	182.7
1978	158.8 + (18.1 + 13.6)	166.0
1979	158.8 + (18.1 + 13.6)	175.9
1980	149.7 + (total of 40.8)	132.0
1981	149.7 + (3 x 13.6)	157.7
1982	145.1 + (2 x 13.6)	106.9
1983	154.2 + (2 x 13.6)	82.0
1984	147.0 + (2 x 13.6)	128.6
1985	157.9 + (18.1 + 9.1)	192.5
1986	158.8 + (2 x 13.6)	228.1
1987	No quota set	248.2
1988	172.4 + (2 x27.2)	267.3
1990	181.4 + (5 x 18.1)	242.3
1991	190.5 + (4 x 18.1)	219.5
1992	190.5 + (4 x 18.1)	221.3
1993	226.8 + (4 x 22.7)	213.3
1994	226.8 + (4 x 22.7)	197.2
1995	213.2 + (3 x 18.1)	196.2
1996	213.2 + (3 x 18.1)	218.0
1997	220.0 + (3 x 15.0)	213.2
1998	210.0 + (3 x 15.0)	238.6

Notes: Data are from various editions of the Commission's *Annual Report* and are in expressed in thousands of metric tons, round weight. Figures in parentheses indicate additional increments of catch allowed after the initial quota was set. For example in 1998, three additions of 15,000 tons each (3 x 15.0) were allowed after the 210,000 limit was reached, for a total allowable catch of 255,000 metric tons of yellowfin. From 1980 to 1997, such additions were approved but not implemented.

develop their fishing industries. However, for the purposes of negotiations, comparative levels of local poverty and the contribution of a potential fishing industry to alleviate misery are difficult to quantify. Translating such concerns into fishing effort redistributions relies more on perception than hard data. Thus sovereignty and economic differences among states pose problems in dividing the fixed, optimal amount of fishing capacity.

A second reason for the slow evolution of the effort regime is to allow for the common legitimacy of final regulations among heterogeneous actors. The effort regulations do not stand alone, but instead are combined with a set of substitutable regulations to create an international regime. Stricter effort limits partially or completely alleviate the need for catch limits and other harvesting regulations. For example, the Commission has calculated the months of fishing that would be allowed under different levels of aggregate effort: purse seining could continue year-round if capacity was limited to 135,000 metric tons (mt.), while the season would have to be cut by two months if effort remained at 166,964 mt. (Inter-American Tropical Tuna Commission, 2000d). Other prohibitions such as restrictions against targeting juvenile tuna or requiring retention of bycatch would allow either longer seasons or greater fishing capacity. Participating states come to the negotiations employing different mixes of potentially prohibited practices whose continued use would be costly to monitor and sanction without the full cooperation of each government and industry. Regime legitimacy decreases future costs of compliance and unites the participants against potential intrusion by non-Commission members.

Beyond mere temporary agreements, international regimes provide principles, norms and decision-making procedures for cooperation in addressing future challenges. Thus, effective regimes survive despite the changing interests of members. The process and results of negotiations lead to common expectations and norms about the behavior of members (Krasner, 1983). Lengthy discussions and negotiations prevent the imposition of rules on unwilling actors, a development that would deter compliance and enforcement. While a complete harmonization of interests is unlikely, the productive participation of each actor improves the efficacy of the eventual regime by allowing actors to identify their input into the rulemaking process and realize the common good produced by the comprehensive set of regulations.

The ETP Tuna Regime consists of one long-established practice, three conservation measures and a resolution on fishing effort. The Commission for three decades has recommended limits on the total allowable catch of yellowfin tuna. More recently, participants in the Commission agreed to three yellowfin conservation measures and later, the Resolution on Tuna Fleet Capacity. These regulations have imposed short-term costs on the purse-seine fleet in the hopes of long-term sustainable resource use and a common vision among members of acceptable behavior in the fishery.

4.1 Yellowfin conservation measures

With growing concern for yellowfin management, members of the Commission have agreed to conservation-minded restrictions that impose short-term adjustment costs on members and seek to exclude fishery vessels from countries that are not members or fail to cooperate with Commission policies.

Beginning with the tuna-dolphin controversy, the use of fish aggregating devices (FADs) increased.[5] While FADs provide an alternative to setting nets on dolphin, they lead to the bycatch of non-tuna species and juvenile yellowfin. Harvesting and processing countries not dependent on the US market opposed FADs, claiming that they led to inferior product and depleted the yellowfin stock. For example, on 15 July 1999, Costa Rica, a processing country selling to Latin American markets, unilaterally prohibited FADs in its EEZ (Instituto Costaricense de Pesca y Acuicultura, 2002). A June 1998 Commission resolution limited the number of FADs that each vessel could carry, prohibited high-seas transshipment of tuna (which discouraged large catches of juvenile fish) and eliminated the use of tender vessels, non-fishing vessels that service FADs (Inter-American Tropical Tuna Commission, 1999a). The regulation represents a compromise between countries that seek to target the US market and those not subject to strict US dolphin regulations.

To allow for greater conservation of tuna stocks, the Commission adopted at its 66th Meeting (12-15 June 2000) a resolution calling for purse-seiners to retain and 'land all bigeye, skipjack and yellowfin tuna caught, except fish considered unfit for human consumption for reasons other than size' (Inter-American Tropical Tuna Commission, 2000a). Allowing the discard of small fish gives incentive to high-grade, wherein fishers dispose of less valuable fish at sea to make room in their holds for those commanding a higher price. On the other hand, if vessels were forced to land the less valuable fish, they have incentive to avoid the catch of juvenile tuna, which would then survive to grow and regenerate. The resolution also requires 'vessels to promptly release, to the extent practicable, all sea turtles, sharks, billfishes, rays, mahi-mahi and other non-target species' (Inter-American Tropical Tuna Commission, 2000a). While the regulation concerning non-tuna species relies on a vague requirement, the tuna regulation is specific and enforceable. The retention requirement represents

---

[5] Fish Aggregating Devices are man-made floating objects around which fish gather. Their use decreases the cost of searching for stocks of fish.

a unanimously adopted measure that poses significant costs to purse-seine vessels yet improves resource management.

The construction of a Regional Vessel Register, to better monitor and enforce the anticipated effort regime, resulted from a resolution at the 66th Meeting. Each member country is required to supply the IATTC with detailed information, and changes hereto, on every vessel in its jurisdiction. Non-member governments with vessels in the EPO were requested to do the same (Inter-American Tropical Tuna Commission, 2000a). Non-member vessels were then targeted through a following resolution, which encouraged their compliance with Commission policies and prohibited non-members from engaging 'in activities that undermine the effectiveness of the (Commission's) conservation and management measures' (Inter-American Tropical Tuna Commission, 2000a). As non-cooperators would not be eligible for an effort allocation, abidance to Commission policy would result in their exclusion from the yellowfin fishery, an action allowed by the FAO Code of Conduct for Responsible Fishing.

Each of the above resolutions imposed costs on either members or non-members. However, the debate allowed compromise among sovereign states, facilitating the task of gaining domestic approval and increasing the chance of compliance. Commission actions on FADs, retention and non-members compliment the centerpiece of the regime, the Resolution on Tuna Fleet Capacity.

4.2 The Resolution on Tuna Fleet Capacity

In seeking to limit the amount of harvesting capacity in the purse-seine yellowfin fishery, Commission members began negotiating national quotas, a process that provoked conflict among states with varied interests. Temporarily successful, attempts to create a strict formula for setting effort limits proved divisive, disadvantaged the poorest of countries, and threatened to detract from the legitimacy of the regime. Unanimous agreement on capacity limits was reached by the exclusion of non-participants and replacement of national quotas by a transferable system representing the status quo.

At the 62nd Meeting, the Commission adopted purse-seine efforts limits in the CYRA for the year 1999. Shown in the third column of Table 3, each member state received a limit on vessel carrying capacity based on '…the catches of national fleets during the period 1985-1998; the amount of catch historically taken within the zones where each state exercises sovereignty or national jurisdiction; the landings of tuna in each nation; the contribution of each state to the IATTC conservation program; including the reduction of dolphin mortality; and other factors' (Inter-American Tropical Tuna

Commission, 1999a). While the purpose of the program is to 'address the potential problem of excess capacity' (Inter-American Tropical Tuna Commission, 1999a), in truth each country received an effort limit equal

Table 7-3. Proposed limits of capacity limitation plans

	Actual 1998[A]	1999 Limit[B]	Proposal 1[C]	Proposal 2[D]	Request 3/02[E]
Belize	4,469	1,877	1,982	0	0
Colombia	5,928	6,608	12,000	8,608	12,000
EC.	9,877[F]	7,785[G]	9,595	9,595	9,595
Costa Rica	752	6,000	14,030	8,000	14,030
Ecuador	34,383	32,203	37,086	34,500	34,500
El Salvador	1,312	1,700	5,000	4,000	4,736
FSM	0	0	1,270	0	0
Guatemala	0	0	6,000	5,050	6,502
Honduras	979	499	1,588	0	0
Mexico	40,230	49,500	49,960	50,048	50,048
Nicaragua	0	2,000	4,500	4,000	7,450
Panama	2,774	3,500	5,645	5,600	5,600
Peru	0	0	12,000	3,500	12,000
USA	8,300	8,969	7,747	8,969	8,969
Vanuatu	11,769	12,121	13,332	12,121	12,121
Venezuela	22,682	25,975	25,976	25,975	25,975
Total	137,946	158,737	207,302	176,466	203,056

Notes: Countries that were not members of or did not apply for membership in the IATTC did not receive effort allocations, except for Belize and Honduras in 1999 and the first proposal limits for Belize, Honduras and the Federated States of Micronesia. While not members, Colombia and Peru were seen as cooperating states and thus received a quota, which is not included in the first four totals. Figures are metric tons of carrying capacity. E.C. stands for the European Community and FSM for the Federated States of Micronesia.

[A] Inter-American Tropical Tuna Commission, 1999a. Total includes vessels of Cyprus and Taiwan, which sought to relocate to Central American countries.
[B] Inter-American Tropical Tuna Commission, 2001b.
[C] Inter-American tropical Tuna Commission, 2000d.
[D] Inter-American Tropical Tuna Commission, 2000c.
[E] Inter-American Tropical Tuna Commission, 2002b.
[F, G] Figure is for Spain.

or greater than its 1998 purse-seine effort. The success of the program depends on the satisfaction of each state at the point of adoption and requests for inflated quotas were seen as the cost of creating an enforceable regime. Thus, the purpose of the limits was merely their establishment, even if that entailed arbitrary distributional inequities or a higher effort limit than that most economically efficient.

Although the above limits were intended to serve for 1999 only, Costa Rica successfully motioned to extend them indefinitely when members could not agree on permanent capacity limits. Seen in columns four and five of Table 7-3, allocated effort of subsequent negotiations changed, reflecting the influence of the rights of developing countries to share the wealth of the resource and actions to exclude non-participants.

While members could not agree on the allocation of effort among themselves, they had a common interest in expelling those countries that did not participate in negotiations. A 2001 draft document of fleet capacity expelled vessels from Belize and Honduras, countries that have not joined or cooperated with the Commission, and required such vessels to transfer to participating members (Inter-American Tropical Tuna Commission, 2000c). The addition of new vessels was prohibited and replacement of lost ships and their capacity quota needed to come from willing transfers by member states. Vessels flagged by countries that did not participate or cooperate with the Commission would be placed on a list of non-cooperating vessels (Inter-American Tropical Tuna Commission, 2001b). This action threatens the financing opportunities and resale of vessels, as demonstrated by a similar and successful effort by the Forum Fisheries Association.

The consensus on limiting harvesting capacity and excluding outsiders paved the way for a final resolution limiting fleet capacity, adopted by at the 69th Meeting on 28 June 2002. The use of national quotas was dropped to prevent zero-sum negotiating among states. Instead, members agreed that only purse-seine vessels listed as of June 28, 2000 on the previously established registrar would be allowed access to the CYRA. Countries that wished to increase their fleet size would need to seek the voluntary transfer of vessels from other members. Exemptions to this rule (found in the last column of Table 3-7) were offered to states seeking further development (El Salvador, Guatemala, Nicaragua and Peru) and to Costa Rica, which would use this right only if its processing industry found future difficulty in procuring raw material. The final resolution met the needs of both states seeking further development and those with established industries, which maintained their status quo capacity.

4.3 Countries and their roles in the negotiations

In creating the regime, the contracting parties fell into three groups with different interests.[6] Along with Colombia and Peru, the poorer Central American states had historically been excluded from the socioeconomic benefits of the fishery and positioned themselves for a greater role. Second,

---

[6] This discussion excludes mention of France (not represented by the European Community), which requested a 4,000 mt. allocation.

countries with a long-established harvesting and/or processing capacity in the region (Costa Rica, Ecuador, Mexico, Vanuatu and Venezuela) expressed strong preferences for the effort regime and some were willing to make concessions towards that end. Third, for the United States and Spain (later the European Community), capacity levels had been declining and transferred to coastal states, notably Ecuador. Representatives from both countries strived to maintain their existing capacity levels while pressing for a reliable agreement that would promote the interests of multinational corporations doing business in Latin America. Figures 7-1 and 7-2 demonstrate the differences and trends in harvesting capacity for the Central American countries and the larger fleets, respectively. While most states jockeyed for larger relative shares, increasing the limit of the most vocal states would accommodate such differences.

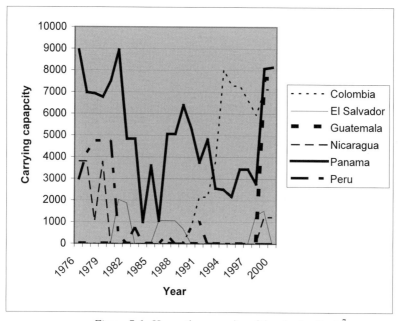

Figure 7-1. Harvesting capacity of developing fleets[7]

Each fleet capacity draft specifically mentions Colombia, Panama, El Salvador, Guatemala, Nicaragua and Peru as countries interested in future capacity limit increases. The latter four countries lacked significant capacity in both the harvesting and processing sectors and feared that a permanent effort regime would exclude them from development. In particular, Guatemala, El Salvador and Nicaragua felt that they needed capacity quotas

---

[7] Data are from Inter-American Tropical Tuna Commission. *Annual Report.* La Jolla, California: IATTC, various years.

significant enough to attract foreign-flagged vessels or foreign direct investment in processing facilities. In the 66th Meeting, El Salvador reminded members of its past inability to attract private investment and hence develop fishing capacity, due to the presence of armed conflict. The representative politely requested an additional 1,700 ton limit but with an intention to request another 1,600 in the 2001. This request found support from Colombia, the European Community, Guatemala, Nicaragua, Peru, Vanuatu and Venezuela(Inter-American Tropical Tuna Commission, 2000d), countries either seeking to develop their own processing industry or those who would benefit from an increase in competition among regional processors. Notably absent in approval are states already possessing both a large harvesting and processing capacity.

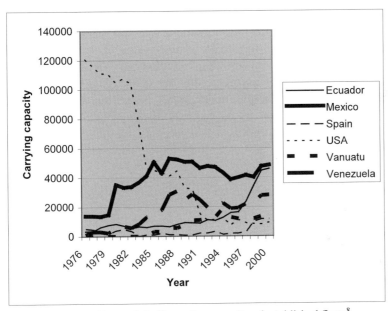

Figure 7-2. Harvesting capacity of established fleets[8]

Central American countries increased their effort quota at the 66th Meeting through a joint statement. El Salvador, Nicaragua and Panama declared their effort limits to be 3,500, 4,000 and 5,600 tons, respectively, based on their sovereign rights and the claim that such action would not pose a threat to the resource (Inter-American Tropical Tuna Commission, 2000a). Furthermore, all Central American countries criticized the 'inflexibility of extraregional countries which have historically exploited these resources,' a

---

[8] Data are from Inter-American Tropical Tuna Commission. *Annual Report*. La Jolla, California: IATTC, various years.

jab at the major fleets that did not offer concession: the US, Vanuatu and Venezuela with four, seven and 15 percent of allocated capacity, respectively (Inter-American Tropical Tuna Commission, 2000c). Panama 'called attention to the fact that Article 5 of the FAO Code of Conduct for Responsible Fishing states that recognition should be given to enhancing the ability of developing countries to develop their fisheries and participate in high-seas fisheries' (Inter-American Tropical Tuna Commission, 1999b). Fulfilling this development norm while not allowing increases in total capacity would be impossible without concessions from the larger fleets.

The two Andean countries did not apply for membership but nonetheless cooperated with the IATTC. Citing its historical participation and processing capacity, Peru reserved a right to a 12,000 mt. capacity fleet (Inter-American Tropical Tuna Commission, 2000a). Representatives of Colombia consistently reminded the Commission that sovereignty and UNCLOS allow the country to determine catches and regulations within Colombia's EEZ, regardless of IATTC resolutions. Declaring that its sovereign rights are 'not negotiable,' Colombia expressed strong disagreement with the process, declaring that the 1998 limits 'do not satisfy the rights and interests of Colombia of having an allocation of 12,000 tons' (Inter-American Tropical Tuna Commission, 2000a). With existing fleets under 6,000 mt., Peru and Colombia shared the development desires of the Central American countries.

In contrast to the other Central American states, Costa Rica had participated in the fishery since the 1930s and was one of the founding members of the Commission. To support its processing industry, Costa Rica sought stable effort levels and restrictions on the catch of lesser-valued, juvenile fish. The country prohibited the use of FADs in its own waters and asked the Commission to do so in the entire CYRA (Inter-American Tropical Tuna Commission, 2000d). While Costa Rica maintained a long-term goal of a 14,030 mt. fleet (Inter-American Tropical Tuna Commission, 2000c), representatives asked only for a "passive capacity allocation" of 8,000 mt. for processing purposes, adding that 'the allocation would not be used as long as its processing industry was adequately supplied' (Inter-American Tropical Tuna Commission, 2000a). Along with Mexico and Ecuador, Costa Rica was willing to make concessions in order to establish an agreement.

From the onset of fleet negotiations, the Mexican delegation sought an immediate, binding agreement. At the 3rd meeting of the Working Group, representatives requested that any document apply not only to the immediate year but also to for future ones. Furthermore, a resolution should be submitted for immediate approval, 'and without waiting for consultations with any government which did not attend' (Inter-American Tropical Tuna Commission, 2000d). Mexico pointed out that some countries had exceeded the limits set at the 62nd meeting (Inter-American Tropical Tuna Commission, 1999b) and that limits should be set for all gear types, not merely purse-seiners (Inter-American Tropical Tuna Commission, 2000d). Initially defending its large fleet, the Mexican representatives expressed disapproval of the rights of developing coastal states to expand their fleets, which would come at the expense of existing fleets or profitable resource management (Inter-American Tropical Tuna Commission, 2000c). However, expressing its immediate desire for a comprehensive plan, Mexico joined Ecuador in offering to reduce their capacity by 10 percent each for transfer to developing states and called on the US, Vanuatu and Venezuela to do the same. Included in this figure is Mexico's proposed transfer of 6,000 mt.. to Central American countries and Ecuador's 3,500 mt. offer to Colombia (Inter-American Tropical Tuna Commission, 2000c).

Costa Rica, Mexico and the US emerged as the principal countries forcing issues of monitoring and compliance. US and Mexican representatives repeated claims that the 1999 limits had been exceeded 'and the consequences of over-capacity were now apparent' (Inter-American Tropical Tuna Commission, 2000a). The countries succeeded in including in the draft resolution that, upon acceptance, the Compliance Working Group would monitor enforcement (Inter-American Tropical Tuna Commission, 2000b).

While the threat of increasing capacity led all actors to the negotiating table, countries with an established industry offered to stabilize or decrease their harvesting capacity while newcomers demanded increases. Among the former group, those countries that experienced both losses from a lack of effort regime and changes in bureaucratic structure proved willing to decrease their national capacity to gain the approval of the developers.

5. EXPLAINING BARGAINING AND SUCCESS IN CREATING THE REGIME

The ETP Tuna Regime is one of the few examples of successful international cooperation to manage common property. In fact, the regime is more ambitious than any other in that it limits the amount of inputs allowed in the fishery, not simply the amount of catch. While the above section detailed what was accomplished and states' interests in negotiations, this section explains the role of international pressures and the leadership of two countries in securing the agreements. The threat of excess harvesting capacity, made more immediate from the anticipated opening of the US market, increased the urgency for an effective regime. Institutional changes within Costa Rica and Mexico allowed both to play a leadership role in international environmental negotiations.

5.1 Pressure from the end of the "dolphin-safe" embargo

The debate over dolphin-safe tuna reflected more than concern over cetacean mortality; the contending sides also argued over the definition of sustainable ecosystem management. As US- and Mexican-led programs [9] drastically reduced the number of dolphin kills from over 400,000 to less than 5,000 per year, an uncomfortable tradeoff emerged. Continuation of purse seining with improved technology and regulations[10] would continue to lead to dolphin mortality, although at much lower levels. The alternative, catching tuna that congregated under logs or FADs, resulted in the bycatch of sea turtles, sharks and billfish. In addition, the latter method led to the harvest of sexually immature tuna, threatening future stock levels as well as profits (Joseph, 1995). The tradeoff led most US environmental groups to endorse the lifting of the embargo, leaving the Earth Island Institute, the Sierra Club and People for the Ethical Treatment of Animals, along with two sympathetic Senators and a federal judge as obstacles to US implementation of the "dolphin-safer" harvesting techniques. A 1992 GATT panel decision, Latin American businesses and governments, other environmental groups such as Greenpeace and the Council for Marine Conservation, most Congressmembers and the Clinton administration pressed for resolution of the conflict.

While the House of Representatives in 1996 and 1997 passed legislation, with Clinton's approval, calling for an end to the embargo, the bill stalled in the Senate. Finally, the upper house passed a compromise measure, whereby the embargo would end on 2 February 2000 but use of the dolphin-safe label would continue. Countries must be contributing members of the IATTC and certified under the Commission's Dolphin Protection Program in order to export tuna to the US. A challenge in the NY Circuit Court of Appeals by the Earth Island Institute to the lifting of the embargo failed (Rechaza, 2000). The continued use of the label would depend on the recommendation by the National Marine Fisheries Service to the Senate that purse seining did

---

[9] Examples include the technological development and educational efforts of the US National Marine Fisheries Service, Mexico's Programa Nacional para el Aprovechamiento del Atún y la Protección de los Delfines (National Program for the Utilization of Tuna and Dolphin Protection, PNAAPD) which has had observers on all vessels since 1977, the 1992 La Jolla Agreement and the 1994 Declaration of Panama.

[10] Improvements include development of the Medina panel, not allowing night sets, requiring personnel to enter nets to release dolphin and a comprehensive observer program.

not have a negative impact on dolphin populations. Some countries (i.e. Costa Rica) received labeling approval.

Throughout the late 1990s, Latin American tuna industries remained optimistic that they would regain access to the US market. Newspaper headlines during this time frequently predicted the lifting of the embargo. Mexican industry leaders acknowledged that the continuation of the dolphin-safe label would hinder sales but exports to the US would gradually increase. The Mexican industry asked for $500 million in loans from the World Bank and Inter-American Development Bank to rebuild harvesting and processing capacity lost during the embargo (Gonzalez, 1997). Industry leaders in Latin America, the US, Europe and Asia rightly believed that, with the reopening of the US market for ETP tuna, prices, investment and harvesting pressure would increase. With the optimism came growing fear of a tragedy of the commons historically held at bay by weather patterns, economic crises and trade disputes.

5.2 Changes in CPUE and profitability

Excess investment in harvesting common property threatens the profits of all participants. A reliable statistic that measures variance in earnings is changes in catch-per-unit-of-effort (CPUE), distinguishable among the different purse-seine fleets in the EPO yellowfin fishery. CPUE, measured here as each country's annual catch divided by total vessel carrying capacity, measures the productivity of each unit of fixed investment. The established harvesters most willing to redistribute effort to developing states have experienced sharp drops in CPUE. In contrast, CPUE for less generous states has remained stable.

Changes in CPUE distinguish the major countries that offered concessions to developing countries from those that held firm on their allocated effort levels.[11] As investment in harvesting equipment continues, the catch of each participant eventually declines as each set of the net yields fewer fish and search costs rise. However, for reasons of technology, geography and other factors, countries vary in their vulnerability to the loss, seen in Figure 7-3. The steady decline of US CPUE reflects the entry of new vessels, mostly Mexican, disputes over fishing access in the 1980s, and the troubles presented by the tuna-dolphin controversy in the 1990s. The Mexican fleet emerged as the most modern and efficient, but regional over-investment led to a dramatic decrease in CPUE starting in 1997. Mexico's

---

[11] The Central American countries and Peru had only a few years of vessel ownership during the 24-year sample. As seen by comparing Figures 7-1 and 7-2, Colombian harvesting participation remained slight compared to major harvester. CPUEs for these countries are therefore not shown.

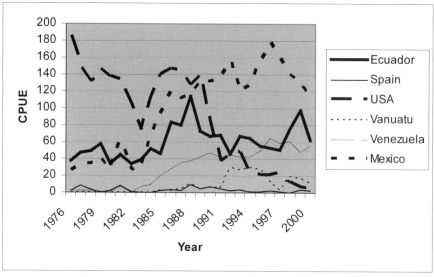

Figure 7-3. Catch-per-unit-of-effort for established fleets[12]

loss appeared to be Ecuador's gain in the next two years, but the latter suffered drops in CPUE beginning in 1999. In perspective, the offer of the two countries to redistribute effort to developing states appears not entirely driven by altruism but also by the opportunity cost of not limiting total effort, namely further declines in harvesting yields per unit of fixed investment.

By contrast, the countries that the Central American countries labeled as most 'inflexible' in redistribution negotiations, Vanuatu and Venezuela, respectively found a smaller or no decline in CPUE. These countries saw no need for dramatic diplomatic gestures while Mexico and Ecuador preferred a stable level of capacity with some redistribution losses to long-term declines in profitability.

5.3 International leadership through changes in administrative structure

The creation of Exclusive Economic Zones provided the initial bureaucratic structure for fisheries development. Diplomatic conflict and the belief in the bonanza of the resource placed fishery regulations and public investment on the agenda of high politics. States empowered several bureaucratic actors to further fishery development, often as a response to

---

[12] Data are from Inter-American Tropical Tuna Commission. *Annual Report*. La Jolla, California: IATTC, various years.

harsh macroeconomic conditions, and later to counter protectionist or environmental policies of the leading tuna markets. Fisheries departments, financing ministries, foreign ministries, government corporations and direct intervention by legislators held major influence at different times in different countries. The pursuance of their goals often sacrificed sustainable fishery objectives to larger political tradeoffs or the nationalistic defense of marine property that resonated with voters at large.

In general, the influence of many of these actors waned over time, increasing the voice of the fishing industry and environmental professionals: 'the approval in the early 1990s of new fisheries laws demonstrated growing local concerns over the sustainability of current fishing practices' (Aguilar, Reid and Thorpe, 2000). The promise of the fishery as a leading source of foreign exchange faded with problems posed by resource volatility, changes in tuna markets, and the emergence of more promising sectors to develop. Trans-boundary stock migration and weather patterns hindered stable harvests. Competition from other tuna producing regions, market gluts, and embargos led to drastic changes in ex-vessel and final product prices. At times, declining prices coincided with national economic crises and rises in dollar-denominated interest rates. For the more advanced economies such as Mexico, Costa Rica and Venezuela, other sectors attracted the attention of state and private investors at the expense of support for the tuna industry. This trend was not uniform however; countries more dependent on primary exports, notably Ecuador, lacked alternatives and continued to support the tuna industry through its changes in harvests and price. Furthermore, those getting a late start on tuna development (Colombia and Central American countries with the exception of Costa Rica) had yet to experience disillusionment with the promise of tuna development. Where foreign ministries and state development banks found other sectors more attractive, support for the tuna fishery faded. The departure of other ministries opened political space for local interests, environmental groups and fishery professionals to gain control over fishery management, a trend especially evident in Mexico and Costa Rica.

In Mexico, the state responded to the extension of marine jurisdiction with ambitious plans for fisheries development. Public investment in fisheries complimented the Portillo administration's (1976-82) national development strategy based on extraction of petroleum and other natural resources, and financing development by borrowing against future revenues. The para-statal Productos Pesqueros Mexicanos (PROPOMEX) received state funds to consolidate marketing and processing tuna and other products. Created in 1980, the state-owned Banco Nacional Pesquero y Portuario (BANPESCA) underwrote loans for fishing cooperatives as well as private vessel owners. Between 1979 and 1982, of the 54 purse-seine vessels financed and built abroad at a cost of $400 million, 31 were financed by

BANPESCA and the remainder had BANPESCA guarantees (Hudgins, 1986). BANPESCA assumed a larger role in the tuna industry after the 1982 peso crash left owners of purse-seiners unable to finance their dollar-denominated loans. In exchange, the government demanded greater tuna catches (and lower wholesale prices) to support Mexico's food policy, processing employment and ambitions as a leader in the global tuna industry.

The indebtedness of private producers, cooperatives and state agencies reached a peak shortly before Carlos Salinas took office in 1988 and state financial support soon diminished. Accelerating the neo-liberal reforms of the previous administration, Salinas privatized PROPOMEX facilities and vessels and closed BANPESCA, the latter receiving only a fraction of the funds loaned out in the previous decade (Aguilar, Reid and Thorpe, 2000). The creation of the National Program for the Development of Fisheries and Resources signaled a significant reduction is state support for fishery development. Resource use was to be rationalized, made able to function without significant government financing. Cooperatives collapsed without state funding, facilitating a legal change that replaced their historic access with permits and concessions favoring the more capital-intensive export sector (Aguilar, Reid and Thorpe, 2000).

The plethora of bureaucratic actors promoting fisheries development was consolidated into a professional core dedicated to sustainable harvests instead of ambitious expansion. In the late-1970s, fisheries Secretary Jorge Casteñeda,[13] who represented the tuna industry in its negotiations with the US, was later promoted to Foreign Minister, pushing development efforts into a more publicized forum. The Subsecretary of Fisheries, upgraded to Ministerial status in 1982, promised to turn the Mexican industry into one of the world's top five fish producers through the National Fisheries Plan (Aguilar, Reid and Thorpe, 2000). Bureaucratic consolidation began with the 1986 Federal Act of the Sea, emphasizing that regulation of marine spaces is a Federal Executive Power, ultimately in the uncontested hands of the President (Special Representative, 1986). Bureaucratic goals changed from industrial expansion to environmental protection with the creation of the Secretaría de Medio Ambiente, Recursos Naturales y Pesca (SEMARP) in 1994.[14]

SEMARP delegated many functions of fisheries management to its National Fisheries Institute (INP),[15] refocusing development activities into a

---

[13] This Jorge Casteñeda served in several posts for the Partido Revolucionario Institutional (the Institutional Revolutionary Party, known by its Spanish acronym PRI) and should not be confused with the widely-read social critic who served under Vicente Fox.

[14] Under the Fox Administration (2000-06), fisheries management was placed under the Ministry of Agriculture. The slow pace of bureaucratic reorganization and replacement of personnel makes unclear the impact of this change on purse-seine effort negotiations.

[15] The INP was originally created in 1971.

less political, more scientific and management-oriented direction. INP objectives include 'looking for methods more agreeable with the environment and the health of marine ecosystem' that will 'create more rational policies for the exploitation of fishery resources.' Over-exploitation, ecological impacts, 'and the overcapitalization of the fleet are the result of longstanding policies that the INP now seeks to reverse' (Sobrexplotación, 1998). Maintaining close contacts with industry groups such as the Cámara Nacional de la Industria Pesquera y Agrícola (Canaipes), the INP broadened its studies to include aquaculture and the protection of marine turtles, gray whales, and dolphins. Currently, more than 400 employees in seven offices conduct biological studies and sustainable planning (Instituto Nacional de la Pesca, 2001). The INP seeks to better not only environmental health but also the profits of fishermen and security of investors. Through the decreased financial role of the state and the refocus of bureaucratic goals, Mexico emerged as an international leader pressing for sustainable fishing practices. SEMARP Secretary Julia Carabias Lillo commented that 'We want to tell the world that Latin America is active and we will to continue working together to unify our interest in sustainable development' (Vigueras, 1997).

Costa Rica, one of the other leaders in the development of an effort regime, underwent similar bureaucratic changes. The Dirección General de Recursos Pesqueros y Agricultura, part of the Ministerio de Agricultura y Ganadería, led post-UNCLOS development efforts. In contrast to Mexico, Costa Rica chose to use foreign harvesters to supply a domestic processing industry. While the country has at times invested in purse-seiners, the government generally has held the belief that the social gains of employment and national receipts in the processing sector outweigh risky investments in yellowfin harvesting. Still in effect, the 1976 Ley Ferreto permitts 'foreign vessels...that land all of their catch to national processors...(to) have the permission to catch without paying (per catch) fees' (Asamblea Legislativa, 1976). Three major companies, Sardimar, Borda Azul S.A. and Enlatodora Nacional Tesoros del Mar currently generate annually $25-32 million in export revenue and directly employ 5000 citizens, predominately in Puntarenas ([1] Instituto Costaricense de Pesca y Acuicultura, 2002). Foreign vessels remained largely unregulated throughout the 1980s and early 1990s.

Controversy over the environmental impact of domestic fishermen and foreign tuna harvesters led to the creation of the Instituto Costaricense de Pesca y Agricultura (INCOPESCA) in 1994. INCOPESCA's three major activities include coordinating and developing national fisheries, achieving rational economic use and preventing pollution from fishing and aquaculture (Quesada, 1998). In addition to proposing stronger domestic legislation, INCOPESCA implemented a dolphin-safe tuna program that gained US approval and prohibited the use of FADs. The president of INCOPESCA serves as Costa Rica's representative at IATTC meetings. The transfer of authority from the Agricultural Ministry to an autonomous fisheries agency led to more stringent regulation of foreign harvesters and an important role for Costa Rica in the IATTC's regulation of purse-seine effort.

In contrast to Mexico and Costa Rica's devolution of fishery management to agencies most concerned with sustainability, representation of the industry in other countries remains in the hands of foreign or commerce ministries. Such agencies express greater interest in economic gain than resource sustainability. At IATTC meetings, Colombia's lead representative is from the Ministry of Foreign Commerce, the Economics Ministry for El Salvador, the Ministry of Agriculture and Ranching for Guatemala and the Minister of Commerce for Venezuela.[16] While these bureaucrats follow the industrialization or export goals of their respective agencies, the empowerment of fishing professionals has allowed Mexico and Costa Rica assume leadership roles. Not only did they offer concessions on capacity quotas yet both countries presented to the IATTC necessary and sometimes technical proposals to limit fishing effort in the ETP.

6. CONCLUSION

Management success in commercial fisheries depends not only on setting sustainable catch rates but also on limiting investments dedicated to harvesting. Although effort-limitation agreements foster greater legitimacy in the system of regulations by eliminating the 'race for fish' and associated factor misallocation, participants find it more difficult to agree on the aggregate limit and distribution of effort. The uncertainty of who will benefit from a total allowable catch may be preferred to the certainty of fixed shares of capacity.

Effort limitation agreements are yet more difficult in international forums. The lack of a coercive authority in global politics requires a regime that satisfies all sovereign actors. Proof of this hurdle can be found in comparing the lack of accords to limit inputs for the exploitation of international common property resources compared to recent, yet limited, advances in domestic legislation or the longstanding success of local, community-based management. The participants in the ETP yellowfin fishery have agreed on total catch limits for the past three decades. The opportunity of market opening, the threat of profit loss from over-investment and changes in bureaucratic structure have led member states to search for a

---

[16] From the list of attendees of various meeting minutes of the IATTC.

deeper solution to common property management, namely a regime to limit fishing effort.

The example and eventual success of international regulation of harvesting capacity in the ETP yellowfin fishery may provide an inspirational example of management directions for other global common properties. If the lessons from the ETP case are transportable to other regions and resources, states can make difficult sacrifices to create regimes for sustainable use. The evolution of agreements and norms find fertile ground when economic pressures threaten the profitability of investments and when leading participants assume leadership roles that initiate short-term sacrifice and press for viable and legitimate measures that protect the resource and those who depend on it.

ACKNOWLEDGMENTS

The author wishes to thank managers and members of the tuna industry and two referees for helpful comments, as well as Mary Casey Kane for providing research assistance. All errors and omissions belong solely to the author.

REFERENCES

Aguilar Ibarra, Alonso, Reid, Chris, and Thorpe, Andy, 2000, The political economy of marine fishery development in Peru, Chile and Mexico, *Journal of Latin American Studies* **32:**503-27.

Asamblea Legislativa de Costa Rica, 1976, Registero de Barcos Atuneros, Ley No 5775, amended by Ley No 6267, 1978.

Gonzalez, Raul Moreno, 1997, Necesario invertir cerca de 500 mmd para reactivar flota atunera, *El Nacional*, Mexico D. F., August 17.

Hardin, Garrett, 1968, The tragedy of the commons, *Science* **162:**1243-1248.

Hudgins, Linda L., 1986, *Development of the Mexican Tuna Industry*, East-West Center, Honolulu.

Instituto Costaricense de Pesca y Agricultura, 2002, *Memoria Institucional: Administración 1998-2002*, INCOPESCA, San José.

Instituto Nacional de la Pesca, 2001, (July 16, 2001);
http://inp.semarnap.gob.mx/institucion_descripcion.htm.

Inter-American Tropical Tuna Commission, 2002a, *Draft Plan for Regional Management of Fishing Capacity*, IATTC, La Jolla, California.

Inter-American Tropical Tuna Commission, 2002b, *Minutes of the 6th Meeting of the Permanent Working Group on Fleet Capacity*, IATTC, La Jolla, California.

Inter-American Tropical Tuna Commission, 2001a, *Draft Plan for Regional Management of Fishing Capacity*, IATTC, La Jolla, California.

Inter-American Tropical Tuna Commission, 2001b, *Minutes of the 68th Meeting*, IATTC, La Jolla, California.
Inter-American Tropical Tuna Commission, 2000a, *Minutes of the 66th Meeting*, IATTC, La Jolla, California.
Inter-American Tropical Tuna Commission, 2000b, *Minutes of the 5th Meeting of the Permanent Working Group on Fleet Capacity*, IATTC, La Jolla, California.
Inter-American Tropical Tuna Commission, 2000c, *Minutes of the 4th Meeting of the Permanent Working Group on Fleet Capacity*, IATTC, La Jolla, California.
Inter-American Tropical Tuna Commission, 2000d, *Minutes of the 3rd Meeting of the Permanent Working Group on Fleet Capacity*, IATTC, La Jolla, California.
Inter-American Tropical Tuna Commission, 1999a, *Annual Report 1998*, IATTC, La Jolla, California.
Inter-American Tropical Tuna Commission, 1999b, *Minutes of the 65th Meeting*, IATTC, La Jolla, California.
Joseph, James, 1995, The tuna-dolphin controversy in the Eastern Pacific Ocean: biological, economic and political impacts, *Ocean Development and International Law* **25**:1-30.
Krasner, Stephen D., 1983, *International Regimes*, Cornell University Press, Ithaca.
Miles, Edward L., et al., 2002, *Environmental Regime Effectiveness: Confronting Theory with Evidence*, MIT Press, Cambridge, MA.
Munro, Gordon R. and Anthony D. Scott, 1985, The economics of fishery management, in: *Handbook of Natural Resource and Energy Economics*, Allen V. Kneese and James L. Sweeney, eds., North-Holland Press, New York.
Quesada Zuñiga, Felicia, 1998, *La Legislación Pesquera Costaricense: Errónea e Inoperante ante la Pesca del Atún*, Universidad de Costa Rica, San José.
Rechaza juez de NY mantener embaro al atún mexicano, *El Economista*, 18 April 2000.
Sobreexplotación de los rescursos pesqueros, 1998, *La Jornada*, Mexico D. F.
Special Representative of the Secretary-General for the Law of the Sea, 1986, *The Law of the Sea: Current Developments in State Practice*, United Nations, New York.
Vigueras, Manuel Noguez, 1997, Explotación de los mares con criterio de sustenabilidad, plantea Julia Carabias, *El Excelsior,* Mexico D. F., August 1.

Chapter 8

ECO-LABELING IN LATIN AMERICA
Providing a scientific foundation for consumer confidence in market-based conservation strategies

Thomas V. Dietsch[1,2]
[1]*University of Michigan, School of Natural Resources and Environment, Ann Arbor, MI 48109;* [2]*Current Address: Smithsonian Migratory Bird Center, National Zoological Park, Washington, DC 20008*

Abstract: Market-based conservation strategies have been proposed to encourage more sustainable and environmentally sensitive management practices in biodiversity rich areas. By paying a price premium, consumers can provide a market signal through their purchases to encourage producers to use less damaging management practices. This approach has been heavily promoted for a wide range of products from the Neotropics. These products are often the focus of sustainable development programs and highlight certification as a means for consumers to recognize the conservation value of the product. Recent research has shown significant benefits for biodiversity from less intensive extraction methods. While these results are a good basis for early optimism, they are not sufficient to ensure long-term conservation success. There is a growing scientific literature on how to monitor ecological integrity in conservation programs. This stronger scientific understanding suggests there are limitations to conservation in managed landscapes. This chapter reviews certification programs thus far proposed and implemented, focusing on shade-grown coffee as an example. Using birds, a highly visible and well-studied taxa, the scientific evidence is reviewed and compared with conservation goals. Based on available evidence, suggestions are made on how certification can enhance current conservation efforts and what research is still needed as programs develop. Overall, a general approach is proposed for monitoring and evaluating environmentally friendly certification programs that may give consumers not only confidence but also an awareness of how their purchase may contribute to conservation.

Key words: eco-labeling; shade-grown coffee; agroecology; international ecosystem management; biodiversity conservation; birds; tropical ecology

1. INTRODUCTION

Building on early success for dolphin-safe tuna and organic agriculture, the number of environmentally friendly certification programs, often referred to as "eco-labeling", has proliferated. These market-based conservation strategies seek to translate a growing environmental movement into an economic mechanism that encourages improved management and harvesting practices. In exchange for these more benign practices, producers theoretically receive a price premium that compensates them for lower yields or higher production costs. These programs often market their goods with images of pristine nature and compassionate farmers living close to the earth. Unfortunately, these same images are used to sell a wide range of other products that have nothing to do with sustainable production. Truly sustainable products are getting harder to distinguish through this flurry of "green-washing" leaving consumers skeptical of all things "green" and wary of products with actual conservation benefits. Survey data shows that many consumers want to choose products associated with causes they care about, but fewer actually make those purchases (Cone and Roper, 1993; Messer et al., 2000). This may be due to skepticism and confusion about how these products actually make a difference. In this chapter, I examine the rationale and potential benefits of certification programs to determine if they may form a basis upon which consumers can support sustainable products.

Certification may help improve sustainable products' recognition and acceptance. Dolphin-safe tuna and organic agriculture have used certification and recognized seals with good success. Organic foods continue to grow rapidly in global agricultural markets (IFOAM, 2002). These programs were pioneering efforts that used markets to reward farmers and fishermen who adopted specified practices. At the time, few products made similar environmentally friendly claims and easy-to-communicate benefits took advantage of a growing awareness of environmental problems among consumers. In the case of organic agriculture, which prohibits agrochemical use, benefits are not just environmental, but also reduce chemical exposure for the consumer. The range of tropical products that available in the green market now includes timber, coffee, cacao, bananas, oranges, and a host of non-timber forest products. For marine systems, sea turtle-safe shrimp has joined dolphin-safe tuna, along with watchlists to steer seafood consumers to sustainable fisheries. In addition, social justice movements have also turned to the marketplace to help impoverished small farmers and indigenous people. This increasingly complex array of buying choices for concerned consumers, however, may cause "label fatigue", especially in competition with a wave of green advertising for products that offer no benefits for the environment (Lohr, 1999).

Nonetheless, challenges for environmentalists continue to grow, requiring innovation to complement established conservation solutions. Deforestation rates in the tropics continue at high levels, with the Americas accounting for nearly half (48%) of the area loss in moist tropical forests for the 1980s (67.4 million hectares) (FOA, 1993; Schwarzman and Kingston, 1997). The consequence of this wave of deforestation and habitat fragmentation will be a dramatically altered landscape (Whitmore, 1997). The full extent of biodiversity loss may never be known, given that huge portions of the Earth's flora and fauna have yet to be catalogued, especially in the tropics (Pimm et al., 1995). Uncertain long-term changes in temperature and precipitation patterns due to climate change may have negative synergistic effects on the distribution and composition of ecosystems (Root & Schneider, 2002). With less than 10% of Neotropical forests in protected areas (Green et al., 1996), there is a need to encourage environmentally friendly land use practices that minimize long-term losses. Conservation biologists can help overcome consumer reluctance by providing a scientific foundation for evaluating programs making claims of conservation benefits. This chapter provides a framework for evaluating market-based conservation solutions, in particular how conservation science can provide the basis for consumer confidence using certified shade-grown coffee as a case study.

2. LIMITS OF TRADITIONAL CONSERVATION

Many criticize traditional conservation of protected areas as inadequate to address the looming crisis in biodiversity loss. Conservation biologists have long recognized that many protected areas were established for reasons other than biodiversity conservation, and thus may be limited in how much biodiversity they can protect. Similarly, even when conservation was the motivation for establishing a park, political and economic realities may limit the area protected below what is necessary to meet conservation goals. In fact, the whole notion of protected areas for wilderness or pristine nature is being challenged as a social construction (Cronon, 1995), and raises social justice issues regarding the treatment of communities inside or near protected areas (Wilshusen et al., 2002). Local people are often excluded from the park management decision-making and on occasion forcibly relocated away from parks. Without adequate resources for guards, subsistence and hunting activities can degrade habitat and exert pressure on wildlife. These criticisms have led to new approaches to address ecological limitations of parks (regional landscape approaches, which include core areas, buffer zones and corridors per Noss, 1983) and social justice concerns (integrated conservation and development plans per Alpert, 1996). These seemingly contrary approaches have much in common, in particular, the inclusion of areas where less-intensive management is encouraged or tolerated. A key challenge for these programs is to determine appropriate land use consistent with conservation goals.

Researchers and policymakers often overlook tropical agriculture when they discuss opportunities for conserving biological diversity (Vandermeer and Perfecto, 1997). While they usually list deforestation for agriculture as the primary threat to tropical biodiversity, they often fail to point out that many areas have been under sustainable agricultural production for extended periods of time, for centuries in some cases. Recent research suggests that these lands may have significant potential to enhance conservation efforts (Perfecto et al., 1996; Greenberg et al., 1997a; Greenberg et al., 1997b). In particular, traditional coffee growing practices that include a heavy shade canopy are thought to act as a buffer against the impacts of deforestation. For example, coffee growing areas in Puerto Rico may have provided a refuge for many species, especially birds and orchids, when much of the island was deforested, and greatly reduced the number of extinctions (Brash, 1987; Nir, 1988). In general, traditional growing practices have undergone dramatic transformation in recent years as agrochemical and plant breeding techniques have increased yields and altered crop management. This intensification often results in a monoculture with a simplified vegetative structure. Unfortunately, overproduction often leaves small farmers with low economic returns due to reduced prices and the higher costs of modern methods. In the case of coffee, prices have fallen to 50-year lows as countries around the world compete to intensify production (ECLAC, 2002). Thus, while agriculture is usually viewed as a cause of deforestation, encouraging less intensive agricultural practices may instead stabilize local economies, thereby reducing pressure on primary forests.

3. CERTIFICATION IN LATIN AMERICA

Modern industrial agriculture has drawn considerable criticism for unsustainable reliance on the use of agrochemicals and damage to tropical ecosystems that results when traditional growing systems are simplified to monocultures (Vandermeer and Perfecto, 1995). Organic and biodynamic (Rudolf Steiner's holistic agricultural system, which is less well-known than organic in the United States, but with stricter practices) programs certify intercropping and alternative agrochemical-free practices that maintain soil fertility and control pests and comprise an alternative sustainable agriculture (Vandermeer, 1995). Higher prices and good international consumer recognition, with biodynamic being more popular and commanding higher prices outside the United States, facilitate the spread of these programs, though certification costs are still a barrier for many farmers (Dudley et al., 1997). In an attempt to reach larger producers, others have developed integrated pest management (IPM) approaches to reduce pesticide use.[1] Rainforest Alliance developed the Sustainable Agriculture Network, which now has certification partners throughout Latin America, to promote and certify best practices, including IPM.[2] This program grew out of their "Eco-OK," better banana program and now also certifies coffee, cacao, and oranges. "Bird friendly™" or shade-grown coffee focuses on enhancing vegetation management practices to benefit associated biodiversity (Perfecto et al., 1996). With fair trade certification, developed in 1988 to improve trading conditions for small farmers (Waridel, 2002), coffee certification programs now incorporate major aspects of sustainability: organizational, economic, production, and conservation.

3.1 Coffee Certification Programs

As the major agricultural commodity legally traded in global markets, coffee is a lead target for certification programs. These programs can be divided into four main approaches: organic, fair-trade, shade-grown, and cause-related. Coffee was first certified biodynamic in Mexico at Finca Irlanda in 1928 (Yussefi and Willer, 2002), and was an important precursor to organic certification, which has now spread throughout Latin America. In fair trade certification, small farmers (farms <5 hectares) are paid a "fair" price for their harvest.[3] These farmers organize into democratic cooperatives and sell directly to buyers from consuming countries. Buyers pay a guaranteed price and consumers pay certification costs.[4] Cause-related coffees usually give a portion of their sale proceeds to aid social or environmental organizations. Good examples of cause-related coffee include the "Coffee Kids" program that uses proceeds for rural development programs in coffee growing areas and Thanksgiving Coffee Company's programs with several conservation organizations.[5] These programs benefit social and environmental issues through organization contributions but do not necessarily have a direct effect on coffee management practices.

---

[1] Integrated pest management programs allow some agrochemical use but focus on eliminating the most hazardous and persistent chemicals, such as DDT.
[2] Rainforest Alliance sustainable agriculture website:
[http://www.rainforestalliance.org/programs/cap/index.html].
[3] Transfair USA website [http://www.transfairusa.org/why/index.html].
[4] The price in December 2002 was $1.26 per pound, $1.41 with organic certification.
[5] Coffee Kids website [http://www.coffeekids.org/].

Shade-grown coffee, known also as "Bird friendly™" and more recently also as "biodiversity friendly", incorporates certification programs with criteria that are designed to provide conservation benefits. The Smithsonian Migratory Bird Center (SMBC) first developed this approach after it recognized the potential benefits of some vegetation management practices for long distance migratory birds. Criteria were first produced following the 1996 First Sustainable Coffee Congress, held in Washington, DC (Greenberg, 1996). In general, shade-grown coffee is designed to encourage environmentally friendly agricultural practices both at the landscape and vegetative levels. No net loss of forested areas is allowed and there are requirements for buffers along streams and rivers to protect riparian vegetation. This approach gets its name from requirements to maintain specific levels of shade in the canopy cover but there are additional criteria regarding the diversity and abundance of shade trees. The canopy must also meet structural requirements that mimic natural forest structure. Epiphytes, plants that grow on the limbs of trees in most tropical forests and are critical habitat for a wide range of tropical diversity, are also encouraged (Nadkarni and Matelson, 1989). Smithsonian Migratory Bird Center and Rainforest Alliance now operate certification programs with similar criteria, though SMBC requires organic certification and has some stricter vegetation criteria, while Rainforest Alliance uses a less stringent IPM approach that allows agrochemical use. SMBC and Rainforest Alliance have joined with Conservation International and Consumers Choice Council to produce general consensus principles for shade-grown coffee (CCC, 2001). Conservation International's (CI) Conservation Coffee program, sold by Starbuck's, does not include certification. This program is better characterized as cause-related with proceeds aiding CI's efforts to encourage the use of shade management practices in the El Triunfo Biosphere reserve in Chiapas, Mexico.[6]

3.2 The Role of Certification

Certification provides a feedback loop that connects consumer preferences and values with particular production options. This is useful when consumers are located at distances too far from producers to know them or to verify their management practices. The certifier, preferably an independent third party, inspects the production system using established criteria. In exchange for paying a price premium, consumers receive an assurance that sustainable land use practices were used. This feedback loop can allow conservation science a mechanism for adaptive management as

---

[6] Conservation International website:
[http://www.conservation.org/xp/CIWEB/newsroom/campaigns/coffee.xml].

used in ecosystem management (Grumbine, 1994). Certification thus provides a unique framework for ecosystem management in an international context. For this approach to be successful, consumers need confidence that this added expense makes a difference.

A well-established scientific basis for certification criteria can build this confidence and if properly conveyed through outreach efforts will both educate consumers and adjust producer behavior. Criteria should be measurable and based on scientifically testable hypotheses allowing producers and vendors to make verifiable claims. Certification organizations and procedures should also be transparent and open for review. Most certification also involves careful tracking of product chain of custody to minimize the risk of cheating or fraud. While consumer confidence and producer participation may depend on a host of unpredictable and difficult to control factors, scientific evidence and testing can help delineate reasonable expectations from certification programs.

3.3 Framework for Conservation in Managed Ecosystems

The certification approach has proven an effective means to improve connections between conscientious consumers and like-minded producers. For organic agriculture, this connection has dramatically reduced pesticide residues on certified produce (Baker et al., 2002). Using certification as an alternative framework for ecosystem management, especially in an international context, will require special attention to the monitoring component to ensure that management practices produce conservation benefits. Direct enforcement is maintained through site visits by certifiers and tracking of chain of custody from producer to consumer. However, monitoring science can provide feedback to consumers through public outreach and education (Figure 8-1). This might serve as an indirect enforcement mechanism. Specific criteria can be evaluated and changes recommended when conservation benefits cannot be verified. Once conservation benefits have been identified they can be provided to land use and protected area management planners to enhance traditional conservation efforts.

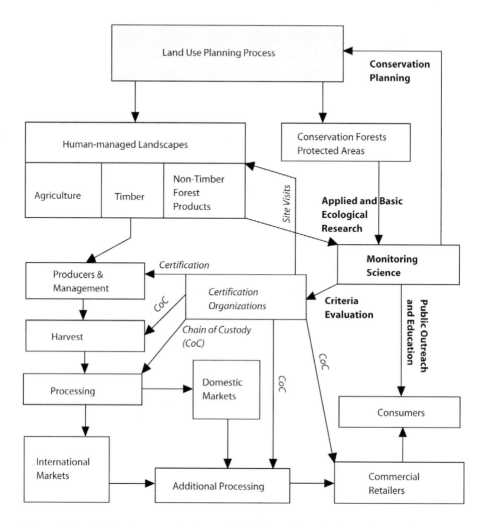

Figure 8-1. Certification organizations (in italics) and monitoring science (in bold) can combine to provide feedback pathways that may improve the conservation potential of managed landscapes (Adapted from Gullison et al., 2001).

Conservation science should clearly define expected benefits from particular management activities that might be certified. A framework for this can be built from basic population and community ecology. One of the first problems identified for endangered species isolated in protected areas was inbreeding depression when population sizes are small (Soulé, 1986). Even minimal improved connectivity between populations can facilitate gene flow (Noss, 1983). However, demographic fluctuation (stochasticity) can also be a significant constraint for isolated populations (Gilpin and Soulé, 1986). Metapopulation theory provides a basis for understanding the dynamic interaction between fluctuating, dispersed populations and landscape context (Hanski, 1999). Populations within suitable habitat patches are separated by less suitable or hostile habitat (Hanski, 1997). Encouraging less intensive land-use can provide conservation benefits if individuals move more freely through this habitat (Doak, 1995). Population models can help define important life history characteristics associated with population persistence. These characteristics include reproductive success, juvenile survival, dispersal, and adult survival. Note that if particular management activities can loosen some life history constraints there can be conservation benefits without addressing each characteristic. Thus even degraded habitat can provide benefits through improved dispersal or juvenile survival if there is adequate vegetative diversity and structure for predator avoidance and resource availability. Of course, community level interactions can be quite complex and lead to situations where management that produces apparently good habitat has unintended negative consequences such as corridors that become predator traps, or facilitate the spread of exotic species or disease (Beier and Noss, 1998).

How can monitoring determine which management practices lead to conservation improvements? There are several approaches to improve conservation within certification programs. In areas where the ecology is well known, monitoring focal-species, those most sensitive to a particular threat, can indicate benefits to other taxa (Lambeck, 1997). In poorly studied ecosystems or where greater complexity puts a broader focus on biodiversity, such as in the tropics, vegetation management can provide a good surrogate. The general hypothesis behind criticisms of modern agriculture argues that simplification of vegetative structure and diversity (planned biodiversity) reduces associated or unplanned biodiversity that can help control pest problems (Swift et al., 1996). Actual biodiversity loss curves can help identify appropriate levels for management criteria (Figure 8-2). A management index can provide a measure of intensification (Mas and Dietsch, 2003). When there are nonlinear responses, levels of management intensity that produce sharp declines in biodiversity measures may indicate where criteria should be established (See Curve IV in Figure 8-2; vegetative characteristics associated with management practices above the threshold of the sharp decline in biodiversity could be used as criteria). An alternative but similar approach is to focus on landscape-level features for restoration and management (Lindenmeyer et al., 2002). Whichever approach is used, appropriate measures of success can be selected as indicators of ecological integrity (Noss, 1990). Monitoring should be based on testing hypotheses related to certification claims and specific criteria. This will allow adaptive management through the adjustment of criteria or conservation planning by defining better management activities that are appropriate for particular landscape uses such as corridors or buffer zones. This approach can provide the ecological basis for realistic expectations of conservation success.

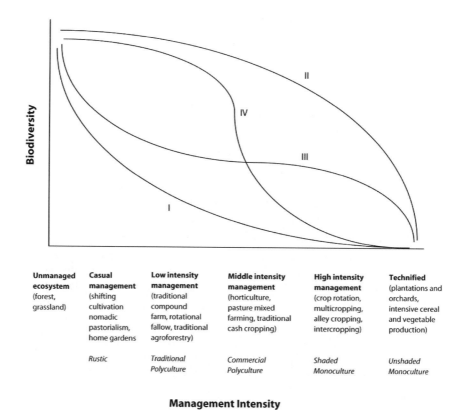

Figure 8-2. Hypothesized curves of biodiversity loss across an intensification gradient (Adapted from Swift et al., 1996). Coffee management categories in italic (Figure 8-3).

Shade gradient in coffee production systems

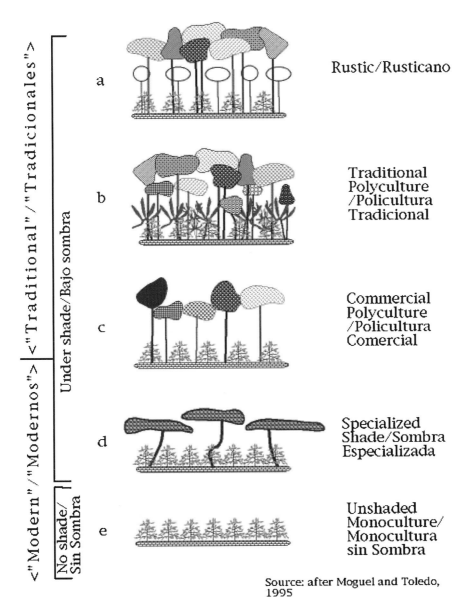

Figure 8-3. "Gestalt" system for classifying coffee agroecosystems, demonstrating coffee intensification gradient (from Robert Rice, Smithsonian Migratory Bird Center based on 1995 Newspaper article by Moguel and Toledo).

4. CURRENT EVIDENCE OF CONSERVATION SUCCESS – THE SHADE-GROWN COFFEE EXAMPLE

4.1 Coffee Agroecosystems

Before cultivation, coffee was a shade tolerant understory shrub in the eastern African highlands of Ethiopia (Wellman, 1961). Consequently, early cultivation methods maintained the shade canopy as an important component of the traditional or rustic growing systems. Coffee first arrived in Latin America in the early 1700's (Pendergrast 1999). Coffee is now well established in most mid-altitude areas (250-1750 m) where there is adequate rainfall. Through a cultivation process extending over centuries, coffee agricultural systems (agroecosystems) have developed a wide range of management practices ranging from integrated low-input agroforestry systems to modern technical high-input coffee monocultures (Perfecto and Snelling, 1995; Moguel and Toledo, 1999). Others have described the conversion process as an intensification gradient (Moguel and Toledo, 1999; Perfecto et al., 1996). Moguel and Toledo (1996) have developed a useful "gestalt" categorization system for describing typical coffee agroecosystems in Mexico that has wide application throughout Latin America (Figure 8-3). However, this is a gradient across multiple interrelated variables and oversimplification may mask important within farm and regional variation in management practices.

Traditional coffee agroecosystems contain a wide diversity of crops and often retain many native overstory trees for shade (Wellman, 1961). This diversity of primary producers combined with habitat structure variability supports a number of invertebrate and vertebrate organisms (Andow, 1991; Karr and Roth, 1971). The increased diversity in habitat structure produces a higher inherent complexity with many benefits for agriculture. Though still poorly studied, structural and vegetative diversity may provide microclimatic conditions and habitat for natural enemies such as predatory ants and birds (Perfecto and Snelling, 1995). Of the 332 Neararctic migratory species that winter in the Neotropics (51% of the U.S. avifauna), 82% include arboreal invertebrates in their diets while only 32% include fruit (Rappole et al., 1983). Many aspects of this more complex environment such as a reduced pest load and improved nutrient cycling may make this a productive long-term system competitive with input intensive monocultures. Interactions within this agricultural flora and fauna can closely resemble native tropical forest systems (Borrero, 1986; Perfecto and Vandermeer, 2002; Mas and Dietsch, 2003).

Current shade-grown coffee certification programs assume that greater vegetative structure and diversity will promote greater biodiversity, a vegetative management approach. Thus far, there is good empirical evidence for this assumption particularly in systems with contrasting levels of intensification. Comparisons between sun and shade coffee plantations have shown this pattern at opposite ends of the coffee production spectrum (Greenberg et al., 1997b; Wunderle and Latta, 1996). While the specific pattern across the intensification gradient for many taxa and mechanisms behind the distributions of particular species remain to be elucidated, these early patterns suggest that birds may make good indicators of the conservation value of agricultural systems such as coffee. However, higher shade density usually reduces coffee productivity (Soto-Pinto et al., 2000), suggesting the need for additional incentives for farmers. This has been the basis and possibly the impetus for the certification of coffee plantations as "Shade-grown", "Bird Friendly™", "Eco-ok", or "Café Amagable".

4.2 Birds and Coffee

Scientific connections between birds and coffee date back to an early expedition in 1691 that returned to England with a coffee branch and Andean Condor feathers (Sloane, 1694). More recently, well-known tropical ornithologists such as Griscom, Phillips, Wagner, Terborgh, and Skutch have studied avian natural history in coffee agroecosystems but not as an explicit component of distinct ecosystems. General public interest in birds has creates a natural focus for coffee certification efforts; much of the early publicity highlights benefits for bird conservation. Fortunately, birds fit the suggested criteria for selecting conservation indicator groups across broad geographic regions. In particular, birds are taxonomically well-known and stable, have better-studied general life histories than many other tropical taxa, are readily surveyed, and occupy a breadth of habitats. The sensitivity of different bird species to management intensity varies considerably, which may be useful for monitoring conservation success. Caution should be used, however. Though birds may be a good general indicator, the patterns of biodiversity loss may be different for other taxa (Perfecto et al., 2003). The general trend of reduction in diversity with intensification has also been shown for ants, spiders, mammals, reptiles, and butterflies (Perfecto et al., 1996; Ibarra-Nuñez, 1990; Mas and Dietsch, 2003; Horvath and Muñoz, unpublished data). However, the shape of the biodiversity loss curve could vary dramatically between taxa (i.e. between curves I and II of Figure 8-2). With this in mind, this review draws on the considerable work on birds in coffee ecosystems and other agroecosystems to establish a theoretical framework for how management practices could have conservation benefits.

4.3 Potential for Conservation of Birds in Coffee

Many coffee growing regions are located in tropical montane areas that also contain many of the world's endemic bird areas (EBA), areas with two or more restricted-range (<50,000 sq. km) bird species entirely confined to them (ICBP, 1992). These endemic species may be of conservation interest because their restricted range may make them more vulnerable to extinction. In 221 EBA's world wide, 26% of the world's species are found in 5% of the global area and 20% are found in 2% of the area (Stattersfield et al., 1998). Many major coffee producing nations also host the highest numbers of range-restricted endemic birds (Table 8-1). Further, most Latin American coffee production nations contain these rare birds, many of which are among 66 bird species of global conservation significance that have already been observed in coffee with a shade component (Dietsch, 2000). In southern Mexico and Central America, endemic bird areas are associated with highland areas surrounded by and including major coffee growing areas (Dietsch, 2000). Additionally, Moguel and Toledo (1999) identified 14 coffee growing areas as hotspots of biodiversity conservation of which, all but two were still dominated by traditional coffee. This suggests that improved conservation in coffee agroecosystems, as part of a comprehensive conservation strategy for some species or groups of species, could enhance established reserves, which tend to be at higher elevations for cloud forests.

8. Eco-Labeling in Latin America

Table 8-1. Countries with high numbers of range-restricted birds and their coffee production (updated from Dietsch, 2000 using data from Stattersfield et al., 1998 and FAO, 2002). Endemic birds are those found only in the country listed. Range-restricted birds are those whose geographic range is less than 50,000 hectares (ICBP, 1992).

Country	# of Range-Restricted Endemic Birds	# of Range-Restricted Birds Occurring	Hectares of Coffee (in 1000's)	Total Coffee Production for 2001, 1000's of mT (Rank)
Indonesia	319	403	891	376 (4)
Peru	101	211	228	158 (13)
Brazil	104	164	2302	1780 (1)
Columbia	63	192	850	560 (3)
Papua New Guinea	81	167	87	84 (18)
Ecuador	30	160	369	146 (14)
Philippines	121	126	136	130 (15)
Venezuela	39	110	220	69 (22)
Mexico	55	83	758	330 (5)
Costa Rica	6	77	106	181 (12)
India	38	74	310	301 (6)
Other Latin American & Caribbean Countries				
Guatemala	0	23	273	276 (8)
Honduras	1	20	217	205 (10)
El Salvador	0	17	162	112 (16)
Nicaragua	0	14	90	78 (19)
Dominican Republic	0	32	139	35 (26)
Haiti	1	33	54	28 (27)
Bolivia	17	61	25	24 (29)
Cuba	6	10	72	15 (35)
Puerto Rico	11	23	30	12.8 (37)
Panama	7	91	21	12.4 (39)
Paraguay	0	6	5.1	3.6 (47)
Jamaica	28	35	5	2.7 (52)
Lesser Antilles	24	33	1	0.6
Trinidad and Tobago	1	2	3.5	0.5 (57)
Guyana	0	20	0.4	0.14 (62)
Central Am & Carib		324	1933	1290
South America		658	4000	2742
Global Totals		2623	10766	7044

4.4 Conservation in Agroecosystems

Conservation benefits from traditional agroecosystems, though not ideal, do exist. According to Robbins et al. (1992), bird populations in cacao and shade coffee resembled those in native broadleaf forests though many ground-foraging migrants were missing. In Guatemala, 38 species of Neotropical migrants utilized an orange grove during 2 consecutive field seasons and 16 species were suggested as likely winter residents (Rogers et al., 1982). Mills and Rogers (1992) found extremely high ratios of migrant birds in Belizean citrus plantations, ranging from 50-80.6% of individuals and 43.5-58% of the species.

The diversification of habitat structure seems to encourage use of agricultural habitats by a wide range of bird species (Mellink, 1991). This habitat use is further enhanced by the proximity and connectedness of intact natural habitats. Estrada et al. (1997) studied anthropogenic landscapes in Los Tuxtlas, Mexico and found higher avian diversity in arboreal agricultural habitats (cacao, coffee, mixed, citrus, and allspice) than non-arboreal (corn, jalapeño, chili pepper, and bananas) with isolating distance and continued disturbance as important variables affecting species richness. Agricultural islands and live fencerows were important elements that reduce physical and biotic isolation among remaining forest fragments. In Illinois, prairie avifauna diversity and abundance was highest in areas adjacent to grasslands with high spatial heterogeneity and connectivity. In contrast, in areas where agricultural practices were intensified severe population declines and extirpations were reported (Warner, 1994).

For migratory songbirds, which by their nature may be more mobile, connectedness may be less important than actual habitat structure. In Belize, migrants were common in mature moist forest but were more abundant in early successional stages, particularly in habitat mosaics created by Kekchi-Mayan milpa agriculture (Kricher and Davis, 1992). Changes in vegetative structure and food availability can also affect distribution and behavior of forest birds. On the wintering grounds, hooded warblers segregate by sex and defend territories in different habitats. Morton et al. (1993) found that primary cues for this habitat selection are based on the relative amounts of vertical and oblique elements in the habitat. So, not only does changing these cues favor one species over another, but it may also favor one sex over another. Unlike conspecifics in primary habitats that are relatively sedentary, disturbance changes behavior of individuals so that in edge and second growth habitats these same species tend to become nomads (Rappole and Morton, 1985). Severe degradation may completely change bird community behavior in an area. Insectivorous birds in forests tend to be solitary, whereas granivorous birds in open areas have larger flock sizes

(Carrascal and Telleria, 1990). After Hurricane Gilbert passed over Jamaica, removal of overstory trees in lowland coffee plantation areas and montane forests caused severe reductions in bird species numbers by removing foraging substrate and changing proximal habitat selection cues (Wunderle et al., 1992).

Conversion of traditional coffee to more intensive "modern" systems seems to be a particularly drastic alteration of local ecology. In Colombia, Borrero (1986) found 99 (13 migratory) bird species in traditional coffee plantations that were being converted to sun coffee. This was significantly higher than other agricultural areas and conversion virtually eliminated this bird fauna. In an Indian integrated coffee system using remnant forest overstory supplemented with planted overstory, Beehler et al. (1987) observed an avifauna nearly as diverse (37 species vs. 43 species) as intact forests. They suggest that integrating coffee and pepper plantations with intact forest systems would preserve much of the native avifauna though they acknowledge large-bodied species were not likely to survive due to illegal hunting. Improved hunting regulation and alternative food sources should also be incorporated into conservation plans.

Agricultural intensification brings an additional hazard to native fauna. Pesticides and other agrochemical use in the Neotropics poses a direct threat to tropical biodiversity (Weir and Schapiro, 1981). Calvo and Blake (1998) noted fauna declines in censuses conducted one week after the application of herbicide (Gramaxone) in a traditional coffee plantation. This could be due either to direct mortality or bird movement related to a reduction in ground cover or food resources. Repeated exposure may cause declines in overall survival. Pesticide levels in migratory bird body fat indicate that contamination on wintering grounds is a serious problem, particularly for predators high in the food chain (Rappole et al., 1983). The primary general threats to Costa Rica's avifauna are environmental contamination and habitat destruction (Stiles, 1989). Agricultural practices contribute to both these problems. Land cleared for industrial farming produces chemical run-off and is unavailable as habitat for most species of birds.

4.5 Recent Research on Birds in Coffee Agroecosystems

The ornithological literature and ecological theory suggest that increased structure in coffee agroecosystems should improve avian species diversity. Though not an easy relationship to elucidate, vegetative structure has long been linked to avian diversity (MacArthur and MacArthur, 1961). Agricultural intensification results in significant declines in the diversity and abundance of Neotropical birds found along a gradient of reduced vegetative structure in coffee agroecosystems (Table 8-2). The management categories in Table 8-2 represent chosen variation on the part of the farmer in the vegetative structure that produces bird habitats. Similarly, within each plantation, management decisions, terrain limitations, temporal changes, and worker efficiency also create variation across the landscape both between and within management categories. This produces a patchwork of more and less intensive production areas as the basic framework for avian habitat selection.

Table 8-2. Bird species richness observed in coffee agroecosystems from published studies. Shade categories are estimated using classification from Moguel and Toledo (1999). Reported values are from point count data only. Greenberg et al. papers report estimated species richness.

Authors	Location	Sun	Shade Monoculture	Commercial Polyculture	Traditional Polyculture	Rustic	Forest
Wunderle & Latta 1996	Dominican Republic	20		24			35
Greenberg et al. 1997a	Guatemala	49.2-58.7	*Gliricidia* 46.3-58.1				72.2-95.3
Greenberg et al. 1997a	Guatemala		*Inga* 55.5-62.1				
Greenberg et al. 1997b	Chiapas, Mexico		64.7			68.7	56.0-68.1
Calvo & Blake 1998	Guatemala		49-51		63-72		

Observed species loss raises questions about the mechanisms that underlie the loss of forest-associated birds as coffee agriculture is intensified. Habitat selection theory suggests that bird abundance for many species is linked to factors associated with vegetative structure (i.e., appropriate nest-site, food, and other resource availability, Cody, 1985). Additionally, source-sink dynamics, a special case of metapopulation theory, suggest that in lower quality habitat patches, populations can be maintained by the influx of individuals from ource populations (Hanski, 1999; Robinson et al., 1995). Though this may mask species loss near habitat fragments, proximity or connectedness to higher quality habitat can facilitate movement between suitable patches (Noss, 1983; Noss and Harris, 1983). Thus for some species, less intensively managed coffee agroecosystems may be utilized as suitable habitat while for others coffee is only a matrix for movement between patches of preferred habitat. Other than the survey work cited above, bird habitat use has not been well studied in coffee or other tropical agroecosystems. More specific data are needed to assess whether coffee agroecosystems represent viable habitat for forest birds or transitory habitat providing short-term resources or dispersal corridors. As a matrix between forest patches, coffee agroecosystems can provide the following habitat use benefits for forest-associated birds:

Improved Connectivity - Forage and cover allow movement of species through the matrix that would otherwise be isolated.

Temporary Resources – Individuals leave preferred habitat for short-term forays (daily or seasonal migration) to forage then return to roost.

Temporary Refuge - Allows some reproduction (perhaps very low and the benefit is reproductive experience) for individuals who would prefer to breed in higher quality habitats but are excluded by members of their own species or prevented from reaching those areas (through dispersal barriers). When able, adults (or the next generation) move to higher quality territory and younger birds replace them.

Permanent Residents - Individuals breed successfully and hold high quality territories. Adults remain on territories; young disperse because habitat is saturated (i.e., when there are no available territories).

Willingness to move between patches and to use agriculture may vary depending on the type of agriculture at the edge (i.e., the sharpness or fuzziness of the edge) with greater use by forest-associated birds at lower contrast edges. In temperate agroecosystems, hedgerows and woodlots have been identified as important habitat for natural enemies, nesting sites for birds, and overwintering habitat for insects. For birds, pest control is limited by sharp differences in habitat between crops and woodlots. Birds rarely venture far from cover even when pest populations are high (Kirk et al., 1996). Certification efforts to raise the quality of the agricultural matrix (i.e., increased density and diversity of shade trees) may help overcome this limitation. Furthermore, at the landscape level, as more farmers adopt certification methods, increases in matrix habitat quality may show threshold effects by improving overall connectivity and reducing fragmentation. Forest-associated species may then be less constrained by dispersal abilities and distance from source populations in reserves.

Resident species seem to be more sensitive than migrants to habitat alteration in general and the intensification of coffee management practices specifically. Wunderle (1999) found that plantation size affected resident species richness but no area effect for migrants. Also among shade plantations, habitat characteristics did not affect migrant species richness while resident species richness was higher in larger older plantations at lower elevations with numerous stems (> 3 dbh), little or no pruning of the overstory, and maximum canopy cover at 12.0-15.0 m. Many migrants make extensive use of habitats with intermediate disturbance such as shade coffee and other arboreal agriculture. Consequently, there may be some conservation trade-offs if migrants reach higher abundances at lower levels of shade and drop off as shade increases to levels that may benefit more residents. Perhaps this could be counteracted by encouraging conservation of riparian corridors and forest fragments within plantations. Petit et al. (1999) suggest that proximity of extensive forest and presence of riparian vegetation may enhance plantation habitat value for many forest-associated species in Panama. They found 43% community similarity to lowland forest fragments and higher proportions of frugivorous and nectivorous species than in native forests. They suggest that assessing conservation value will require an analysis not only of species richness, but also type of species supported, reproductive output, and survival rates.

Calvo and Blake (1998) suggest that flowers and fruits of Inga sp. were important resources in both shade monoculture (modern) and polyculture (traditional) coffee agroecosystems. Proportion of migrants (23-32%) did not differ, but total abundance was higher on the traditional farm. Temporal patterns varied between the two systems; insectivores and omnivores declined during the wet season in the traditional farm but increased on the modern farm. More species had higher abundances on the traditional farm (11 species) during the wet season than on the modernized farm (2 species) and 9 species were equally abundant. Though less pronounced, there was a similar pattern during the dry season. In these two coffee agroecosystems, habitat preferences for individual species seem not only to vary seasonally but are also more pronounced during the wet season (an elevated period of resident breeding). Phenology of fruiting and flowering in the shade canopy may be a key determinant of bird abundance in coffee plantations as birds select suitable habitat within the landscape matrix.

Thus, diet provides an important consideration in limiting distribution of birds across a landscape, and the availability of particular food items may constrain how some species use coffee agroecosystems. Resident birds may be more specialized for particular components of the insect fauna (e.g. larger size categories), especially during the breeding season (Greenberg, 1995). Comparing available arthropod taxa proportions with the proportion in stomach contents may determine diet selectivity and degree of specialization among species (Poulin and Lefebvre, 1996). A better understanding of resource use and availability produced by coffee management practices may suggest regional recommendations or criteria changes for certification programs. The SMBC has done this for areas of Peru by recommending the use of native tree species that provide important seasonal fruit and flower resources for the diverse frugivore and nectivore community (Greenberg, personal communication).

Since migratory birds spend less than half of their life cycle on breeding grounds, an improved understanding of mortality factors through the wintering season is critical for successful conservation efforts. Bird community structure on wintering grounds is not well understood. Some factors that may be important are habitat association and migrant-resident interactions. Variation in foraging and territorial behavior needs to be better understood especially within the context of managed ecosystems. Relationships between migrants and residents inside and outside of agricultural settings are poorly understood.

5. RESEARCH NEEDS

Because tropical ecosystems are poorly studied relative to temperate areas, there is a general need for basic ecological research. This chapter has focused on birds as a central taxon in the development of shade coffee certification programs. However, birds may have limits as focal species, as other taxa may be more sensitive to management intensity (Perfecto et al., 2003). This suggests that vegetative management and landscape restoration approaches may provide better general biodiversity conservation benefits. The current "gestalt" system of categorizing coffee management systems is useful from an educational standpoint but may have scientific limits due to variation within each category. To aid monitoring, a vegetation management index similar to one designed by Mas and Dietsch (2003) based on the range of variation found across management categories can provide the basis for a refined and more general index connected with certification criteria. At the landscape level, there are many confounding effects that must be considered when comparing farms or management systems. More research is needed to identify and quantify conservation value from landscape features in coffee management areas. In particular, care must be taken to distinguish conservation benefits that derive from vegetation management practices from those that result from the landscape context of the farm (Mas and Dietsch, 2004).

Thus far, species richness has been used to compare different coffee management systems. However, the presence of a species may not be an adequate criterion with which to assess or provide conservation benefits (Van Horne, 1983; Gordon and Ornelas, 2000). Existing alternative indices and indicators can improve initial assessments of conservation success but need to be evaluated (Magurran, 1988). With most mid-altitude forests already converted to coffee, understanding the role of small reserves and forest fragments in contributing to landscape diversity will be an important part of establishing a conservation baseline to provide a more fixed assessment of conservation benefits (Schelhas and Greenberg, 1996). This baseline can be connected with focused studies to understand how sensitive taxa and species select and use habitat. In particular, knowledge of survival and reproductive rates under different management practices will be necessary to understand the population ecology of species in managed landscapes. Apparently poor habitat may provide some population benefits (Foppen et al., 2000; Murphy, 2001). Thus, metapopulation dynamics should be evaluated at the landscape level and combine fecundity and mortality estimates to predict patch persistence. Though shade-grown coffee may benefit species that prefer closed canopy forests, there may be conservation trade-offs with those that prefer more open habitats. More research is needed to identify how management activities affect each species and identify characteristics associated with those adversely affected. This will involve an assessment of species benefits (resources) and hazards (predators and disease) within management systems.

The two main certification programs differ in whether they allow agrochemical use. Agrochemicals hazards need to be better evaluated for tropical agroecosystems. Along these lines, research should focus on the concerns of coffee producers as an important component of sustainability (Lélé and Norgaard, 1996). This can be combined with work to assess the ecological role of biodiversity in coffee agroecosystems. Depending on ecological interactions within a management system, there may be additional benefits or costs for growers.

Thus far, shade coffee research and certification efforts have focused on Latin America, however, since coffee is a global commodity, research is needed to explore geographic differences in production systems and potential conservation benefits. Shade coffee is an important test case for the implementation of eco-labeling in tropical agroecosystems. Other agricultural products may present opportunities for similar efforts in the future. Finally, evaluation of certification criteria with the goal to produce a process that is easily implemented and understood by farmers while maximizing conservation benefits will have a greater probability of acceptance by producers (Schelhas, 1994; Bray et al., 2002).

6. CERTIFICATION OF BIRD FRIENDLY™ AND SUSTAINABLE COFFEE

Recent efforts to identify and establish sustainable growing practices have brought the role of birds into a new focus. This coincides with an increasing interest in improving conservation of biodiversity in coffee agroecosystems. A number of certification efforts have been proposed,[7] but only the Bird Friendly™ program from SMBC and Sustainable Coffee program from Rainforest Alliance are using specific published criteria to certify shade-grown coffee. These programs distinguish between management systems of different intensities. Mas and Dietsch (2004) show that the Smithsonian program would certify Rustic or Traditional Polyculture systems, while the Rainforest Alliance program would also certify Commercial Polycultures. While most management systems they evaluate were organic and contained some shade, some organic practices were too intensive for shade-grown certification (Mas and Dietsch, 2004). In other words, not all shade is created equal. In associated studies, Rustic and Traditional Polyculture practices also produced significantly greater diversity for ants, birds, and butterflies (Mas and Dietsch, 2003; Perfecto and Vandermeer, 2002; Dietsch, 2003). By certifying particular characteristics that enhance bird or biodiversity conservation, coffee can be marketed to conscientious consumers at a higher price to compensate for lost production. In many cases, farmers already use growing practices that provide benefits to associated biodiversity. In Mexico, coffee grown under a shade canopy that would be categorized as Rustic or Traditional Polyculture constitutes 39% of the total area grown, Commercial Polyculture accounts for another 10% (Moguel and Toledo, 1999). Rewarding these growing practices can reduce incentives to intensify production or to switch to different agricultural systems (such as pasture systems) that are more damaging.

While certification enables consumers to identify products that are more environmentally responsible, diffuse program goals can still cause confusion and thereby reduce the effectiveness of all programs. While better than intensive modern methods, organic practices may not be enough to provide adequate conservation benefits. This problem was the basis for a conference held March 29-30, 2000 by the NAFTA-created tri-national Commission on Environmental Cooperation in Oaxaca, Mexico. Though participants, ranging from producers to roasters, had difficulty reaching consensus on specifics, they recognized tremendous potential in harmonizing standards for certification and linking programs, possibly through a unified seal to improve consumer recognition (TES, 2000). Sustainable coffee was proposed as a term for coffee that is tri-certified for the 3 main certification efforts: organic, fair trade, and shade-grown (Rice and McLean, 1999). One obstacle to unification has been the general impression that economic viability for small farmers and conservation goals are at odds. This has not

---

[7] North American Commission for Environmental Cooperation coffee certification program online database:
[http://www.cec.org/programs_projects/trade_environ_econ/sustain_agriculture/databases/index.cfm?varlan=english].

been the case in Mexico, where despite heavy initial investment costs for organic certification, small farmer cooperatives have had good success receiving certification with a corresponding improvement in price (Bray, 2002). This is one reason Mexico is now the global leader in organic coffee production with a growing domestic market (IFOAM, 2002; Bray, 2002). Throughout Latin America, promoting small farmer organizations provides a good model not only for linking shade-grown coffee with the better established organic and fair trade movements, but also as a basis for encouraging general eco-labeling efforts.

7. CONCLUSIONS

Coffee marketed as shade-grown is now readily available, suggesting that consumer education efforts have begun to pay off. Unfortunately, most of this coffee is not currently certified and not all shade is created equal in terms of biodiversity conservation. Consumers need to be directed toward certified shade-grown coffee to encourage those practices that scientific evidence suggests provide conservation benefits. This can increase demand for certified shade coffee to help ensure that farmers entering the certification process receive adequate compensation. While this chapter focuses on shade-grown coffee, a similar tension between increasing production and strong incentives to reduce management intensity exists in other environmentally friendly certification efforts. Efforts need to balance optimal conservation and economic viability. If conservation needs outweigh what is economically viable, then alternative and perhaps more traditional conservation strategies may be necessary. Advertisers may be guilty of over-selling in some cases and fraud in others, however, conservation biologists need to question these images and provide an alternative and realistic perspective on how managed landscapes can contribute to conservation goals. The shade-grown coffee example suggests that consumers can be confident that science-based certification programs constitute an effective conservation tool.

Market-based conservation strategies address social justice concerns of traditional conservation in protected areas by identifying compatible land use practices and rewarding producers that use them. Certification should require specific actions from producers, for which they deserve fair compensation. These specific actions provide the basis for monitoring ecological integrity. Rewarding those current practices that produce measurable conservation benefits can build credibility among producers and lay the foundation for restoration of intensively managed areas. Encouraging less intensive management practices near protected areas may help stem the current biodiversity crisis, but long-term success will come when intensive agriculture is replaced with a diverse and environmentally friendly managed mosaic. Environmentally friendly certification can provide economic incentives for this large-scale restoration project.

REFERENCES

Alpert, P., 1996, Integrated conservation and development projects: Examples from Africa, *BioScience* **46**:845-855.

Andow, D., 1991, Vegetational diversity and arthropod population response, *Annual Review of Entomology* **39**:561-586.

Baker, B.P., Benbrook, C.M., Groth III, E., Lutz Benbrook, K., 2002, Pesticide residues in conventional, integrated pest management (IPM)-grown and organic foods: insights from three US data sets, *Food Additives and Contaminants Journal* **19**:427-446.

Beehler, B.M., Krishna Raju, K.S.R., and Ali, S., 1987, Avian use of man-disturbed forest habitats in the Eastern Ghats, India, *Ibis* **129**:197-211.

Beier, P., and Noss, R.F., 1998, Do habitat corridors provide connectivity?, *Conservation Biology* **12**:1241-1252.

Borrero, J.I.H., 1986, La substitucion de cafetales de sombrio por caturrales y su efecto negativo sobre la fauna de vertebrados, *Caladasia* **15**(71-75):725-732.

Brash, A.R., 1987, The History of Avian Extinction and Forest Conversion on Puerto Rico, *Biological Conservation* **39**:97-111.

Bray, D.B., Sanchez, J.L.P., Murphy, E.C., 2002, Social dimensions of organic coffee production in Mexico: Lessons for eco-labeling initiatives, *Society and Natural Resources*, **15**:429-446.

Calvo, L., and Blake. J., 1998, Bird diversity and abundance on two different shade coffee plantations in Guatemala. *Bird Conservation International* **8**:297-308.

Carrascal, L.M., and Telleria, J.L., 1990, Flock size of birds wintering in a cultivated area: Influence of vegetation structure and type of diet, *Ekologia Polska* **38**:201-210.

Cody, M.L., 1985, An introduction to habitat selection in birds, in *Habitat Selection in Birds*, ed. M.L. Cody, San Diego: Academic Press.

Cone and Roper, 1993, *Benchmark Study on Cause-Related Marketing*, cited on Transfair USA website: [http://www.transfairusa.org/why/coffee.html], May 2002.

Cronon, W. 1995, The trouble with wilderness: or getting back to the wrong nature, in *Uncommon Ground: toward reinventing nature*, ed. W. Cronon, New York: W. W. Norton, pp. 23-90.

CCC (Consumers Choice Council), 2001, *Conservation Principles for Coffee Production*, Consumers Choice Council, downloaded from [http://www.consumerscouncil.org/coffee/principles_eng.pdf], May 2002.

Dietsch, T.V., 2000, Assessing the conservation value of shade-grown coffee: a biological perspective using Neotropical birds, *Endangered Species Update* **17**: 22-124.

Dietsch, T.V., 2003, *Avian Ecology and Conservation in the Coffee Agroecosystems of Chiapas, Mexico*. PhD dissertation. School of Natural Resources & Environment, University of Michigan, Ann Arbor, MI.

Doak, D., 1995, Source-sink models and the problem of habitat degradation: General models and applications to the Yellowstone Grizzly, *Conservation Biology*, **9**:1370-1379.

Dudley, N., Elliott, C., and Stolton, S., 1997, A framework for environmental labeling, *Environment* **39**:16-20, 42-45.

ECLAC (United Nations Economic Commission for Latin America and the Caribbean), 2002, *Centroamérica: El impacto de la caída de los precios del café*, downloaded from [http://www.eclac.org/publicaciones/Mexico/7/LCMEXL517/L517.pdf], May 2002.

Estrada, A., Coates-Estrada, R., Meritt, Jr., D.A., 1997, Anthropogenic landscape changes and avian diversity at Los Tuxtlas, Mexico, *Biodiversity and Conservation* **6**:19-43.

FAO (Food and agriculture organization of the United Nations), 1993, *Forest resources assessment 1990: Tropical countries*, FAO Forestry Paper 112, Rome: FAO.

FAO, 2002, *FAOSTAT Database Collections*, Rome: FAO, downloaded from [http://apps.fao.org/page/collections?subset=agriculture], May 2002.

Foppen, R.P.B., Chardon, J.P., and Liefveld, W., 2000, Understanding the role of sink patches in source-sink metapopulations: Reed Warbler in an agricultural landscape, *Conservation Biology* **14**:1881-1892.

Gilpin, M.E. and Soulé, M.E., 1986, Minimum viable populations: Processes of species extinction, in *Conservation biology: The science of scarcity and diversity*, ed. M.E. Soulé, Sunderland: Sinauer Associates, pp. 19-34.

Gordon, C.E., and Ornelas, J.F., 2000, Comparing endemism and habitat restriction in Mesoamerican tropical deciduous forest birds: implications for biodiversity conservation planning, *Bird Conservation International* **10**:289-303.

Green, M.J.B., Murray, M.G., Bunting, G.C., Paine, J.R., 1996, *WCMC Biodiversity Bulletin 1: Priorities for biodiversity conservation in the tropics*, Cambridge, UK: World conservation Monitoring Centre, available at [http://www.unep-wcmc.org/].

Greenberg, R., 1995, Insectivorous migratory birds in tropical ecosystems - the breeding currency hypothesis, *Journal of Avian Biology* **26**:260-264.

Greenberg, R., 1996, *Criteria Working Group Thought Paper*, Smithsonian Migratory Bird Center (SMBC), [http://natzoo.si.edu/smbc/], downloaded May 2002.

Greenberg, R., Bichier, P., Angon, A.C., and Reitsma, R., 1997a, Bird populations in shade and sun coffee plantations in Central Guatemala, *Conservation Biology* **11**:448-459.

Greenberg, R., Bichier, P., and Sterling, J., 1997b, Bird populations in rustic and planted shade coffee plantations of eastern Chiapas, Mexico, *Biotropica* **29**:501-514.

Grumbine, R.E., 1994, What is ecosystem management?, *Conservation Biology* **8**:27-38.

Gullison, T., Melnyk, M., and Wong, C., 2001, *Logging Off: Mechanisms to stop or prevent industrial logging in forests of high conservation value*, Cambridge, USA: Union of Concerned Scientists, 6.

Hanksi, I, 1997, Predictive and practical metapopulation models: The incidence function approach, in *Spatial Ecology: The role of space in population dynamics and interspecific interactions*, eds. D. Tilman and P. Kareiva, Princeton: Princeton University Press.

Hanski, I., 1999, *Metapopulation Ecology*, New York: Oxford University Press.

ICBP (International Council for Bird Preservation), 1992, *Putting Biodiversity On the Map: Priority areas for global conservation*, Cambridge: ICBP.

IFOAM (International Federation of Organic Agriculture Movements), 2002, *Reports on Organic Agriculture Worldwide*. [http://www.ifoam.org/orgagri/oaworld.html], downloaded May 2002.

Karr, J.R., and Roth, R.R., 1971, Vegetation structure and avian diversity in several New World Areas, *The American Naturalist* **105**:423-435.

Kirk, D.A., Evenden, M.D., and Mineau, P., 1996, Past and current attempts to evaluate the role of birds as predators of insect pests in temperate agriculture, in *Current Ornithology Vol. 13*, ed. V. Nolan Jr. and E. D. Ketterson, New York, Plenum Press, pp. 175-269.

Kricher, J.C., and Davis, W.E., 1992, Patterns of avian species richness in disturbed and undisturbed habitats in Belize, in *Ecology and Conservation of Neotropical Migrant*

Landbirds, ed. Hagan, J.M. and Johnston, D.W., Washington DC: Smithsonian Institution Press, pp. 240-246.

Lambeck, R.J., 1997, Focal species: A multi-species umbrella for nature conservation, *Conservation Biology* **11**:849-856.

Lélé, S., and Norgaard, R.B., 1996, Sustainability and the scientist's burden, *Conservation Biology* **10**:354-365.

Lindenmayer, D.B., Manning, A.D., Smith, P.L., Possingham, H.P., Fischer, J., Oliver, I., and McCarthy, M.A., 2002, The focal-species approach and landscape restoration: A critique, *Conservation Biology* **16**:338-345.

Lohr, L., 1999, Welfare effects of eco-label proliferation: Too much of a good thing?, presented at Food and Agricultural Marketing Consortium 1999 Annual Meeting, Alexandria, VA.

MacArthur, R.H., and MacArthur, J.W., 1961, On bird species diversity, *Ecology* **42**:594-599.

Magurran, A.E., 1988, *Ecological Diversity and Its Measurement*, Princeton: Princeton University Press.

Mas, A.H., and Dietsch, T.V., 2003, An index of management intensity for coffee agroecosystems to evaluate butterfly species richness, *Ecological Applications* **13**:1491-1501.

Mas, A.H., and Dietsch, T.V., 2004, Linking shade coffee certification programs to biodiversity conservation: Butterflies and birds in Chiapas, Mexico, *Ecological Applications* **14**:642-654.

Mellink, E., 1991, Bird communities associated with three traditional agroecosystems in the San Luis Potosi Plateau, Mexico, *Agriculture, Ecosystems and Environment* **36**:37-50.

Messer, K.D., Kotchen, M.J., and Moore, M.R., 2000, Can shade-grown coffee help tropical biodiversity? A Market Perspective, *Endangered Species Update* **17**:125-131.

Mills, E.D., and Rogers, Jr., D.T., 1982, Ratios of Neotropical migrant and Neotropical resident birds in winter in a citrus plantation in central Belize, *Journal of Field Ornithology* **63**:109-116.

Moguel, P., and Toledo, V.M., 1996, El café en México: ecología, cultura indígena y sustentabilidad, *Ciencias* **43**:40-51.

Moguel, P., and Toledo, V.M., 1999, Biodiversity conservation in traditional coffee systems in Mexico, *Conservation Biology* **12**:1-11.

Morton, E.S., Van der Voort, M. and Greenberg, R., 1993, How a warbler chooses its habitat: field support for laboratory experiments, *Animal Behaviour* **46**:47-53.

Murphy, M.T., 2001, Source-sink dynamics of a declining Eastern Kingbird population and the value of sink habitats, *Conservation Biology*, **15**:737-748.

Nadkarni, N.M., and Matelson, T.J., 1989, Bird use of epiphyte resources in Neotropical trees, *Condor* **91**:891-907.

Nir, M.A., 1988, The survivors: orchids on a Puerto Rican coffee finca, *American Orchid Society Bulletin* **57**:989-995.

Noss, R.F., 1983, A regional landscape approach to maintain diversity, *BioScience* **33**:700-706.

Noss, R.F., 1990, Indicators for monitoring biodiversity: a hierarchical approach, *Conservation Biology* **4**:355-364.

Noss, R.F., and Harris, L.D., 1986, Nodes, networks, and MUMs: preserving diversity at all scales, *Environmental Management* **10**:299-309.

Pendergrast, M., 1999, *Uncommon Grounds: The history of coffee and how it transformed our world*, New York: Basic Books.

Perfecto, I., and Vandermeer, J., 2002, Quality of agroecological matrix in a tropical montane landscape: Ants in coffee plantations in southern Mexico, *ConservationBiology* **16**:174-182.

Perfecto, I., and Snelling, R., 1995, Biodiversity and the transformation of a tropical agroecosystem: ants in coffee plantations, *Ecological Applications* **5**:1084-1097.

Perfecto, I., Rice, R.A., Greenberg, R., and Van der Voort, M.E., 1996, Shade coffee: a disappearing refuge for biodiversity. *BioScience* **46**:598-608.

Perfecto, I., Mas, A., Dietsch, T., and Vandermeer, J., 2003, Conservation of biodiversity in coffee agroecosystems: A tri-taxa comparison in southern Mexico, *Biodiversity and Conservation*, **12**:1239-1252.

Petit, L.J., Petit, D.R., Christian, D.G., and Powell, H.D.W., 1999, Bird communities of natural and modified habitats in Panama, *Ecography* **22**:292-304.

Pimm, S.L., Russell, G.J., Gittleman, J.L., and Brooks, T.M., 1995, The future of biodiversity, *Science* **269**:347-350.

Poulin, B. and Lefebvre, G., 1996, Dietary relationships of migrant and resident birds from a humid forest in central Panama, *Auk* **113**:277-287.

Rappole, J.H., Morton, E.S., Lovejoy, T.E., and Ruos, J.L., 1983, *Neararctic Avian Migrants in the Neotropics*, Washington DC: U.S. Department of Interior, Fish and Wildlife Service, U.S. Government Printing Office.

Rappole, J.H. and Morton, E.S., 1985, Effects of habitat alteration on a tropical avian forest community in *Neotropical Ornithology*, ed. P.A. Buckley, M.S. Foster, E.S. Morton, R.S. Ridgely, and F.G. Buckley, *Ornithological Monographs* **36**:1013-1036.

Rice, P.D., and McLean, J., 1999, *Sustainable Coffee at the Crossroads*, Washington DC: Consumer's Choice Council, downloaded from [http://www.consumerscouncil.org/coffee/coffeebook/coffee.pdf], May 2002.

Robbins, C.S., Dowell, B.A., Dawson, D.K., Colon, J.A., Estrada, R., Sutton, A., Sutton, R. and Weyer, D., 1992, Comparison of Neotropical migrant landbird populations wintering in tropical forest, isolated forest fragments, and agricultural habitats, in *Ecology and Conservation of Neotropical Migrant Landbird*, ed. J.M. Hagan and D.W. Johnston, Washington DC: Smithsonian Institution Press, pp. 207-220.

Robinson, S. K., Thompson, F.R., Donovan, T.M., Whitehead, D.R., and Faaborg, J., 1995, Regional forest fragmentation and the nesting success of migratory birds, *Science* **267**:1987-1990.

Rogers, D.T., Jr., Hicks, D.L., Wischusen, E.W., and Parrish, J.R., 1982, Repeats, returns, and estimated flight ranges of some North American migrants in Guatemala, *Journal of Field Ornithology* 53:133-138.

Root, T.L. and Schneider, S.H., 2002 Climate change: Overview and implication for wildlife, in *Wildlife Responses to Climate Change: North American Case Studies*, ed. S.H. Schneider and T.L. Root, Washington, DC: Island Press, pp. 1-56.

Schelhas, J., 1994, Building sustainable land use on existing practices: Smallholder land-use mosaics in tropical lowland Costa Rica, *Society and Natural Resources*, **7**:67-84.

Schelhas, J. and Greenberg, R., 1996, Introduction: The value of forest patches, in *Forest Patches in Tropical Landscapes*, J. Schelhas and R. Greenberg, eds., Washington DC: Island Press.

Schwarzman, S. and Kingston, M., 1997, *Global Deforestation, Timber and the Struggle for Sustainability: Making the label stick*, New York: Environmental Defense Fund.

Sloane, H. 1694. An Account of a Prodigiously Large Feather of the Bird Cuntur, Brought from Chili, and Supposed to be a Kind of Vultur; and of the Coffee-Shrub. By Hans Sloane, M. D. S. R. S. (in Number 208), *Philosophical Transactions (1683-1775)* **18**:61-64.

Soulé, M.E., 1986, The fitness and viability of populations, in *Conservation Biology: The science of scarcity and diversity*, ed. M.E. Soulé, Sunderland: Sinauer Associates, pp. 13-18

Stattersfield, A.J., Crosby, M.J., Long, A.J., and Wege, D.C., 1998, *Endemic Bird Areas of the World: Priorities for biodiversity conservation*, Cambridge: Birdlife International.

Stiles, F.G., Skutch, A.F., and Gardner, D., 1989, *A Guide to the Birds of Costa Rica*, Ithaca: Cornell University Press.

Swift, M.J., Vandermeer, J., Ramakrishnan, P.S., Anderson, J.M., Ong, C.K., and Hawkins, B.A., 1996, Biodiversity and agroecosystem function, in *Functional Roles of Biodiversity: A global perspective*, ed. H.A. Mooney, J.H. Cushman, E. Medina, O.E. Sala and E.D. Schulze, Chichester, UK: John Wiley & Sons Ltd.

TES (TerraChoice Environmental Services), 2000, *Environmental and Other Labeling of Coffee: The role of mutual recognition*, Montréal: CEC, downloaded from [http://www.cec.org/programs_projects/trade_environ_econ/pdfs/Terra-e.pdf], May 2002.

Vandermeer, J., 1995, The ecological basis of alternative agriculture, *Annual Review of Ecology and Systematics* **26**:201-224.

Vandermeer, J. and Perfecto, I., 1995, *Breakfast of Biodiversity: The truth about rain forest destruction*, Oakland: Institute for Food and Development Policy.

Vandermeer, J. and I. Perfecto, 1997, The agroecosystem: A need for the conservation biologist's lens, *Conservation Biology* **11**:591-592.

Van Horne, B., 1983, Density as a misleading indicator of habitat quality, *Journal of Wildlife Management* **47**:893-901.

Waridel, L., 2002, *Coffee with Pleasure: Just java and world trade*, Tonawanda: Black Rose Books.

Warner, R.E., 1994, Agricultural land use and grassland habitat in Illinois: Future shock for Midwestern birds?, *Conservation Biology* **8**:147-156.

Weir, D. and Schapiro, M., 1981, *Circle of Poison: Pesticides and People in a Hungry World*, Oakland, CA: Food First Books.

Wellman, F.L., 1961, *Coffee: Botany, Cultivation, and Utilization*, New York, NY: Interscience Publishers, Inc.

Whitmore, T.C., 1997, Tropical forest disturbance, disappearance, and species loss, in *Tropical Forest Remnants: Ecology, management, and conservation of fragmented communities*, ed. W.F. Laurance and R.O. Bierregaard, Jr., Chicago: University of Chicago Press.

Wilshusen, P.R., Brechin, S.R., Fortwangler, C.L., and West, P.C., 2002 Reinventing a square wheel: Critique of a resurgent "Protection Paradigm" in international biodiversity conservation, *Society and Natural Resources* **15**:17-40.

Wunderle, J.M., 1999, Avian distribution in Dominican shade coffee plantations: Area and habitat relationships, *Journal of Field Ornithology* **70**:58-70.

Wunderle, J.M., Lodge, D.J., and Waide, R.B., 1992, Short-term effects of Hurricane Gilbert on terrestrial bird populations on Jamaica, *The Auk* **109**:148-166.

Wunderle, Jr., J.M., and Latta, S.C., 1996, Avian abundance in sun and shade coffee plantations and remnant pine forest in the Cordillera Central, Dominican Republic, *Ornitologia Neotropical* 7:19-34.

Yussefi, M., and Willer, H., 2002, *Organic Agriculture Worldwide 2002: Statistics and future prospects*. Sponsored by BIOFACH, in collaboration with IFOAM, Bad Dürkheim, Nederlands: Stiftung Ökologie & Landbau (SÖL), downloaded from [http://www.soel.de/inhalte/publikationen/s_74_03.pdf], May 2002.

PART 4: PUBLIC PARTICIPATION AND JUSTICE SYSTEMS

Chapter 9

PUBLIC PROSECUTORS AND ENVIRONMENTAL PROTECTION IN BRAZIL

Lesley K. McAllister[1]
[1]*Ph.D. Candidate, Energy and Resources Group, 310 Barrows Hall, University of California at Berkeley, Berkeley, California 94720.*

Abstract: Since the passage of the Brazilian Federal Constitution of 1988, the public prosecutors of the Brazilian Ministério Público have become significant actors in environmental protection through their use of investigative and legal instruments to impose civil and criminal liability for environmental harms. The Brazilian Ministério Público's work is a response to the 'non-enforcement problem' of Brazilian environmental law in which laws tend to be strong 'on-the-books' but weak in practice. This chapter describes and analyzes the involvement of the Brazilian Ministério Público in environmental protection. It presents the instruments that prosecutors use to defend environmental interests; describes and provides an explanation for the legal and institutional changes in the 1980s that allowed the Ministério Público to become a significant actor in environmental protection; and assesses the effectiveness of the prosecutorial enforcement of environmental laws in Brazil.

Key words: enforcement; prosecution; environmental law; environmental policy; Latin America; Ministério Público; public interest; public civil action

1. INTRODUCTION

A new model of environmental regulation is being forged in Brazil that may stem the chronic under-enforcement of environmental laws. The public prosecutors of the Ministério Público have become active in environmental protection, enforcing Brazil's environmental laws against private as well as public actors. Transformed by the Federal Constitution of 1988 from an institution that was dedicated primarily to defending the interests of the state in criminal cases into one that is also responsible for defending environmental and other public interests, the Ministério Público has changed the face of environmental law enforcement in Brazil.

Many Latin American countries have substantial but under-enforced legal frameworks for environmental protection. Like Brazil, other countries wrote new constitutions guaranteeing a wide array of social and political rights, including the right to a healthy environment, to mark their transitions from authoritarian rule to democratic rule in the 1980s and early 1990s.[1] Several of these, including Argentina, Venezuela and Colombia, have reformed their laws to enable citizen groups and the Ministério Público or a similar institution to legally defend these new rights.[2] Among these, Brazil's Ministério Público has been the most active in defending environmental interests.[3]

Brazilian environmental legal experts have often acclaimed the strength of the Brazilian constitution and environmental laws while criticizing their lack of enforcement.[4] As stated by one scholar, "Brazil has an advanced legal framework that is one of the best in the world. What is lacking is compliance."[5] Discussing environmental laws in Latin America generally, he explains, "In many cases, the law is a mere document, written for the sake of national, and lately, international public opinion" (Benjamin, 1995). In other words, environmental law tends to suffer from a 'non-enforcement problem,' wherein strong laws are passed but they are not adequately implemented or enforced.

In Brazil, however, a new type of environmental enforcement – prosecutorial enforcement – is reshaping environmental protection. Both at the federal and state levels, the prosecutors of the Ministério Público have

---

[1] The constitutions of Argentina (1994), Chile (1980 with amendments), Colombia (1991 with amendments), Costa Rica (1949 with amendments), Ecuador (1998), Nicaragua (1987 with amendments) and Paraguay (1992) all guarantee the right to a healthy environment. Georgetown University and Organización de Estados Americanos, 1998, *Base de Datos Políticos de las Américas*; http://www.georgetown.edu/pdba/Comp/Ambiente/derecho.html.

[2] Argentina's Constitution of 1994 (Section 43) gives citizens and environmental groups the right to sue for any act or omission of the public authorities, private entities or individuals affecting or threatening to affect environmental rights. Venezuela's 1991 Constitution (Article 88, Section 1) provides for citizen actions to defend collective rights and interests such as environment, property, public health, safety, and public morals. Colombia's Constitution of 1991 (Article 277) charges the Ministério Público with defending 'collective interests, especially the environment.'

[3] While this study focuses on environmental interests, the Brazilian Ministério Público became similarly active in the defense of other public interests such as consumer rights, children's rights, housing and urban problems, and anti-corruption.

[4] The English term 'enforcement' lacks a precise translation in Portuguese. The most common translations are *implementação da lei, aplicação da lei* and *fazer cumprir a lei*. *Fiscalização* and *controle* are also used, particularly in the administrative sphere. As such, the English word has come into usage in Portuguese (see, e.g., Ferraz and Ferraz, 1997).

[5] Quote by Antonio Herman Benjamin, published in "Judiciárias: Aplicação da lei ambiental foi tema em debate na semana do MP," *A Tarde*, Salvador, Bahia (December 12, 1998).

become involved in enforcing environmental laws.[6] Throughout Brazil, there were about 10,000 public prosecutors in 2003 (Sadek and Cavalcanti, 2003). Of these, over 2,000 were responsible for environmental protection, and perhaps about 200 worked in this area exclusively.

This chapter describes and analyzes the involvement of the Brazilian Ministério Público in environmental protection.[7] It first discusses the instruments used by prosecutors to enforce environmental laws and presents data on prosecutorial enforcement from the Brazilian state of São Paulo. It then describes and explains the rise of prosecutorial enforcement in Brazil. Finally, it assesses the effectiveness of prosecutorial enforcement in reducing the 'non-enforcement' problem of Brazilian environmental law.

2. PROSECUTORIAL ENFORCEMENT IN THE STATE OF SÃO PAULO

The state of São Paulo's Ministério Público is the largest and most active in environmental protection. While federal prosecutors and state prosecutors in other states are also involved in environmental protection, the São Paulo Ministério Público was the first to become involved in environmental protection and it is viewed as a model by other Brazilian prosecutorial institutions. After introducing the instruments used by prosecutors to enforcement environmental laws, this section presents data on the São Paulo Ministério Público's environmental enforcement activity, analyzes the types of problems addressed, and discusses several 'high impact' cases.

2.1 Instruments of prosecutorial enforcement

Prosecutors primarily use three tools in their environmental protection work: the investigation, the public civil action, and the conduct adjustment

---

[6] The Brazilian Ministério Público is actually 27 separate institutions – the Federal Ministério Público and a Ministério Público in each of Brazil's 26 states. While their general organization and functions are governed by the same constitutional provisions and federal laws, they are autonomous institutions. The Federal Ministério Público is headquartered in the national capital, Brasilia, and has federal prosecutors in each state. Each state Ministério Público has a headquarters in the state capital and one or more state prosecutors in each judicial district (*comarca*) in the state.

[7] This chapter is based on fieldwork conducted in Brazil by the author from June to August 2000 and from September 2001 to June 2002. Research methods included participant observation in the Ministério Público and in the environmental agencies; semi-structured interviews with prosecutors, agency officials, environmental group leaders, scholars, and private lawyers; analysis of institutional databases; and archival research in academic and institutional library collections.

agreement. In addition, prosecutors may criminally prosecute environmental harms that result from intentional or negligent actions.[8]

Prosecutors generally open environmental investigations because of either a public complaint or a notification from another governmental agency regarding a potential environmental harm. A prosecutor may also open an environmental investigation based on personal observation or knowledge. The investigation seeks to verify the environmental harm, determine its extent and severity, and identify the responsible party. The investigation may be a regular civil investigation (*inquérito civil*) or it may be a preparatory investigation (*procedimento preparatório do inquérito civil*), used to determine whether a regular investigation is warranted.

When conducting an investigation, prosecutors have the legal authority to demand information from private and public entities. Often the prosecutor demands relevant documents or a technical report about a certain problem from the environmental agency. If the agency fails to comply with the prosecutor's requests, the prosecutor can hold agency officials criminally liable. In cases where the agency itself may be responsible for the environmental harm, the prosecutor may seek a technical opinion from an expert outside the agency.

Investigations are designed to facilitate the collection of evidence for the eventual filing of a public civil action (*ação civil pública*). Passed in 1985, the Public Civil Action Law (*Lei de Ação Civil Pública*) is a procedural law that enables interests considered to be "diffuse" or "collective" such as environmental interests to be defended in court.[9] The decision to file a public civil action is considered by many prosecutors to be non-discretionary: where there is proof of environmental harm and the case is not resolved extra-judicially as described below, the prosecutor has a duty to file a public civil action (Mazzilli, 2001).[10]

Public prosecutors can use the public civil action both to impose liability when environmental harm has already occurred and to prevent environment

---

[8] Under Brazilian law, a single environmental offense may result in administrative, civil, and criminal liability. Generally, environmental agencies issue administrative sanctions while the Ministério Público pursues civil and criminal sanctions.

[9] Federal Law 7,347 of July 24, 1985. Federal Law 8,078 of September 11, 1990 (the Consumer Defense Code, *Codigo da Defesa do Consumidor*) amended the Public Civil Action Law to enable the judicial defense of all diffuse, collective, and shared individual interests (*interesses difusos, interesses coletivos* and *interesses individuais homogeneos*). Diffuse interests are those held by an undetermined or undeterminable number of people, such as the interest in urban air quality. Collective interests are those held by a certain group, category, or class of people, such as the protection of the lands of a certain Indian tribe. Shared individual interests are those that are held by an identifiable group and that can be indemnified individually, such as recovery of damages by purchasers of an automobile with a manufacturing defect (Mazzilli, 2001; Silva, 2001).

[10] Other prosecutors argue that prosecutors should exercise discretion so as to be able to dedicate greater resources to larger or more important cases (Proença, 2001).

harm. In cases where the harm has already occurred, public prosecutors use the public civil action to obtain a judicial order imposing civil liability. If found liable by the court, the responsible party may be ordered to pay money damages to a state environmental fund or it may be enjoined to take an action (*obrigação de fazer*) or to abstain from taking an action (*obrigação de não fazer*) to remedy the harm.[11] Except where the environmental harm is considered irreparable, prosecutors generally seek to enjoin the party rather than to collect money damages. The public civil action may also be used to prevent environmental harm. Where environmental harm is threatened by a party's activities, a prosecutor may request the court to issue a preliminary injunction (*mandado liminar*) to prevent the harm.[12]

Unlike the power to open an investigation, the power to bring a public civil action is not exclusive to the Ministério Público. Environmental public civil actions can also be filed by a variety of other governmental and non-governmental actors, including environmental agencies and environmental organization.[13] The Ministério Público, however, has brought over ninety percent of public civil actions in Brazil (Cappelli, 2000). Environmental agencies generally choose to enforce the law using administrative sanctions - warnings, fines, and facility closures. Environmental organizations often make complaints to the Ministério Público rather than filing legal action themselves.

Once the prosecutor has sufficient information about the problem being investigated, he generally tries settle the case extra-judicially. The majority of the Ministério Público's environmental investigations result in settlements, referred to as conduct adjustment agreements (*termos de ajustamento de conduta*). Negotiated with the responsible party, these agreements determine the actions that the party must take to remedy the harm and the monetary penalties to be applied if the agreement is not complied with. In the case of noncompliance, the prosecutor may judicially enforce the agreement. Such lawsuits are won almost automatically, given that the conduct adjustment agreement represents the responsible party's admission to causing the harm and acceptance of the stated penalty. In addition to being faster and less costly, extra-judicial resolution carries other advantages including the party's voluntary assumption of responsibility and the greater potential to customize the remedy (Fink, 2001).

While criminal sanctions for some environmentally harmful activities have existed in Brazil for many years, the Environmental Crimes Law of 1998 (*Lei de Crimes Ambientais*) consolidated and enhanced the penalties for a variety of crimes including mistreatment or illegal killing of animals,

---

[11] Federal Law 7,347, Article 3 and Article 13.
[12] To be granted a preliminary injunction, the prosecutor must prove the likelihood of success on the merits (*fumus boni iuris*) and the risk of irreparable harm (*periculum in mora*).
[13] Federal Law 7,347, Article 5.

deforestation, pollution, and destruction of historical preservation sites.[14] Under the environmental crimes law, it is also a crime for a public official to state false or incomplete information in the environmental permitting process or to issue an environmental permit that does not comply with environmental regulations.[15] Both individuals and legal entities may incur criminal liability with sanctions including prison, house arrest, community service, disqualification from the receipt of public contracts or subsidies, and monetary fines of up to 50 million Brazilian *reais*.[16]

2.2 Prosecutorial enforcement activity in São Paulo

Using investigations, public civil actions, and conduct adjustment agreements, the São Paulo Ministério Público became a key actor in environmental enforcement in the state of São Paulo. The Ministério Público opened over 1,300 regular civil investigations and 2,500 preparatory investigations in the year 2000. As shown in *Figure 9-1*, the annual number of regular and preparatory investigations generally grew between 1985 and 2000, with particularly precipitous growth beginning in 1995. In the year 2000, the São Paulo Ministério Público also filed 229 environmental public civil actions. Between 1985 and 2000, there were over 3,000 environmental public civil actions filed. As can be observed in *Figure 9-1*, the annual number of public civil actions stayed fairly constant in the 1990s, averaging 240 per year from 1991 to 2000. In São Paulo, there were about 1,000 environmental conduct adjustment agreements signed in the year 2000 – four times the number of public civil actions.[17]

The government, whether federal, state, or municipal, was a defendant in many of the São Paulo Ministério Público's public civil actions. Municipal governments were sued for failing to treat sewage or handle solid wastes in accordance with environmental laws. State and federal governments were sued for environmental harms caused by infrastructure projects such as roads, dams, and energy facilities. In some cases, the environmental agency was sued for failing to implement and enforce environmental laws. A governmental entity was named as a defendant in almost one quarter of the environmental public civil actions filed by the Ministério Público of São Paulo between 1985 and 1997.

---

[14] Federal Law 9,605 of February 12, 1998.
[15] Federal Law 9,605, Articles 66 and 67.
[16] Decree 3,179 of September 21, 1999. Fifty million Brazilian *reais* (R$) was equivalent to US$ 21.7 million using the commercial exchange rate of January 1, 2002.
[17] Antônio Herman Benjamin, Conselho Superior of the Ministério Público of São Paulo, pers. comm., March 2002.

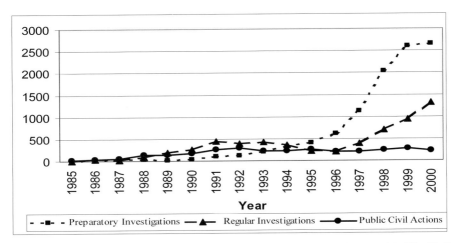

Figure 9-1. Number of Preparatory Investigations, Regular Investigations, and Public Civil Actions by the Ministério Público of the State of São Paulo, 1985 to 2000.[18]

While criminal sanctions are available for environmental enforcement, the São Paulo Ministério Público's environmental work has been concentrated in the civil sphere. In addition to the novelty of the environmental crimes law, there are several reasons that the Ministério Público relies more heavily on civil instruments. In civil cases, the Ministério Público does not have to prove that the party acted intentionally or negligently. As stated by a São Paulo environmental prosecutor, "In the civil sphere, we have a much more powerful tool: strict liability. We do not have to argue about guilt – it doesn't matter if the party acted intentionally or not. In the criminal sphere, you have to prove a guilty mental state."[19]

A second reason is that the civil instruments allow the possibility of preventing environmental harm rather than just punishing it. In most environmental public civil actions, the São Paulo Ministério Público seeks an injunction to force the party to cease the harmful activity. Even when punishment is the goal, civil liability – making the party either undertake an environmental recuperation project or pay money damages valued according to the environmental harm – is often considered a more appropriate tool than criminal liability.

Organizational aspects, specifically the traditional division of the Ministério Público's civil and criminal work, provide a third reason. São Paulo prosecutors that specialize in environmental protection are classified as civil prosecutors and do not have the authority to act in the criminal

---

[18] Pers. comm., Centro de Apoio Operacional das Promotorias de Justiça do Meio Ambiente, Ministério Público de São Paulo (November 2001).
[19] Ibid.

sphere. Instead, the environmental prosecutor refers the case to a criminal prosecutor, who also handles other types of criminal cases and may not view environmental cases as a priority.[20] Moreover, it is worth noting that, unlike civil investigations, criminal investigations are conducted by the police rather than by the Ministério Público. This lack of control over the investigative process also contributes to the lower usage of criminal process.

While criminal environmental cases are uncommon in São Paulo, it is likely that the possibility of criminal sanctions deters many potential offenders. In addition, the possibility of criminal charges is likely to make parties that are under civil investigation by the Ministério Público more willing to submit to a conduct adjustment agreement.

2.3 Types of environmental cases

The Ministério Público handles cases involving a large range of different environmental problems. *Figure 9-2* shows a breakdown of the types of problems dealt with in the 2,197 public civil actions that were filed in the state of São Paulo from 1984 to 1997. The most common types of environmental public civil actions involve deforestation and other violations of Brazilian forestry laws. Actions based on such violations account for 35% of all actions filed by the Ministério Público of São Paulo between 1984 and 1997.[21] Deforestation and devegetation in areas of the Atlantic Coastal Rainforest (*Mata Atlântica*) and in erosion-prone areas are extremely restricted under Brazilian forestry code and other laws, requiring a special permit that can only be obtained for works of public utility or social interest.[22]

Other issues commonly dealt with by the São Paulo Ministério Público include water quality (13%), land use and construction (12%), air quality (9%), solid wastes (9%) and mineral extraction (8%). While most of the water quality cases involved the lack of domestic sewage treatment, others dealt with industrial water pollution, oil spills, fish kills, and drinking water

---

[20] In some states, including Rio Grande do Sul, specialized environmental prosecutors can also bring criminal cases.
[21] Despite being most prevalent, cases concerning deforestation and devegetation are not considered to be the most significant cases by many prosecutors because they often involve small areas and few trees. The data indicate that 35% of cases in the deforestation and erosion categories concerned areas of less than 0.5 hectare or 50 trees and 50% concerned less than 1 hectare or 100 trees. The prevalence of these lawsuits is partly explained by the fact that prosecutors are routinely informed of such violations by the São Paulo Environmental Police (*Comando de Policiamento Ambiental*), and thus these are the problems about which prosecutors hear most and have the information available to file a public civil action
[22] See, e.g., Forestry Code (*Código Florestal*), Federal Law 4,771 of September 15, 1965 and Atlantic Coastal Rainforest Law, Federal Decree 750 of February 10, 1993.

9. Public Prosecutors and Environmental Protection

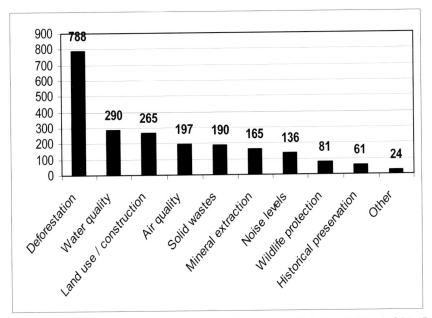

Figure 9-2. Number of Public Civil Actions filed by the Ministério Público of São Paulo according to Type of Environmental Harm, 1985 – 1997.[23]

treatment. Land use and construction cases involved environmental impacts and permitting irregularities in the construction of subdivisions, buildings, and roads. The majority of air quality cases pertained to the practice of burning of sugar cane fields to facilitate harvest. Solid waste cases generally dealt with illegal dumping and the impacts of unprotected landfills. Most mineral extraction cases involved the extraction of sand and other construction materials.

The remainder of the cases involved noise levels (6%), wildlife and animal protection (4%), historical preservation (3%), and other assorted issues (1%). Noise pollution cases targeted factories, music clubs, evangelical churches and other establishments that produced illegally high noise levels. Wildlife and animal protection cases generally concerned illegal hunting and animal mistreatment. Historical preservation cases sought protection of historical buildings such as churches, railroad stations, and plazas in urban areas.[24]

---

[23] Pers. comm., Centro de Apoio Operacional das Promotorias de Justiça do Meio Ambiente, Ministério Público de São Paulo (November 2001).
[24] In a study of environmental public interests filed by the Ministério Público of the state of Rio de Janeiro between 1985 and 1991, Fuks (1999) finds a similar breakdown in the subject of cases. The five problems most prevalent in cases were deforestation (25%); noise levels (12%); water pollution (10%); mineral extraction (9%); and air pollution (7%).

2.4 High impact cases

While cases involving large actual or potential environmental harms are not the majority of the Ministério Público's cases, they are the cases that are most remembered and commented upon by prosecutors and others. The Brazilian Ministério Público has received national attention for its role in many headline environmental cases in the 1990s involving, for example, the imposition of significant monetary fines on the national oil company for oil spills and challenges to governmental approval of many large infrastructure projects such as highways, power plants, and canals.[25] This section describes two cases filed by the São Paulo Ministério Público that are considered by prosecutors and others to have been 'high impact' cases.

In the well-publicized Cubatão case, the São Paulo Ministério Público sought to impose liability for large-scale environmental harms caused by the cluster of petrochemical companies operating in Brazil's largest and most important industrial district, Cubatão.[26] In the early 1980s, the area surrounding Cubatão became known as the 'valley of death' (*vale da morte*) because of high levels of industrial pollution suspected of causing serious birth defects and illnesses. Environmental damage to the nearby coastal mountain range, moreover, had increased the risk of major landslides.

In 1986, the Ministério Público and a private environmental organization jointly filed an environmental public civil action against the 24 national and multinational companies located in Cubatão to impose civil liability for an estimated US $800 million of environmental damages (Milaré, 1992). The legal action requested that the court hold the companies liable for the cost of stabilizing the mountainsides; decontaminating the soils; cleaning and recuperating local streams and rivers; and reforesting with native species. After surviving many years of challenges on procedural grounds by the companies, the case was still in the phase of collecting expert opinions as of 2001.[27]

In another well-publicized case, the São Paulo Ministério Público filed a public civil action alleging potential environmental harm from the proposed construction of a Volkswagen auto assembly plant in an environmentally sensitive area near the city of São Carlos. The 1996 suit was filed against Volkswagen of Brazil, the city of São Carlos, the state of São Paulo, and the

---

[25] See, e.g., "MP quer que Petrobrás pague R$ 100 milhões de multa," *O Estado de São Paulo* (January 24, 2002); "Procuradores querem ser ouvidos sobre Belo Monte," *O Liberal*, Belém, Pará (May 31, 2002); "Cai contrato para hidrovia de Marajó," *O Liberal*, Belém, Pará (January 29, 2002).

[26] In 1983, Cubatão generated $1 billion in exports and produced 47% of Brazil's nitrogen, 40% of its steel, 38% of its fertilizers, 32% of its phosphoric acid, 30% of its polyethylene, 25% of its chlorosoda, 18% of its bottled gas, and 12% of its gasoline (Findley, 1988).

[27] "Caso tramita há 15 anos na Justiça de SP," *A Folha de São Paulo* (October 14, 2001), p. C2.

São Paulo state environmental agency. The state environmental agency had previously determined there was no risk of significant environmental harm from the plant and an environmental impact study (EIS) was not required. The public civil action sought a court determination that an EIS was necessary. The action also requested a preliminary injunction that would enjoin construction of the plant and invalidate the environmental permit.

The preliminary injunction was not granted based on the court's reasoning that the company would have the resources to compensate for any eventual environmental harm. With the denial of the injunction, the construction continued and the plant became operational in October 1996.[28] The Ministério Público, however, ultimately won the lawsuit in January 2000 when the judge ruled that an EIS was necessary and that no new activities, including the planned construction of a test drive area, could be developed in the area without the preparation of an EIS.[29]

3. THE RISE OF PROSECUTORIAL ENFORCEMENT

The role of the Brazilian Ministério Público in the defense of environmental and other public interests was the result of a series of legal and institutional reforms in the 1980s. This section reviews the changes that so significantly altered the Brazilian Ministério Público and led to its importance in environmental protection. This transformation is best explained as resulting from a combination of the Ministério Público's own institutional initiative and the receptive political context present in Brazil during the period of democratization in the 1980s.

3.1 Legal and institutional changes

Historically, the Ministério Público's primary function was as a prosecutor in criminal actions. In the civil sphere, it was generally limited to intervening in cases between two private parties when the interests of minors, absentees, or otherwise legally incompetents parties were at stake.[30] Legal reforms in the 1980s, however, gave the institution a new role in civil litigation, to file lawsuits on behalf of public interests, understood as the interests of societal groups or society as a whole. As proclaimed in a 1991 meeting of Brazilian

---

[28] "Volks inaugura fábrica em S. Carlos," Folha de São Paulo (October 11, 1996), p. 2-2.
[29] Judicial decision of the 3rd Vara Civil of São Carlos in favor of Ministério Público on 13 January 2000, on file with author.
[30] Its role in civil matters included writing legal advisory opinions to the judge in cases involving, for example, marriage and property records, bankruptcies, and the interests of minors (Ferraz and Guimaraes Júnior, 1999).

environmental prosecutors, "What happened, above all, is a remarkable transformation which places Brazil as one of the most pioneering countries in the world in terms of this new function of the Ministério Público, making it the most qualified institution to protect social, diffuse, and collective interests" (Milaré, 1992). This section traces the legal and institutional changes in the 1980s and 1990s that enabled the Ministério Público to become an important actor in environmental protection, highlighting how the Ministério Público contributed to its own transformation.

The first legal reference to the Ministério Público's role in environmental protection was contained in the National Environmental Policy Act of 1981 (*Lei da Política Nacional do Meio Ambiente*), Brazil's first comprehensive federal environmental protection law. The law authorizes the Federal and the State Ministério Público to bring civil and criminal liability actions for damages caused to the environment.[31] This provision was proposed by a São Paulo prosecutor, Paulo Affonso Leme Machado, who participated in the commission that drafted the law.[32]

After the enactment of the National Environmental Policy Act, several environmental disasters occurred in the state of São Paulo that prompted the Ministério Público to begin working in environmental protection. In October 1983, dynamiting in a stone quarry caused a rockslide that ruptured an oil pipeline. The spill affected mangroves, waterways, and nearby beaches and was considered to be the worst ecological disaster that had occurred in the country (Ferraz et al., 1984). The attorney general appointed a prosecutor, Édis Milaré, to work exclusively on this case and other environmental cases. In November 1983, Milaré filed the first judicial action under the National Environmental Policy Act seeking to impose civil liability for the oil spill (Ferraz et al., 1984). In early 1984, the attorney general created a permanent environmental prosecutor's office in the city of São Paulo, and Milaré was appointed to be the coordinator and lead prosecutor. The office was responsible for carrying out investigations, filing lawsuits, and coordinating the institution's environmental work throughout the state (Ferraz, 2000).

These early efforts were greatly boosted by the passage of the Public Civil Action Law of 1985, a law authored by Milaré and two other São Paulo prosecutors. The Public Civil Action Law created a new type of lawsuit that could be used to defend environmental, consumer, and other public interests in court. The idea of a procedural law that would authorize and govern the judicial resolution of conflicts involving diffuse and collective interests developed in legal circles in the late 1970s and early 1980s. Drawing inspiration from the work of the Italian legal comparativist Mauro Cappelletti (1978; 1979) and other European legal scholars on the need for access to justice for public interests, four São Paulo jurists drafted a

---

[31] Federal Law 6,938 of 31 August 1981, Article 14, Chap. IV, §1.
[32] Pers. comm., Paulo Affonso Leme Machado, December 13, 2001.

bill designed primarily to enable civil society organizations to bring civil actions on behalf of environmental interests as well as artistic, aesthetic, historical, and scenic values. The bill was introduced in the Brazilian legislature in early 1984.[33]

Prompted by the jurists' work, Milaré and several colleagues in the São Paulo Ministério Público initiated an institutional discussion about the role of the Ministério Público in this new area and the ways that the jurists' bill might be modified to enable the Ministério Público to play a more significant role in environmental protection and the defense of other public interests. With the endorsement of the institution, they authored a new bill based on the one drafted by the jurists that strengthened the Ministério Público's role. While both bills gave the Ministério Público, other governmental entities, and civil society organizations authority to file public civil actions, the Ministério Público's version included several innovations that favored the institution. Most importantly, it created a powerful investigative instrument, the civil investigation, which the Ministério Público could use to collect evidence in preparation for a public civil action (Arantes, 2002). Introduced into Congress by the Ministry of Justice in late 1984, it received higher priority than the jurists' bill and was passed into law as the Public Civil Action Law in July 1985.[34]

The establishment of environment rights and the ability of the Ministério Público to defend them were consolidated in the Federal Constitution of 1988.[35] The environmental article of the Constitution ensures that "everyone has the right to an ecologically equilibrated environment, a good used in common by all citizens and essential to a healthy quality of life, imposing a duty on the government and the community to defend and preserve it for present and future generations."[36] Commenting that the Brazilian constitution was the most advanced in the world in terms of its environmental clauses, Milaré states that it "can very well be called 'green,' considering the emphasis… that it gives to the environment" (Milaré, 2001). Environmental prosecutors in São Paulo and other states actively participated in writing and lobbying on behalf of the environmental article.[37]

The Brazilian constitution also features very strong provisions with respect to the Ministério Público. It proclaimed the Ministério Público to be "a permanent institution, essential to the judicial function of the State, responsible for the defense of the legal order, the democratic regime and the

---

[33] Proposed Law 3034/84, drafted by Ada P. Grinover, Candido Dinamarco, Kazuo Watanabe, and Waldemar Mariz de Oliveira Júnior and introduced by Congressman Flávio Bierrenbach (Arantes, 2002).
[34] Federal Law 7,347 of June 24, 1985.
[35] The Constitution of the Federal Republic of Brazil was promulgated on October 5, 1988.
[36] Federal Constitution of 1988, Article 225.
[37] Édis Milaré, pers. comm., October 2001.

indispensable interests of society and individuals."[38] It elevated its role of protecting the public interest to the constitutional level, including among the institution's primary functions "carrying out civil investigations and filing public civil actions to protect the public and social heritage, the environment, and other diffuse and collective interests."[39]

As set forth in the Brazilian constitution, the Ministério Público has functional as well as administrative and financial independence from the three traditional branches of government. Each prosecutor enjoys functional autonomy – he chooses how to conduct his investigations and judicial actions. To protect this independence, the constitution provides guarantees of life tenure, as well as protection from demotion, transfer, and salary reduction. As an institution, the Ministério Público enjoys administrative and financial autonomy. It creates its own job openings, writes its own budget, and determines its own salary levels. While the governor appoints the attorney general, he is restricted to choosing from a list of three prosecutors elected by the other prosecutors.[40] Having emerged from the constitutional process as a powerful and independent institution, the Ministério Público is often referred to as the 'fourth branch' of government (*o quarto poder*) (Mazzilli, 1991).

These gains were, in large part, the result of a strong and organized lobby of the Ministério Público throughout the country, led by the Ministério Público of São Paulo (Mazzilli, 1989; Nunes, 1999; Arantes, 2002). For several years before the passage of the Federal Constitution of 1988, meetings were organized to discuss the future of the institution and determine what the institution would seek in terms of its rights and responsibilities in the new constitution (Mazzilli, 1989). In a national meeting in 1986, prosecutors from throughout the country endorsed a draft of the constitutional provisions relating to the Ministério Público (Mazzilli, 1989). While the proposal was somewhat modified in the constitutional assembly, the Ministério Público basically emerged with the profile that it had sought – a largely autonomous institution responsible not only for traditional criminal prosecution but also for the legal defense of a wide range of public interests.

After the passage of the Federal Constitution of 1988, the Ministério Público restructured itself to fulfill its new institutional mission of defending environmental and other public interests. These changes were visible in institutional growth, increasing specialization, and decentralization of its environmental protection work. While only data from the São Paulo Ministério Público are presented, a similar process occurred in the institution throughout the country at both the state and federal levels.

---

[38] Federal Constitution of 1988, Article 127.
[39] Federal Constitution of 1988, Article 129, III.
[40] Federal Constitution of 1988, Articles 127 and 128 (see also Mazzilli, 1991).

The most notable change was a marked growth in the number of prosecutors. Between 1985 and 2001, the number of prosecutors in São Paulo almost doubled, increasing from 849 to 1,620 (see *Figure 9-3*).[41] To accommodate this growth, the Ministério Público began acquiring its own buildings and hiring administrative support staff. Previously, prosecutors had usually worked in the courthouse and had little or no administrative support. To pay for these changes, the budget of the Ministério Público of São Paulo grew significantly, particularly in the early and mid-1990s (Pozzo, 1993; Fabbrini, 2002).

With growth came increasing specialization.[42] The city of São Paulo gained five specialized environmental prosecutor positions; other large cities gained two or three. Even in smaller towns, it was often possible to have at least two prosecutors, allowing one to specialize in criminal matters and the other to specialize in civil matters. Prosecutors were encouraged to make their work less bureaucratic and more oriented toward societal concerns by becoming involved in their local communities and coordinating their activities with other local prosecutors (Pozzo, 1990).

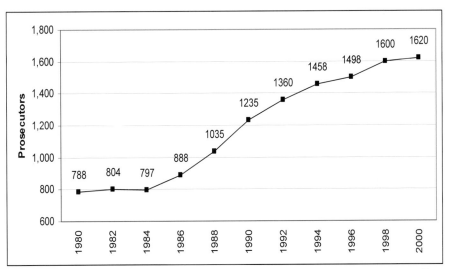

Figure 9-3. Number of Prosecutors in the São Paulo Ministério Público, 1980 – 2000.[43]

---

[41] Prosecutors are hired through a competitive civil service exam (*concurso público*).
[42] After a two-year period of training, prosecutors are usually assigned to a rural judicial district where they serve as the only prosecutor. As a prosecutor receives promotions, he moves to more populated judicial districts where there are more prosecutors and he may increasingly specialize in the criminal or civil area. After several promotions to larger cities and towns, he may be promoted to a position in the capital where he may completely specialize in one area of law such as environmental protection or prosecution of homicides.
[43] Data collected from the "Lista de Antiguidade", the official list of prosecutors published annually in the São Paulo Ministério Público's journal, *Justitia*.

In addition, the environmental protection work of the Ministério Público was decentralized, ultimately allowing greater participation and activity by a larger number of prosecutors. Until the early 1990s, environmental prosecution for the whole state had been centralized in the environmental prosecutor's office in São Paulo city. In 1993, this office was restructured into an Environmental Prosecution Support Center (*Centro de Apoio Operacional das Promotorias de Justiça do Meio Ambiente*).[44] The Center's role was to help unspecialized prosecutors conduct environmental investigations and file environmental public civil actions on their own.

3.2   Explaining the rise of prosecutorial enforcement

The environmental activity of the Brazilian Ministério Público raises the question of why a legal institution traditionally dedicated to criminal prosecution became involved in environmental protection and other public interest areas. The explanation is found in a combination of the institutional initiative of the Ministério Público and a receptive political context.

The most significant factor in the Ministério Público's transformation into a key actor in environmental protection was arguably the Ministério Público's own institutional initiative and mobilization.[45] As shown above, members of the Ministério Público, particularly the São Paulo Ministério Público, took an active role in redefining the institution's role through contributing to major legal developments such as the Public Civil Action Law and the Federal Constitution of 1988, restructuring the institution to fulfill its new role, and actively using the instruments of prosecutorial enforcement to defend environmental interests.

The Ministério Público's initiative in defending environmental and other public interests stemmed from both the idealism of its leadership and its long-held institutional objective of increasing its autonomy from the executive branch. Those that pioneered the Ministério Público's environmental protection work were idealistic in thinking about the public service role of the institution. In 1981, soon after the passage of the National Environmental Policy Act, a prosecutor called for the institution to act upon its new authority in the enforcement of environmental laws: "The task of protecting the [environment], for the good of mankind, has been left to the most promising of the institutions of the modern state, the Ministério Público – defender of the public interest" (Guimaraes Júnior, 1981). In 1985, the attorney general wrote, "The Ministério Público has been evolving to

---

[44] This reform was instituted by article 33 of Federal Law 8,625, of February 12, 1993.

[45] In his work on the Brazilian Ministério Público, Arantes (2002) argues that its institutional reconstruction and the legal changes that enabled its reconstruction were intentionally pursued by the members of the institution.

become, among other state institutions, the one that is committed to the special task of fighting on behalf of society's interests in the judicial arena" (Frontini et al., 1985).

This idealistic vision of the Ministério Público, however, also favored the institution and its prosecutors in several ways. Becoming the defender of society implied the need for independence and autonomy from the rest of the government, particularly the executive branch. As explained by Arantes (2002), the Ministério Público went to the Constitutional Assembly "with the discourse that its political independence was essential to the future constitutional democracy and a benefit more for society than for the institution itself." With these arguments, the Ministério Público won a strong set of constitutional guarantees making it almost completely autonomous of the other three branches of government and giving prosecutors job securities similar to those of judges.

As the defender of the public interest, the Ministério Público grew markedly in political stature. Its new role involved prosecutors not only in environmental protection but also in many other areas of social and political relevance such as governmental corruption, consumer defense, urban development, public health, and education. As explained by one prosecutor, "through its innumerable investigations and lawsuits, [the Ministério Público] began to question a series of extremely relevant practices involving large economic interests of private groups as well as the government itself. The Ministério Público's actions began to directly affect public policies and social interests" (Macedo Júnior, 1999). In sum, the Ministério Público became a "new and important political actor" (Macedo Júnior, 1999).

The political context of the 1980s also contributed to the Ministério Público's new role in environmental protection. In the wake of the military dictatorship, there was growing media and public attention to environmental problems and a surge in the Brazilian environmental movement.[46] As early as 1978, environmentalists endorsed the idea that the Ministério Público should become active in environmental protection. In the First National Ecological Symposium in 1978, symposium participants adopted a São Paulo prosecutor's proposal that the "Ministério Público become active, and that it be given broad powers to prevent, punish, and seek reparations from those that harm the environment" as one of its legislative recommendations (Guimaraes Júnior, 1981). In the same year, a professor of ecology gave a speech to a group of prosecutors, including the attorney general, asserting that "the active participation of the Ministério Público is lacking in the process of creating a national environmental conscience" and calling for the creation of a new job title: Prosecutor for the Protection of Human Health and the Environment (Tomazi, 1981).

---

[46] Viola and Nickel (1994) estimate that there were 40 environmental organizations in 1980, 400 in 1985, and 900 in 1991.

More generally, the period of democratization in the early 1980s was a time for questioning the authority of the government and exposing the ways that it had violated citizen rights during the dictatorship. There was a general lack of confidence in government, particularly in the executive branch that had dominated and subdued the judicial and legislative branches during the military dictatorship. The Ministério Público, despite being part of the government, was able to attract praise rather than criticism from social movements during this period. While still formally linked to the executive branch, the Ministério Público aligned itself with societal interests and proclaimed itself the legal representative of environmental and other social interests.[47]

Given the post-dictatorship fervor for democracy, the idea of having the Ministério Público become politically independent of the executive branch and assume the role of defending the democratic legal order and societal rights was well-received in the 1988 Constitutional Assembly.[48] As Arantes (2002) points out, a historically common notion in Brazilian political thought was that Brazilian society was fragile, disorganized, and incapable of defending its fundamental rights. "From this basic notion emerged abundant complaints about the artificiality of our political institutions, especially representative institutions, and criticisms that our tradition lacked mechanisms to enforce laws and the constitution. At times, this point of departure led to a desire for a neutral power, external to the world of politics and with sufficient autonomy to protect and lead society." The Ministério Público's proposal to the Constitutional Assembly responded to this idea by exhibiting a strong willingness to become such a "neutral power" and defend societal interests.

The political environment of the 1980s, marked by a growth in social movements and widespread support for democratization, thus proved very receptive to the Ministério Público's institutional mobilization to redefine and expand its institutional responsibilities. As stated by a member of the Constitutional Assembly before the passage of the Federal Constitution of 1988, "With the chapter on the Ministério Público, I have the impression that we are taking – if it is accepted by the Assembly - a historic step. It is a

---

[47] Arantes (2002) discusses the paradox that the Ministério Público, a state institution, was able to expand its role in this new area despite the strong "anti-state" sentiment of the new social movements emergent during the period of democratization. As he explains, "the paradoxical ascension of the Ministério Público in a context marked by anti-state ideologies was able to occur only because during the redemocratization, the Ministério Público fought to disassociate itself from the executive branch and to construct the image of itself as an agent of society capable of overseeing the state, despite its being a part of the state" (Arantes, 2002).

[48] The National Constitutional Assembly (*Asembléia Nacional Constituinte*) was composed of 559 elected representatives from diverse political parties. The work of the assembly took 1 year and 8 months and included the preparation of 8 drafts as well as the final constitutional text. There was significant public participation in the process and more than 65,000 amendments were offered (Nunes, 1999).

historic step, and one with very important theoretical repercussions, because we are creating an organ outside the scheme of the three powers. It is an organ of enforcement that does not fit in any of the branches of Montesquieu's scheme. Why are we proposing financial, political, and administrative autonomy for this organ? Because we want a strong agent of legal enforcement."[49]

4. THE EFFECTIVENESS OF PROSECUTORIAL ENFORCEMENT

The involvement of the Brazilian Ministério Público in environmental protection provides a partial remedy to the 'non-enforcement problem' of Brazilian environmental law. The Ministério Público contributes significantly to the development and usage of Brazilian environmental law. It exercises oversight of environmental agencies through its investigations and lawsuits and thereby enhances the accountability of environmental agency officials. It also facilitates access to the courts, often referred to as 'access to justice,' for environmental problems. In all these ways, prosecutorial enforcement is effective in promoting and furthering environmental enforcement and compliance. This section offers preliminary observations about the effectiveness of the Ministério Público as an actor in environmental protection, highlighting its positive impacts as well as potential limitations and inefficiencies.

4.1 Impacts on the usage and development of environmental law

The Ministério Público has had a significant impact on the development and usage of environmental law in Brazil. Its many investigations, public civil actions and conduct adjustment agreements contribute directly to the implementation and enforcement of the law. The case of Cubatão, discussed above, provides an example. Despite the lack of a judicial ruling, the lead prosecutor stated in 1995, "Even with the delay and even without any decision, this case has partially accomplished its mission. It made the 24 companies install pollution control mechanisms. Cubatão today is much better than it was in 1985."[50] In a statement about the case several years later, he opined that it had served its pedagogical and historical purpose: "It

---

[49] Plínio de Arruda Sampaio, member of the Constitutional Assembly, cited in Nunes (1999).
[50] "Ação ambiental tem 'efeito educativo' para empresas," A Folha de São Paulo (October 21, 1995), p. 3-2.

was because of this case that many companies began to worry about the environment."[51]

Although the investigations and lawsuits directly contribute to environmental law enforcement, the Ministério Público has not developed mechanisms to prioritize and coordinate its environmental protection activities. São Paulo environmental prosecutors often express concern that they spend a disproportionate amount of time on relatively insignificant problems because they are required to act upon all complaints and notices that they receive.[52] Moreover, because local prosecutors act independently of each other, there is little consistency in the responses of prosecutors to similar environmental problems. As a result, a local prosecutor of one community may address part of an environmental problem while a larger part of the problem may remain unaddressed by the prosecutor of a neighboring community. The lack of prioritization and coordination among prosecutors compromises the effectiveness of the institution's environmental work.

4.2 Impacts on environmental agencies

The Ministério Público also contributes to the implementation and enforcement of environmental law in Brazil through its impact on environmental agencies. Environmental agency officials often acknowledge the importance of the Ministério Público's work. They observe that the Ministério Público's power to bring civil and criminal lawsuits against violators provokes fear of punishment and thus deters illegal behavior. Agency officials also emphasize that the Ministério Público's involvement in environmental protection has led to greater resistance within the agency to corruption and inappropriate political influence.

While these effects imply an increase in agency accountability, the conflictual and adversarial nature of interactions between environmental agencies and prosecutors signals that the impact may not be wholly positive. Indeed, a dynamic of institutional conflict has arisen wherein prosecutors accuse environmental agencies of being inept and working against rather than for environmental protection. Moreover, prosecutors make large numbers of demands on environmental agencies for technical information and assistance in their investigations, often overburdening environmental agencies that suffer from lack of budgetary and staff resources. In sum, while both agencies and the Ministério Público purport to work for

---

[51] "Caso tramita há 15 anos na Justiça de SP," *A Folha de São Paulo* (October 14, 2001), p. C2.
[52] As expressed by an environmental prosecutor in the city of São Paulo, "There are two types of problems in São Paulo, the insignificant and the unsolvable."

environmental protection, the dynamic of institutional conflict may compromise their efforts.

4.3 Impacts on access to justice

The Ministério Público has significantly facilitated access to the courts or 'access to justice' for environmental problems.[53] As stated by one legal scholar, "Thanks to public civil actions, the judiciary abandoned its frequently distant and remote stance, and became a protagonist in large national controversies" (Grinover, 1999). An environmental prosecutor in São Paulo writes, "Through the public civil action, the individual and private tradition of the national legal culture was broken, considerably expanding the access to justice to innumerable citizens" (Macedo Júnior, 1999).

While the Ministério Público has increased access to justice for environmental problems in Brazil, the involvement of the judiciary in environmental protection carries with it certain problems. Brazilian prosecutors often complain that courts are very slow in deciding cases and that judges are unfamiliar with environmental law. Both the Cubatão case and Volkswagen cases discussed above provide evidence of the delay and of the difficulties of judicial resolution. More fundamentally, judicial resolution may be inappropriate and inefficient for resolution of many environmental problems. Research on regulatory enforcement in industrialized countries has suggested that a legalistic approach to environmental enforcement may lead to legal uncertainty, high compliance-related expenses, and defensiveness among regulated firms (Kagan 2001). To the extent such inefficiencies arise in Brazil, the presumably positive effect of access to justice in improving environmental protection may be compromised.

5. CONCLUSION

This chapter described and analyzed the role of the Brazilian Ministério Público in environmental protection. It presented the instruments that prosecutors use in their environmental work, explained the legal and institutional changes that led to the rise of prosecutorial enforcement of environmental law in Brazil, and provided a preliminary assessment of the effectiveness of prosecutorial enforcement.

As argued in this chapter, prosecutorial enforcement provides a partial solution to the non-enforcement problem that characterizes Brazilian

---

[53] On access to justice, see Cappelletti and Garth (1978); Cappelletti and Weisner (1978). On access to justice in the Brazilian context, see Cavalcanti (1999); Associação dos Magistrados Brasileiros (1996).

environmental law. The Ministério Público's environmental work has had significant impacts on the development and usage of environmental law, the accountability of environmental agencies, and access to justice for environmental problems. While several inefficiencies of prosecutorial enforcement were also observed, the environmental protection work of the Brazilian Ministério Público serves as a model for the many other Latin American countries that suffer from the non-enforcement problem of environmental law.

ACKNOWLEDGMENTS

The author would like to thank the prosecutors and staff of the Ministério Público of São Paulo for their generous access to data and other information. The author would also like to thank her academic advisors at the University of California at Berkeley: Robert A. Kagan, Peter B. Evans and Richard B. Norgaard. Research for this chapter was supported through grants from the National Science Foundation Law and Social Science Program, the Environmental Protection Agency 'Science to Achieve Results' (STAR) Fellowship program, and the Organization of American States.

REFERENCES

Arantes, R. B., 2002, *Ministério Público e Política no Brasil*, Educ; Editora Sumaré; Fapesp, São Paulo.
Associação dos Magistrados Brasileiros, ed. 1996, *Justiça: Promessa e Realidade*, Rio de Janeiro, Nova Fronteira.
Benjamin, A. H., 1995, A proteção do meio ambiente nos paises menos desenvolvidos: O caso da América Latina, *Revista de Direito Ambiental* **0**:83-105.
Cappelletti, M., 1978, Governmental and private advocates for the public interest in civil litigation: a comparative study, in: *Access to Justice, Volume II, Promising Institutions*, M. Cappelletti and J. Weisner, Sijthoff and Noordhoff, Amsterdam, pp. 769-865.
Cappelletti, M., 1979, Vindicating the public interest through the courts: a comparativist's contribution, in: *Access to Justice, Volume I: A World Survey*, M. Cappelletti and B. Garth, Sijthoff and Noordhoff, Amsterdam, pp.
Cappelletti, M. and B. Garth, eds., 1978, *Access to Justice, Volume I, A World Survey*, Amsterdam, Sijthoff and Noordhoff.
Cappelletti, M. and J. Weisner, eds., 1978, *Access to Justice: Volume II, Promising Institutions*, Amsterdam, Sijthoff and Noordhoff.
Cappelli, S., 2000, Novos rumos do direito ambiental, in: *Temas de Direito Ambiental: Uma Visão Interdisciplinar*, E. C. Hausen, O. P. B. Teixeira and P. B. Alvares, AEBA, APESP, Porto Alegre, RS, pp. 53-78.
Cavalcanti, R. B., 1999, *Cidadania e Acesso a Justiça*, Sumaré/Idesp, São Paulo.
Fabbrini, R. N., ed. 2002, *O MP e a Crise Orçamentária*, São Paulo, Associação Paulista do Ministério Publico.

Ferraz, A. A. M. d. C., 2000, Primórdios da ação civil pública, presented at: *15 Anos de Ação Civil Público: Polemicas e controvérsias, evolução legislativa e tendências jurisprudenciais*, São Paulo.
Ferraz, A. A. M. d. C. and P. A. d. C. Ferraz, 1997, Ministério Público e enforcement, in: *Ministério Público e Afirmação da Cidadania*, A. A. M. d. C. Ferraz, by author, São Paulo, pp. 114-119.
Ferraz, A. A. M. d. C. and J. L. Guimaraes Júnior, 1999, A necessária elaboração de uma nova doutrina de Ministério Público, compatível com seu atual perfil constitucional, in: *Ministério Público: Instituição e Processo*, A. A. M. d. C. Ferraz, Atlas S.A., São Paulo, pp. 19-35.
Ferraz, A. A. M. d. C., É. Milaré and N. Nery Júnior, 1984, *A Ação Civil Público e a Tutela Jurisdicional dos Interesses Difusos*, Saraiva, São Paulo.
Findley, R. W., 1988, Pollution control in Brazil, *Ecology Law Quarterly* **15**(1):1-68.
Fink, D. R., 2001, Alternativa a ação civil público ambiental (reflexões sobre as vantagens do termo de ajustamento de conduta), in: *Ação Civil Pública, Lei 7347/1985 - 15 anos*, É. Milaré, Revista dos Tribunais, São Paulo, pp. 113-139.
Frontini, P. S., É. Milaré and A. A. M. d. C. Ferraz, 1985, Ministério Público, ação civil pública, e defesa dos interesses difusos, *Justitia* **47**(131):263-278.
Fuks, M., 1999, *Arenas de Ação e Debate Públicos: os Conflitos do Meio Ambiente enquanto Problema Social no Rio de Janeiro (1985-1992)*, Tribunal de Contas, Instituto Serzedello Correa, Rio de Janeiro.
Grinover, A. P., 1999, A ação civil pública refém do autoritarismo, presented at: *3rd International Environmental Law Conference (3o Congresso Internacional de Direito Ambiental)*, São Paulo, Brazil, IMESP.
Guimaraes Júnior, R., 1981, O futuro do Ministério Público como guardião do meio ambiente e a história do direito ecológico, *Justitia* **43**(113):151-192.
Macedo Júnior, R. P., 1999, Ministério Público brasileiro: um novo ator político, in: *Ministério Público II: Democracia*, J. M. M. Vigliar and R. P. Macedo Júnior, Atlas S.A., São Paulo, pp. 103-114.
Mazzilli, H. N., 1989, *O Ministério Público na Constituição de 1988*, Saraiva, São Paulo.
Mazzilli, H. N., 1991, *Manual do Promotor de Justiça*, Saraiva, São Paulo.
Mazzilli, H. N., 2001, *A Defesa dos Interesses Difusos em Juízo*, Saraiva, São Paulo.
Milaré, É. ed. 1992, *O Ministério Público e a defesa do meio ambiente*, Revista do Ministério Público, Rio Grande do Sul.
Milaré, É. 2001, *Direito do Ambiente*, Revista dos Tribunais, São Paulo.
Nunes, F. J. K., 1999, O Ministério Público e a constituinte de 1987/88, in: *O Sistema de Justiça*, M. T. Sadek, Idesp, Sumaré, São Paulo, pp. 61-77.
Pozzo, A. A. F. D., 1990, Reunião Geral dos Membros do Ministério Público, *Justitia* **52**(152):364-377.
Pozzo, A. A. F. d., 1993, 1990-1993: Um relatório de reformas, *Justitia* **55**(164):163-189.
Proença, L. R., 2001, *Inquérito Civil*, Revista dos Tribunais, São Paulo.
Sadek, M. T. and R. B. Cavalcanti, 2003, The new Brazilian Public Prosecution: an agent of accountability, in: *Democratic Accountability in Latin America*, S. Mainwaring and C. Welna, Oxford University Press, Oxford ; New York, pp. 201-227.
Silva, C. A., 2001, *Justiça em Jogo: Novas Facetas da Atuação dos Promotores da Justiça*, USP, São Paulo.
Tomazi, L. R., 1981, Ministério Público e defesa do meio ambiente, *Justitia* **43**(113):135-142.
Viola, E. and J. W. Nickel, 1994, Integrando a defesa dos direitos humanos e do meio ambiente: lições do Brasil, *Novos Estudos* **40**:171-184.

Chapter 10

DEMOCRACY BY PROXY
Environmental NGOs and Policy Change in Mexico

Raul Pacheco-Vega[1]
[1] *Institute for Resources, Environment and Sustainability, The University of British Columbia, 214-1924 West Mall, Vancouver, B.C. V6T 1Z4, Canada*

Abstract: Citizen participation in policy-making has been touted as the cornerstone of every democratic regime. Encouraging society members to participate in political life allows for a more meaningful, inclusive approach to policy agenda setting, design and implementation. As a result, innovative processes that encourage citizen input in policy-making also have the potential to provide solid foundations to a vigorous, strong democracy. Despite this need for citizen participation, it should also be acknowledged that, for society at large, it is almost impossible to participate in every forum of the environmental policy-making arena. As a result, the number and variety of environmental non-governmental organizations (ENGOs) has increased almost exponentially in the last few years. ENGOs take upon themselves the role of representing the interests of the people (thus encouraging 'democracy by proxy'). This paper traces ENGO involvement in effecting policy change in Mexico by documenting instances where their influence proved to be effective. I also outline the different strategies used by these ENGOs, with a strong focus on NGO coalition-building and network formation. Building on an interdisciplinary theoretical framework, I analyse empirical evidence of environmental policy change and then I trace the extent to which this change can be attributed to ENGO influence. Theoretical and empirical implications of my research are also discussed.

Key words: Citizen participation; pollution control; voluntary instruments; information dissemination

1. INTRODUCTION

Despite the assertion by Chasek that "states are responsible for adopting national and international policies that directly and indirectly affect the environment" (Chasek, 2000), it is also true that non-state actors play a key role in shaping how the nation-state designs and implements policy. The design, planning and implementation of environmental policy at the national level are not tasks that remain solely in the hands of nation-states. Non-state actors also play a significant, as yet unquantified, role and exert a certain degree of influence (Risse-Kappen, 1995). Therefore, we need to assess their impact on environmental policy-making. This paper examines how domestic (Mexican) and international ENGOs have exerted influence on the design and implementation of Mexican environmental policies. Specifically, I use the Mexican *Registro de Emisiones y Transferencia de Contaminantes* (RETC) as an example of a policy instrument whose formulation, design and implementation has been significantly influenced by coalitions of ENGOs.

The RETC is the Mexican version of one of the newest approaches to pollution reduction, the Pollutant Release and Transfer Registry (PRTR). PRTRs are based on the so called 'right-to-know' approach, which arises in response to Principle # 10 of the United Nation's Commission on Environment and Development Agenda 21, indicating that states should facilitate and encourage public participation through disseminating information. This raises public awareness and mobilizes interest groups to influence industrial firms to reduce pollution (Pacheco and Nemetz, 2001). These registries include detailed data and information on the types, locations and amounts of substances of concern released on-site and transferred off-site by industrial facilities. Ideally, governments compile these data from each industry and make it available to the public through the Internet and print. This is an effective strategy when individual citizens or interest groups make use of the information and influence firms to reduce the impact on the environment.

While the US and Canada have accumulated much more experience with PRTRs (for example, the US Toxics Release Inventory, TRI and Canada's National Pollutant Release Inventory, NPRI), Mexico's program just started in 1996. Increasing comparability between the three North American PRTR is proving to be a difficult task because the Canadian and American systems are based on mandatory reporting while the Mexican system is entirely voluntary. There is also disagreement on the types of substances or even the level of data aggregation used to report on pollution. Furthermore, while the US and Canada have been working towards making NPRI and TRI more comparable for a number of years, Mexico is still at the developmental stages.

ENGOs in all three North American countries have lobbied to harmonise the Mexican system with that used in Canada and the U.S. They argue that mandatory reporting is a necessary condition to achieve transparency and accountability in polluters. This lobbying has raised a number of questions regarding the appropriate design and implementation of the RETC and involves significant policy changes. One of the most important changes for RETC took place in the year 2001. On June 29th, 2001, Mr. Victor

10. Democracy by Proxy: Environmental NGOs in Mexico 233

Lichtinger, Secretary of Environment and Natural Resources in Mexico, announced that his ministry (SEMARNAT) would put forward a motion to the Congress to make RETC mandatory. And in December of 2001, Mr. Lichtinger's motion was passed by the Mexican Congress. As the RETC is the youngest and least developed PRTR, this is a bold move.[1] This case, therefore, provides an opportunity to understand the relationships that have evolved throughout the development of this information-based instrument. This is also an opportunity to showcase how domestic and international environmental policies intertwine. The need to assess the impact of non-state actors on environmental policy-making, thus becomes critical and it is addressed within this case study.

The purpose of this chapter is threefold. First, it documents the developments of the last few months in the development of a comparable Mexican PRTR. Second, it examines the role that domestic and international ENGOs have played in changing the reporting mode of RETC. Third, and more broadly, it examines the degree of influence that can be ascribed to environmental NGOs in changing policy objectives. In this chapter, I am particularly interested in the specific strategies and tactics that ENGOs have used to increase pressure on the Mexican government and what has been the net effect of these tactics. I argue that Mexican, Canadian and American ENGOs have formed a coalition to increase pressure on the Mexican government to effect a change in the mode of reporting of the RETC, from voluntary to mandatory. In characterising NGO influences, I outline two types of pressure transmission mechanisms (Harrison and Antweiler, 2001). *First-order mechanisms* are those where the influencing actor has a direct link with the target actor, and *second-order mechanisms* are those where the influencing actor seeks the intervention of an intermediate actor to exert pressure on a target actor. I will describe these mechanisms in more detail in further sections.[2] I also argue that the Mexican, US and Canadian ENGO coalition used 'second-order' mechanisms to bring the North American Commission on Environmental Cooperation (NACEC) into the RETC negotiations and increase NGO leverage over the Mexican environmental ministry. Using direct lobbying strategies, disseminating information on the

---

[1] At the time of writing this chapter, the motion to change the reporting mode of RETC had already been sent to the Congress for approval. It was approved on December 6, 2001.

[2] The nomenclature of 'first-order' and 'second-order' is borrowed from Kathryn Harrison, whom I thank for comments and suggestions for this paper. Also, the first-order mechanisms would be equivalent to Wright's (2000) direct methods of influence, and the second-order mechanisms would be equivalent to his indirect methods of influence. I prefer to use first-order and second-order to emphasise the ordinal nature of the decision-making process of ENGOs: if a first order mechanism doesn't work, they'll probably seek to use a second order mechanism. For Harrison's nomenclature, please see Harrison, K. and W. Antweiler (2001). Environmental Regulation vs. Environmental Information: A View from Canada's National Pollution Release Inventory. Annual Meeting of the Association for Public Policy Analysis and Management, Washington, D.C., APPAM.

potential hazards that toxics pose to human health and participating in the meetings of the North American PRTR project, organised by NACEC, these ENGOs have exerted pressure on Mexico, both in a direct and an indirect manner.

I argue that ENGO pressure has changed the way the Mexican government proceeds with the design and implementation of RETC. However, ENGO pressure has not been the only driver for this change. Instead, preliminary evidence indicates that participation of international environmental institutions also played a major catalytic role. I will address this issue in further sections. The nature of this paper is exploratory and therefore caution should be exercised to draw implications at this early stage.

The paper is developed as follows: First, I outline the theoretical framework that I use for this case study. I draw from works in comparative politics, international relations, social movement theory and organisation theory to inform the case study. Second, I describe the main characteristics of information-based instruments, paying particular attention to pollutant release inventories. I also describe the main features of the North American PRTRs. In the third section I outline my research questions. I am specifically interested in the interplay of two key factors and their influence on environmental policy change: transnational ENGO involvement and the influence of international environmental institutions. I also describe strategies and tactics used by ENGOs. I examine historical developments of the RETC as well as ENGO involvement in those events. I analyse how ENGOs used coalition-formation strategies to increase pressure on the Mexican government and the pressure transmission mechanisms used by these ENGOs. Finally, I discuss the caveats and next steps necessary for this research.

2. THEORETICAL FRAMEWORK

In explaining this environmental policy change, I examine two specific aspects: First, I analyse what mechanisms ENGOs use to exert pressure on national governments. Second, I describe one of the strategies that ENGOs use to increase pressure on governments. By coupling strategies of coalition-formation with second-order mechanisms (whereby ENGOs appeal to international or intergovernmental bodies to increase the pressure on target national governments), I argue that ENGOs are able to effect policy change.[3] Some authors assert that ENGOs are unable to exact policy change (since

---

[3] It is quite categorical to say this without examining cases where ENGOs have in fact used coalition-building strategies AND second-order mechanisms and this failed. However, I am more interested in the case when it worked, and I will look for cases where these strategies didn't work further down the road.

policy formation is a responsibility of nation-states). Others, however, state that ENGOs influence policy changes and evolution (Edwards and Sen, 2000; Wright, 2000). In seeking to influence nation-states, ENGOs make use of a variety of strategies and tactics. Not only do they seek to influence the process of governmental agenda-setting but also the different stages of policy-making: design, implementation and evaluation (Breitmeier and Rittberger, 2000). ENGOs disseminate information, educate the public about environmental challenges and issues, formulate policy options, and lobby governments to further their interests.

3. MECHANISMS OF PRESSURE TRANSMISSION

In harnessing the above-mentioned forms of power, ENGOs use a variety of pressure transmission mechanisms to influence state-behaviour. I use Wright's (Wright, 2000) definition of 'influence': an actor A influences an actor B through information transmission, in an effort to alter actor B's behaviour. The trajectory of influence (or the order or choice of influence mechanism) is important because ENGOs transmit pressure on the basis of previous successes and failures. I define two types of pressure transmission mechanisms. *First-order mechanisms* include (but are not limited to) direct government lobby activities[4] (personal interactions with government policymakers), networking and socializing, participation in decision-making roundtables and intergovernmental bodies, environmental regulation monitoring and enforcement (watchdog activities). When an ENGO uses a first-order mechanism, the relationship between the ENGO and the target group (individual, state or organisation) is more direct. For example, in this case study, I argue that ENGOs lobbying the Mexican government were using first-order pressure transmission mechanisms to exert influence on policy-makers.

Second-order mechanisms include (but are not limited to) public education through information, raising awareness, targeting international environmental institutions in lobbying campaigns, etc. For example, when ENGOs coalitions did not find the results of their lobbying activities satisfying, they found a propitious forum to voice their concerns when representatives of a trinational intergovernmental body were seated at the same table. By actively seeking the involvement of NACEC in the

---

[4] I believe my definition of first-order mechanisms would fit well with Wright's (2000) definition of direct methods, and second-order mechanisms would be equivalent to his indirect methods.

development process of RETC, the ENGO coalition effectively exerted pressure on the Mexican government, but in an indirect manner.[5]

4. ENGO STRATEGIES AND COALITION FORMATION

Wapner argues that one of the most striking (and perhaps underrated) characteristics of ENGOs is their ability to establish networks and build ties between organisations in different geographical locations. Coalition building is without doubt a common strategy that ENGOs use to increase their influence. In building coalitions, ENGOs exchange resources, information, and harness the power of each other through the consolidation of a common front. By sharing expertise and bringing together participants from different organisations with strong knowledge bases in a particular area, ENGOs that belong to these networks are able to increase their leverage and further their demands when exerting pressure on nation-states. NGO coalitions may be found in the realms of human rights advocacy (Keck and Sikkink, 1999), social movements to resist free-trade (Legler, 2000), environmental protection activities (Princen and Finger, 1994), amongst other examples. Definitions of *coalition* abound in the literature and there is ample range for disagreement on what constitutes a coalition. For Sikkink, a coalition is a transnational advocacy network that organises itself around a shared campaign and therefore, shares stronger ties and reaches a higher level of coordination.[6] Keck and Sikkink define a transnational advocacy network (TAN) as a group of actors that "includes those actors working internationally on an issue, who are bound together by shared values, a common discourse and dense exchanges of information and services".(Keck and Sikkink, 1998). Wright (2000) argues that "coalitions are distinct from issue networks and alliances because ENGOs work together on a single joint campaign. Coalitions are held together by shared goals and understandings, shared political experience, finances, expertise, and joint participation in international forums". Jonathan Fox arrives (through a different pathway) to a similar definition as the one Sikkink suggests. He argues that coalitions are networks that have a higher degree of relationship density and cohesion (although he rightly points out that these conceptual distinctions often overlap with each other) (Fox, 2002). For Fox, the main purpose of the distinction is to understand the power relations and varying degrees of cohesion and coordination between the different actors, rather than assuming

---

[5] This strategy is analogous to Keck and Sikkink's boomerang effect. See Nelson (1996) and Keck and Sikkink (1998).
[6] Kathryn Sikkink, e-mail communication, August 8, 2001.

that all are equal and agree on everything.[7] For Hajer (Hajer, 1993), coalitions share a common 'discourse' (along the same lines defined by Keck and Sikkink). Thus, a consensus seems to appear with respect to shared values and beliefs as a characteristic of coalitions.

While I agree that coordination levels have to be higher for a coalition to work, I fail to see where shared values and beliefs lie. Keck and Sikkinks usefully conceptualize a factor that increases cohesiveness in the coalition, but fail to demonstrate how NGOs in a transnational advocacy network share beliefs and values.[8] As a result, and in an effort to simplify this analysis, I propose another definition of coalition. For the purposes of this paper, I define a coalition as a network of organisations and individuals that is held together in a cohesive manner by sharing the same focus and targeting the same problem. As a result, this collaborative network allows these organisations to exchange complementary information, resources, thus complementing each other's core competencies. In this definition, I borrow concepts from the complementarity perspective to analyse coalition-formation strategies (Pacheco-Vega, 1998). I argue that coalitions form when ENGOs 'coalesce' around a central issue and those organisations share resources that become complementary to each other. In further sections of this case study, I show how ENGOs complement each other and share information and key resources to increase their influence on Mexican environmental policy.

5. BACKGROUND

Since the publication of Agenda 21 in 1992, the United Nations Commission on Environment and Development (UNCED) has encouraged nation-states to adopt these instruments. Efforts to increase access to information are on the rise, and, in 1996, the Organisation for Economic Cooperation and Development (OECD) published a guide to help governments implement PRTRs. As a result, several countries, including Czech Republic, the United Kingdom, Canada, the US and more recently, Mexico, have adopted these instruments and started to create their own PRTR.[9]

Mexico has needed to speed up the design and implementation process of the PRTR for two main reasons. First, as a member of the North American

---

[7] Jonathan Fox, personal communication via email, March 2, 2001.
[8] Although I confess that I have the same conceptual problem with Peter Haas's work with epistemic communities, it is easier (I believe) to identify and track shared causal knowledge (and scientific knowledge) than values and beliefs.
[9] Arguably, since the US TRI started earlier than 1992, and given Canada's proximity to the US, the possibility that Canada's NRPI arose as a response to the emergence of the US TRI is quite plausible.

Free Trade Agreement (NAFTA) and a signatory of the North American Agreement on Environmental Cooperation, NAAEC, Mexico agreed to Resolution 97-04, which encourages Mexico, the United States, and Canada to work toward adopting more comparable PRTRs. This means that the Mexican PRTR, RETC, must become similar to the US and Canadian equivalents. Second, in 1994, Mexico became a member of the OECD. The OECD nation-state members have also agreed to harmonize existing PRTRs. As a result, the Mexican government has been pressured to develop a comparable PRTR.

The US Environmental Protection Agency (EPA) Toxics Release Inventory (TRI), the first information dissemination program, was established under the Emergency Planning and Community Right-to-Know Act of 1986 (EPCRA). Considered by the literature on policy instruments as an information-dissemination tool, TRI is a publicly accessible toxic chemical database developed and maintained by the US EPA. Its primary aim is to provide public access to valuable information on how much toxic material is released by manufacturing activities. The provision and release of information to the government body is mandatory although reductions in pollutant releases are expected to be voluntary. Section 313 of EPCRA specifically requires manufacturers to report releases of a number (over 6000) of designated toxic chemicals to the environment. These reports are then submitted to the US EPA and state governments. EPA compiles the data and publishes it online.[10]

> Any manufacturing facility with 10 or more full time employees that manufactures or processes 25,000 pounds (or otherwise uses 10,000 pounds) of a toxic chemical during a given calendar year must file EPA Form R for each chemical processed (Bunge, Cohen-Rosenthal et al., 1996).

The Canadian version of TRI, the National Pollutant Release and Transfer Inventory (NPRI) was created in 1992 and launched in 1993, mostly as a result of policy transfer and learning from the US experience. It aims to "provide Canadians with information on pollutants released to their environment."[11]

Mexico started tracking data on emissions and pollutant releases in 1996, albeit on a voluntary basis. A case has been made that Mexico should adopt a mandatory reporting scheme, because at the moment, industrial firms in Mexico do not have to report to the RETC; they do so on a voluntary basis. This situation is viewed as problematic, because the non-mandatory nature of the Mexican RETC does not allow for cross-national comparisons, clearly

---

[10] Data are available through the EPA website: http://www.epa.gov/tri/general.htm.
[11] Environment Canada's NPRI website: http://www.ec.gc.ca/pdb/npri.

an objective of OECD. In 1994, representatives of 38 governmental, ENG, industrial and academic organizations established a National Coordinating Group (*Grupo Coordinador Nacional, GNC*).[12] This group met over a period of two years under the umbrella co-ordination of a United Nations Institute for Training and Research (UNITAR) grant, to discuss, analyse and design the administrative and overarching framework to establish a Mexican PRTR. In the period of 1995-1996, a pilot project that involved voluntary reporting by 80 facilities was conducted in the State of Querétaro. Only 51 of the 80 facilities that were invited completed the forms, and the quality of information received varied considerably. In April of 1997, the Mexican environmental secretariat proposed an integrated system that would allow the development of the RETC in a more meaningful manner. Data to be reported in the RETC would come from Section V of the *Cedula de Operación Unica (COA)*. An important issue that was also raised was the fact that facilities would not be mandated to fill Section V (they are still not mandated to do so).

The year 1997 marked was the first of voluntary reporting by industrial facilities, and the first report on Pollutant Releases and Transfers was issued in December 1999 by the Mexican government. Results do not paint a very promising picture, although it could be argued that the lack of results was due mainly to inexperience with the PRTR system. The second year of voluntary reporting (1998) also produced results that were apparently discouraging (only 50 facilities reported in Section V, which forms the basis for the RETC and is also optional). However, the RETC has continued to develop.

As indicated by documentary evidence, one of the most important (and yet unresolved) issues that was not agreed upon (within the National Coordinating Group) was the type of access to information and availability of information on specific facilities. Also, the fact that reporting in RETC is voluntary has been highly debated and continued to be the most important issue of relevance in the discussions held in March 5-6, 2001.

In June 2000, the NACEC council signed Resolution 00-08, which establishes the criteria that are considered as key to making a PRTR effective. Table 1 compares these criteria against the current situation in Mexico (at the time of writing).

In December of 2000, the Mexican government approved a voluntary standard (*Norma Voluntaria Mexicana, NMX*) that established the RETC system. While this proposed NMX has been highly debated (not only

---

[12] Portions of the material presented here are taken from the "Background Paper for Roundtable Discussion on PRTR Reporting in Mexico", publicly available from the CEC. This background paper was provided to participants in the March 5 and 6, 2001 meetings. More detailed discussions can be found there. The section above is only intended to set the stage for my discussions on ENGO involvement.

because it establishes voluntary reporting but also because the NMX itself is not mandatory, as opposed to the NOMs or *Norma Oficial Mexicana*), it has undergone approval already.

Table 10-1. Current status of PRTRs in North America

Criteria	TRI	NPRI	RETC
Reporting on individual substances	Yes	Yes	Not yet implemented
Facility-specific reporting	Yes	Yes	Not yet implemented
Periodic reporting (annual)	Yes	Yes	Yes
Public disclosure of reported data on a facility- and chemical-specific basis	Yes	Yes	Not yet implemented (unclear)
Table 1. (continued)			
Limited data confidentiality	Yes	Yes	Unclear - Still not implemented
Standardized database structure to facilitate electronic reporting, data collection, analysis and dissemination	Yes	Yes	Developmental stages

Source: Constructed from Taking Stock 1995, 1996 reports and website references.

ENGO involvement in the RETC development process started in the early 1990s when the GNC first initiated steps towards developing the concept of RETC. However, the presence of ENGOs was felt more strongly since the early months of 2000, when the proposal for a voluntary standard was first drafted and put forward. Interestingly enough, a Mexican ENGO (Programa LaNeta, S.C.) initiated the project "Proyecto Emisiones" in 1997, specifically targeted at disseminating information about toxic releases.[13] While the role of this ENGO in disseminating information on toxic releases is very important, its impact seemed relatively insignificant until early March of 2001, when efforts to build a coalition to fight for a mandatory reporting in RETC began in earnest.

6. RESEARCH QUESTIONS

The research questions I pose in this case study are: *What drove the Mexican government to change the RETC mode of reporting? What factors have driven this policy change?* I hypothesise that two factors explain the

---

[13] Personal communication via email, Ms. Azucena Franco, Proyecto Emisiones, May 7, 2001.

outcome. I use policy change as the dependent variable and ENGO involvement and the influence of international environmental institutions as independent variables. I hypothesise that Mexico changed the mode of reporting to the RETC as a result of external pressures from international and domestic ENGOs working in coalition, as well as the influence of international environmental institutions (e.g. NACEC).

Moreover, I hypothesise that the intervention of international environmental institutions becomes a catalyst in increasing NGO influence on environmental policy-making. This is what I call a 'second-order' mechanism, where an ENGO coalition brings an international institution (international organisation or intergovernmental organisation) to the decision-making process, therefore increasing the pressure on a target national government (Nelson, 1996; Keck and Sikkink, 1999). I will illustrate how second-order mechanisms come into play in the next sections.[14]

7. METHODOLOGY

I began to follow the development of the North American PRTR project in early June of 2000. I used document analysis (letters, communications, press releases and official reports) as well as semi-structured informal talks with government officials, intergovernmental panels, ENGOs and academics. I traced ENGO involvement by documenting their activities and inferentially determining the extent to which civil society pressures have effected changes in this particular environmental policy context. Using semi-structured questionnaire techniques, I informally interviewed key informants and sustained several discussions with ENGO representatives, government representatives and industry associations over the period from March 3, 2001 to November 25, 2001. I also had informal discussions with the participants in the March 5th and 6th meetings of the Consultative Group for the North American PRTR project.[15] I actively participated in and observed the discussions held over these two days and observed the involvement of people in the dialogues. I tracked down those people who were more vocal and seemed to be highly involved with the development of RETC. I followed the participation of environmental NGOs in list-servs, and exchanged electronic mail communications with several NGOs and

---

[14] Other potential explanations would include policy convergence between US, Mexico and Canada, favourable domestic politics towards a mandatory RETC and policy learning within the Mexican government. However, these hypotheses are not supported by the evidence presented here.
[15] NACEC provided funding for many participants in these meetings, including academics, ENGOs and government officials.

government officials, academics and industry representatives from March 2001 to November 2001. I also held exploratory discussions in the period of October 1 and 2, 2001, in Mexico City.

8. EMPIRICAL EVIDENCE

In this research, I show that the change from voluntary reporting to mandatory reporting may be explained by arguing that NACEC has acted as a catalyst to ENGO involvement in this policy change. By providing Mexican, American and Canadian ENGOs with spaces, forums and venues where they can share information, expertise and organise pressure campaigns, NACEC played a catalytic role that has strengthened (albeit perhaps involuntarily) the demands of civil society organisations. Cooperation amongst these ENGOs by working in coalition (both at the domestic and international levels) has solidified their role and further their interests. I argue that the change in Mexican environmental policy with respect to toxic release inventory reporting has arisen (at least partially) as a result of pressure from ENGOs (both domestic and international), working in coalitions, and the influence of international environmental institutions within which Mexico is embedded. In studying how ENGOs alter state behaviour, I look for explicit interactions between government officials of the target state (Mexico) and activists from ENGOs involved in the case study.

While ENGOs became involved in the National Coordinating Committee since it started working to develop a proposal for a Mexican PRTR, it only became apparent until March of 2001 that their efforts to influence the Mexican government would not be effective if they did not build a coalition. While this coalition formation exercise does not appear to have been explicit, two factors might have influenced the formation of a coalition. First, Mexican ENGOs recognised early in 2001 that they needed to cooperate with other Mexican ENGOs and with Canadian and US ENGOs to increase pressure intensity on Mexican government officials. ENGOs realised that no matter how much they lobbied the Mexican government, their claims would not be heard (or prioritised) if they did not form a coalition that brought these issues to the table. Second, the number of meetings and interactions among ENGO representatives dealing with RETC increased within the last two years. Therefore, exchanges of information and resources also increased and ENGOs developed cooperative networks that would allow effectively influence Mexican environmental policy-making.

Also, the influence of three ENGOs seems to have been instrumental both in promoting a mandatory RETC and in building a coalition. Information dissemination efforts of *Emisiones: Espacio Virtual*, a project of

10. Democracy by Proxy: Environmental NGOs in Mexico 243

Programa LaNeta raised awareness in the environmental activist community, thus leading to the formation of a more-or-less defined coalition of ENGOs. Emisiones was a project that first started operating in 1997, as a sub-program of Programa LaNeta (a Mexican non-governmental organisation). Programa LaNeta was founded in 1993, with five full-time members (two of them work in the project Emisiones).[16] Emisiones worked on a non-regular basis with other ENGOs such as Greenpeace Mexico, Colectivo Ecologista Jalisco, Proyecto Fronterizo de Educación Ambiental, Enlace Ecologico, etc. However, there were no specific ties built with other organisations (until 2000).

Under the leadership of Colectivo Ecologista Jalisco, a national campaign to promote appropriate reporting of the Cedula de Operación Anual was launched. This campaign is a formal attempt to strengthen comparability between RETC, TRI and NPRI.[17] The third ENGO that played an influential role, Presencia Ciudadana Mexicana, led by Martha Delgado-Peralta, produced a handbook on Access and Use of Information on Pollutant Emissions aimed to educate and inform the public about potential health hazards posed by chemical contaminants being released to the atmosphere, water, and soil.[18] This handbook promotes the development of a mandatory RETC, where information by facility and by substance could be easily accessed and monitored. Ms. Delgado-Peralta is also the head of a national coalition of ENGOs named "Union de Grupos Ambientalistas" (UGA). UGA was previously headed by Regina Barba (who was also heavily involved in the development of RETC, formerly as an activist and since January of 2001 as a government official in charge of transparency and citizen participation at SEMARNAT). UGA has also worked intensely on lobbying the Mexican government to implement a PRTR that has a high degree of comparability with TRI and NPRI.

Other groups supported these three ENGOs through information exchange and collaboration in lobbying and letter-signing campaigns. For example, Fronteras Comunes (a northern Mexico-based ENGO) has submitted several requests of support to Mexican ENGOs in an effort to increase leverage and strengthen pressure levels on government officials. Also, Fronterizo de Educación Ambiental (led by Mrs. Laura Silvan de Durazo, who now sits on the Joint Public Advisory Council of NACEC) published a number of documents that encourage civil society to increase

---

[16] Personal communication via e-mail with Ms. Azucena Franco, Emisiones - Programa LaNeta, May 7, 2001. I thank Ms. Franco, and Emisiones for keeping me updated with information about the development of the RETC.
[17] More information about this campaign can be found at the following website: http://www.laneta.apc.org/emis/reporte.htm. This webpage is hosted by Programa LaNeta.
[18] The handbook can be accessed online through the following website link: http://presenciaciudadana.org.mx/accesoantecedentes.html.

citizen participation through information-dissemination strategies.[19] Acting as a coalition, these ENGOs have pressured the Mexican government to change the reporting mode of RETC.

Until 2000, the work of these ENGOs was scattered and there was little visible coordination. However, since early 2001, six ENGOs appeared to intensify their exchanges and build cooperative efforts to propose a mandatory RETC.[20] These organisations (Emisiones, Fronterizo de Educacion Ambiental, Colectivo Ecologista Jalisco, Presencia Ciudadana, Greenpeace Mexico and Fronteras Comunes) have been vocal and influential in their lobbying efforts. At the March 5, 2001 meeting organised by the NACEC to request input from the public on its document *Taking Stock '99*, representatives from all these organisations attended the round-table discussions and voiced their concerns. The Mexican ENGOs, bolstered by Canadian and US ENGOs participating in the coalition used this opportunity to bring relevant issues to the table.[21] For example, the leader of Colectivo Ecologista Jalisco (Maite Garcia-Lozano) outlined the work her organisation had been carrying out with the Mexican government, NACEC and the ENGOs mentioned above through the program "Campaña Nacional para el Reporte Acertado y Completo de la Cédula de Operación Anual y la Norma Mexicana - Registro de Emisiones y Transferencia de Contaminantes". Garcia-Lozano also demanded that the Mexican government commit to a mandatory RETC. Joining her in these demands were Laura Durazo of Fronterizo de Educación Ambiental, Paul Orum of the Working Group on Community Right to Know (in the US), Michael Gregory of the Arizona Toxics Information group (in the US), and several other representatives of Canadian ENGOs.

ENGOs repeatedly asserted that RETC must be changed to a mandatory reporting system, and called for placing such discussion on the agenda of the meetings. ENGOs felt that their concerns had not been properly taken into account, that since 1997-98 progress on the PRTR had stopped, and that the nongovernmental groups had been shut out of the process. They felt the real issue was "the need to generate political will to move the RETC forward by giving legal authority to the RETC" (CEC, 2001). As evidenced in the Summary of the Consultative Meeting on the Development of the *Taking Stock 1999* report, "many participants voiced their support for a mandatory PRTR in Mexico" (CEC, 2001). Again, this demonstrates that ENGOs used the meetings to voice their concerns and increase pressure on the target

---

[19] See http://www.laneta.apc.org/emis/sustanci/retc/particip.htm

[20] While Presencia Ciudadana, Emisiones and Colectivo Ecologista Jalisco seemed to be instrumental to start coalition-building efforts, they have also been supported by the other mentioned ENGOs.

[21] US ENGOs included Arizona Toxics Group, the DC-based Working Group on Community-Right-to-Know. And Canadian ENGOs included Pollution Probe and the Canadian Institute for Environmental Law and Policy.

nation-state (Mexico). At the end of the meetings, government officials admitted that there was a real need for a mandatory RETC and said that they would commit to work towards that goal. As a result, on March 6, 29 representatives of 15 ENGOs from all three countries submitted a letter to Mr. Lichtinger praising him for making this commitment and asking for a formal calendar with milestones and deliverables that would fast-track the development of a comparable RETC.[22] Included in their demands were the creation of a Technical Secretariat for the GNC, and explicit mechanisms for civil society participation in decision-making with respect to RETC.

On June 7, 2001, Ms. Laura Silvan de Durazo, director of the Proyecto Fronterizo de Educación Ambiental, A.C. in Tijuana, Baja California, was appointed to the 15 member committee of the NACEC's Joint Public Advisory Committee (JPAC). Fronterizo de Educacion Ambiental is "an organization dedicated to strengthening citizen participation in environmental management by means of public information, education and consultation as well as through the formation of local, national, border and international networks".[23] Ms. Durazo's participation in JPAC might have also played a role in increasing NGO leverage at the trinational level, since Ms. Durazo's lobbying and information-dissemination activities were mainly focused on the 'community right-to-know' area. Her presence in JPAC would probably help raise awareness on the need for a mandatory RETC.

On April 18, 2001, SEMARNAT issued a voluntary standard (NMX-AA-118-SCFI-2001) for the RETC, something that ENGOs strongly opposed. Because NMXs are voluntary as opposed to NOMS (which are mandatory), ENGOs felt that this voluntary standard would work against the purpose of RETC and the commitment agreement reached in the March meetings. However, on repeated occasions (June 5 and June 29th 2001), Lichtinger supported the idea of a mandatory RETC. On June 4th, 2001, Emisiones and Fronteras Comunes sent a new letter to SEMARNAT asking for an update on the status of RETC. On June 13, 2001, representatives of Greenpeace Mexico, Fronteras Comunes, Emisiones and Presencia Ciudadana met with Juan Barrera and Regina Barba from SEMARNAT, to discuss the RETC mandatory versus voluntary situation. In that meeting it became clear that some issues (such as the wording of the suggested modification to the General Law of Ecological Equilibrium and Environmental Protection, LGEEPA) had to be resolved before RETC could become mandatory. Concerned with the outcome of this meeting, the attending ENGOs wrote another letter to Mr. Lichtinger in which they stressed the need for RETC to

---

[22] Letter submitted on March 6, 2001 to Mr. Victor Lichtinger, Secretary of Environment and Natural Resources of Mexico. A copy of the letter is available from the author. Ms. Martha Delgado-Peralta was designated as the coordinator for this campaign.
[23] As quoted in the June 7, 2001 press release of the NACEC.

become mandatory and for the information to be transparent and publicly available.[24]

On June 29, 2001, during the regular session of NACEC's Council, Mr. Lichtinger announced that he would submit a motion to the Mexican Congress to transform RETC from a voluntary instrument to a mandatory one. This event was considered a huge success in improving the comparability of PRTRs across North America. As evidence indicates, ENGO involvement played a key role in raising awareness, influencing government officials and bringing umbrella international organisations to the table to move their demands forward. It is important to recognise, however, that this influence might not have had such a strong impact if there were not for the catalytic intervention of NACEC as an overarching international environmental institution.

On December 6, 2001, the ammendments to Article 109 Bis of the LGEEPA were approved in the Chamber of Deputies of Mexico, and on December 13, 2001, the Senate also approved these amendments.[25] With these approvals, changes in the Mexican Right-To-Know law will allow RETC to be a mandatory reporting policy instrument, thus enhancing comparability with NPRI and TRI.

9. CAVEATS AND ISSUES TO ANALYSE FOR FUTURE RESEARCH

Scholars in the policy sciences field would argue that to study policy change, one would have to study cases for a period of time long enough to allow for a meaningful evaluation of whether policy change has in fact, taken place. For Sabatier and Jenkins-Smith, who designed one of the most influential theoretical frameworks for analysing policy change (the advocacy coalition framework or ACF), "understanding the process of policy change - and the role of technical information therein - requires a time perspective of a decade or more" (Sabatier and Jenkins-Smith, 1999). I acknowledge that in my study of this case, the time frame has not been a complete decade. Future steps for this research will include following the process for the next few years in an effort to capture the richness of the case. Thus, this will allow us to analyse not only policy change but also policy stability. Also, it would be

---

[24] Emisiones, personal communication via e-mail, June 25, 2001.
[25] It is worth noting that members of the Mexican Congress had previously received a letter signed by Greenpeace and all other Mexican ENGOs. This letter was basically lobbying for a mandatory RETC, and it was sent to all members of the Congress. This might have proved an important event that influenced the outcome of this process.

important to examine similar cases in Europe and other non-North American contexts.

Also, this case study calls for a search on comparable cases where transnational civil society was involved and no policy change took effect (in order to find a counterfactual). As I have argued elsewhere, two of the most difficult methodological issues in evaluating cases of public involvement, citizen participation and social movements are causality tracing and counterfactual assessing. It is very hard to find cases where the influence of a variable is not present in order to compare relative effects of the same variable.[26]

Finally, it is worth noting that unpacking the exact degree of influence of each variable (the power of international environmental institutions and the pressure from transnational civil society coalitions) is very difficult and poses a challenge to anyone doing research on this area. Many other factors may influence the outcome and thus should be analysed to increase the level of rigour at later stages of the research[27].

10. CONCLUSIONS AND EPILOGUE

In this case study, I have shown how ENGO involvement has influenced the Mexican government to modify the design and specifications of RETC. Future events will likely involve continued and strong ENGOs lobbying efforts to achieve policy change. As this case study has shown, while the role of the nation-state remains central (regardless of the theoretical orientation of researchers), non-state actors have increasingly taken a more pre-eminent role in domestic and international environmental policy-making. By influencing processes of agenda-setting, decision-making and policy implementation, non-state actors are able to bridge the local-global gap through information dissemination, policy diffusion, and transfer and strengthening processes. Using the metaphor of watering a tree, by acting at the grassroots level (watering the roots), ENGOs are able to exact larger

---

[26] Two very good examples of theoretical and empirical articles that try to deal with exactly these problems are Betsill, M. M. and E. Corell (2001). "NGO Influence in International Environmental Negotiations: A Framework for Analysis." Global Environmental Politics 1(4): 65-85, and Corell, E. and M. M. Betsill (2001). "A Comparative Look at NGO Influence in International Environmental Negotiations: Desertification and Climate Change." Global Environmental Politics 1(4): 86-107.

[27] I have undertaken further work on the development of RETC since this chapter was first written. Due to space limitations, I cannot describe that work in detail here. Suffice it to say that I found substantial evidence that NACEC also played a significant role in shaping RETC. Furthermore, the Registro is already mandatory but the regulation that governs its operation has not been yet published (Feb 2004); declarations of victory on the part of ENGOs may be premature.

scale effects on policy making (at the branch level). This case also highlighted the interplay of domestic and international environmental politics. While influence was felt through horizontal policy mechanisms (through direct contact with other countries' organisations), vertical policy coordination was also relevant (influenced by international environmental institutions). Also, while this case did not aim to provide a comparison between horizontal and vertical environmental policies, it clearly shows that the policy-making process is always the result (to a certain extent) of the interplay between these two schemes of policy coordination. The challenge to make the most of this intertwined relationship still remains fertile ground for future research.

ACKNOWLEDGMENTS

I thank Les Lavkulich, Evan Fraser, Sigfrido Pacheco-Vega, Obdulia Vega-Lopez, Dorothee Schreiber, Kathryn Harrison, Patricia Keen, Brian G. Wright, and Timothy Walls for editorial help and extremely insightful comments on earlier drafts of this chapter. I also thank Manuel Velazquez for formatting help. Tom Legler, Jonathan Fox, and Kathryn Harrison also provided helpful comments on earlier drafts of the chapter. I gratefully acknowledge the North American Commission on Environmental Cooperation (NACEC) for financial support to attend the PRTR project meeting in March 2001, October 2002 and 2003, and CONACyT for financial support from 1999 to 2003. All errors, of course, remain mine. The views presented in this chapter are of the author alone and should not be deemed as representing the position or view of the NACEC or any other organisation or individual mentioned in the paper. None of the people or organisations mentioned in this work are, in any way, responsible for the way I have interpreted their ideas and suggestions.

REFERENCES

Breitmeier, H. and V. Rittberger, 2000, Environmental NGOs in an emerging global civil society. In: *The Global Environment in the Twenty-First Century: Prospects for International Cooperation.* P. S. Chasek. Tokyo, United Nations University Press, pp. 130-163.

Bunge, J., *et al.*, 1996, "Employee participation in pollution reduction: Preliminary analysis of the Toxics Release Inventory." *Journal of Cleaner Production* 4(1): 9-16.

CEC, 2001, Summary of the Consultative Meeting on the Development of the *Taking Stock 1999* Report on North American Pollutant Releases and Transfers. Montreal PQ, North American Commission on Environmental Cooperation, p. 5.

Chasek, P., 2000, Conclusion: The global environment in the twenty-first Century: Prospects for international cooperation. In: *The Global Environment in the Twenty-First Century:*

Prospects for International Cooperation. P. S. Chasek. Tokyo, United Nations University Press, pp. 427-441.

Edwards, M. and G. Sen, 2000, "NGOs, social change and the transformation of human relationships: A 21st-century civic agenda." *Third World Quarterly* **21**(4): 605-616.

Fox, J., 2002, Lessons from Mexico-U.S. Civil Society Coalitions. In: *Cross-Border Dialogues. U.S.-Mexico Social Movement Networking.* D. Brooks and J. Fox. La Jolla, CA, Center for U.S. - Mexican Studies, University of California, San Diego, pp. 341-418.

Hajer, M. A., 1993, Discourse coalitions and the institutionalization of practices. The case of acid rain in Britain. In: *The Argumentative Turn in Policy Analysis and Planning.* F. Fischer and J. Forester. Durham, N.C., Duke University Press, pp. 43-76.

Harrison, K. and W. Antweiler, 2001, Environmental regulation vs. environmental information: A view from Canada's National Pollution Release Inventory. Annual Meeting of the Association for Public Policy Analysis and Management, Washington, D.C., APPAM.

Keck, M. E. and K. Sikkink, 1999, "Transnational advocacy networks in international and regional politics." *International Social Science Journal* **51**(159): 89-101.

Legler, T., 2000, Transnational coalition-building in the Americas: The case of the Hemispheric Social Alliance. Summer Institute on "Building the New Agenda: Hemispheric Integration and Social Cohesion", Robarts Centre for Canadian Studies, York University, Toronto, Ontario.

Nelson, P. J., 1996, "Internationalising economic and environmental policy: Transnational NGO networks and the World Bank's expanding influence." *Millenium: Journal of International Studies* **25**(3): 605-633.

Pacheco, R. and P. N. Nemetz, 2001, Business-not-as-usual: Alternative policy instruments for environmental management. 5th IRE Annual Workshop: Addressing the Knowledge Crisis in Water and Energy: Linking Local and Global Communities, Vancouver, B.C., Institute for Resources and Environment, UBC.

Pacheco-Vega, H. R., 1998, A proposed theoretical model for the construction of strategic alliances in the biotechnology industry. 1998 R&D Management Conference: Technology Strategy and Strategic Alliances, Avila, Spain, R&D Management.

Princen, T. and M. Finger, 1994, *Environmental NGOs in World Politics : Linking the Local and the Global.* London ; New York, Routledge.

Risse-Kappen, T., 1995, *Bringing Transnational Relations Back In : Non-State Actors, Domestic Structures, and International Institutions.* New York, Cambridge University Press.

Sabatier, P. A. and H. C. Jenkins-Smith, 1999, The Advocacy Coalition Framework. An assessment. In: *Theories of the Policy Process.* P. A. Sabatier. Boulder, CO, Westview Press, pp. 289.

Wright, B. G., 2000, "Environmental NGOs and the dolphin-tuna case." *Environmental Politics* **9**(4): 82-103.

PART 5: THE EFFECTS OF TRADE AND DEVELOPMENT POLICIES ON THE ENVIRONMENT

Chapter 11

ECONOMIC PROGRESS IN THE COUNTRYSIDE, FORESTS, AND PUBLIC POLICY:
Some Lessons from Ecuador and Chile

Douglas Southgate,[1] Boris Bravo-Ureta,[2] and Morris Whitaker[3]
[1]*Dept. of Ag., Env., & Dev. Econ., Ohio State Univ., Columbus, OH 43210;* [2]*Office of Int. Affairs and Dept. of Ag. & Res. Econ., Univ. of Connecticut, Storrs, CT 06269;* [3]*Dept. of Econ. and Office of the Provost, Utah State Univ., Logan UT 84322*

Abstract: Recent contributions to the literature on the causes of tropical deforestation indicate that, under certain circumstances, agricultural intensification (i.e., raising crop and livestock yields) can accelerate farmers and ranchers' encroachment on tree-covered land. We contend, however, that the linkage between intensification and habitat conservation is generally positive. Chilean experience during the 1980s and 1990s is a good example. Domestic and foreign demand for agricultural output increased substantially. But because of productivity improvements, agricultural land use fell; the forested portion of the national territory rose from 19 percent in 1990 to 24 percent in 1998. In contrast, large tracts of forests have been converted into cropland and pasture in Ecuador during the last two decades. Productivity-enhancing investment having been very deficient, agricultural land use increased at a 2 percent annual rate during the 1980s and continued to expand in the 1990s.

Key words: agricultural development; tropical deforestation; Latin America

1. INTRODUCTION

In one of the best surveys available of the literature on tropical deforestation, David Kaimowitz and Arild Angelsen, two economists at the International Center for Forestry Research (CIFOR), have examined more than 150 economic models addressing the causes of encroachment on species-rich habitats in the developing world. The general lessons to be drawn from these studies, they argue, are limited. Avoiding endorsement of views and prescriptions that are overly simplistic, Kaimowitz and Angelsen (1998) have rekindled an old debate about whether biodiverse ecosystems near the equator are threatened or safeguarded by agricultural development.

The two economists concede that the application of improved crop-production technology in areas already under cultivation is likely to ease human pressure on natural habitats. However, they point out that there are various circumstances under which the introduction of better farming methods actually leads to deforestation. Even if the higher yields resulting from technological change cause commodity prices to fall, Kaimowitz and Angelsen (1998) argue that the ultimate impact on forests is indeterminate.

In part, this sort of ambivalence has to do with deforestation's great variability. There may be certain motivations that cause one set of farmers to encroach on tree-covered habitats in one setting and others that cause another group to clear forests somewhere else. Given this variability, Kaimowitz and Angelsen's call for careful analysis makes perfect sense, as does their warning against over-simplification. However, excessive ambivalence is also problematical. The pace of demographic, economic, and environmental change is far too rapid to allow for the deliberate and exhaustive resolution of all empirical questions before a strategy is fashioned to reconcile economic progress and conservation of natural habitats in the developing world. In particular, there is ample reason to conclude that agricultural intensification – increasing crop yields – is an essential element of such a strategy in many places.

How agricultural intensification can help to alleviate deforestation is discussed in general terms in the next section of this paper. We then turn to an analysis of what has happened to natural habitats in Ecuador, where crop yields stagnated during the latter part of the twentieth century. Examined next is the situation in Chile, where productivity growth has coincided with diminished agricultural land use. Lessons to be learned from the experiences of these two countries are summarized at the end of the paper.

2. THE IMPACTS OF AGRICULTURAL DEVELOPMENT

To be sure, the development of new technology for raising crops in places where trees and other natural vegetation have been growing creates an impulse for agricultural extensification – an impulse, that is, for agriculture's geographic expansion. Increased soybean production along the eastern and southern fringes of the Brazilian Amazon and around Santa Cruz, Bolivia following the introduction of soybean varieties that flourish in warm, tropical settings is a case in point (Kaimowitz and Smith, 2001).

Otherwise, disagreements over linkages between agricultural intensification and deforestation have a lot to do with geographic scale. At a global level, the benefits of technological change in the agricultural sector are undeniable. By and large, consumption of farm products is not highly sensitive to changes in the prices of those products; in other words, demand is price-inelastic. This means that technological improvement, which lowers production costs and therefore increases supplies, drives down commodity prices. In remote areas traversed by agricultural frontiers, price declines outweigh the costs savings for farmers created by the use of better technology. Thus, land-clearing is discouraged. The same outcome is observed as the years pass and increases in population and living standards cause the demand for food to increase. As long as the supply growth resulting from agricultural development exceeds demand growth, then prices fall and incentives for agricultural land-clearing in frontier settings weaken (Southgate, 1998).

This fortunate result is not just theoretical conjecture. Since the 1960s, yield increases have more than matched global demand growth. This has kept the rate of growth in agricultural land use at very low levels – less than 0.2 percent per annum, to be specific (Tweeten, 1998). Were it not for the Green Revolution, we would now be talking about tropical forests mainly in the past tense. Every spare patch of ground would have to be used to feed a desperately hungry world population.

But just because an inverse relationship exists between intensification and deforestation at a global scale does not prove that it also applies at a more limited geographical scale. Consider a small, open economy that enjoys a comparative advantage in farming. For such a country, prices of exported commodities, which are determined in international markets, are affected very little by its exports. This is true of agricultural production in Guatemala, for example. How much or how little its farmers produce has virtually no effect on global supplies and international prices. From Guatemala's standpoint, demand for its exported output is price-elastic. This means that intensification, which raises output, does not cause major price declines. Since it also lowers farmers' costs, forest loss accelerates.

Even in small, open economies experiencing agricultural intensification, habitat loss is not inevitable. Where production of crops and livestock is continually becoming more efficient, other changes are bound to be taking place that ease pressure on tree-covered hinterlands. Among these changes is increased non-agricultural employment, which results from improved educational attainment (which is an important part of human capital formation) as well as better access to labor markets for rural people. As Kaimowitz and Angelsen (1998) emphasize, a rise in non-agricultural employment diminishes the number of people who are attracted by the prospect of carving an impoverished farm out of a primary forest. Another change is the development of institutions – courts that can be counted on to enforce contracts and property rights, for example – required for the functioning of a robust and competitive market economy. A recently published econometric study, in which a sample of countries from around the world was used, indicates that this sort of institutional development promotes forest conservation (Bohn and Deacon, 2000).

Furthermore, simple economic growth, caused by liberalized trade and the accumulation of human, institutional, and other resources, can have beneficial environmental consequences. For one thing, demand for environmental services increases as incomes rise. At the same time, improved standards of living allow for the development of mechanisms needed to satisfy this demand. For example, demand for visits to places that are scenic and untouched goes up in a society that is growing wealthier. Meanwhile, increased affluence facilitates the establishment of a system of parks, which helps to satisfy this demand.

Recognizing that improved environmental quality can go hand in hand with economic growth in an affluent setting, various economists have posited a so-called Environmental Kuznets Curve (EKC), which relates living standards to environmental quality (Dasgupta *et al.*, 2002). Shaped like an inverted U, this function is consistent with the idea that pressure on natural resources increases as a society emerges from poverty but eventually passes a threshold beyond which economic expansion and environmental quality begin to complement one another. To pass this peak of the EKC for tropical forests, activities like agricultural extensification need to be curtailed. As documented in the rest of the paper, this task has been accomplished in some parts of Latin America but not in others.

3. THE CAUSES OF HABITAT LOSS IN ECUADOR

The environmental price paid where public policy does not encourage movement past the peak of the EKC is clear in Ecuador. In that country, trade distortions and a lack of productivity-enhancing investment have contributed to accelerated deforestation in recent decades.

A pivotal event in the country's recent economic history was the initiation of petroleum production, nearly thirty years ago. During the 1970s, the inflow of petrodollars gave the country's central bank the means to maintain a fixed exchange rate even though inflation within Ecuador should have induced devaluations. The currency over-valuation resulting from this intervention made imports artificially cheap for urban-based consumers, including the politically influential middle class. However, this policy was disadvantageous for farmers and other producers of tradable commodities, including goods competing with imports as well as exports.

In spite of the disincentives for agriculture created by currency over-valuation, deforestation happened at a very rapid rate. In part, this was the result of an ambitious program of road construction in previously inaccessible areas, which was made possible by oil wealth. Migrating along these roads were thousands of farmers from other parts of Ecuador, who promptly set about clearing their new properties (Bromley, 1981). By and large newly deforested land was dedicated to cattle ranching. Livestock products were less tradable than crops. Accordingly, demand for domestically produced output, which grew substantially as incomes rose during the 1970s, was not dampened by the central bank's interventions (Wunder, 2000).

With the devaluations that occurred in the wake of the debt crisis, in 1982, incentives for crop production improved. However, Ecuadorian farmers responded to demand growth less by raising yields and more by using additional land. This is contrary to general patterns of agricultural development, in which intensification eventually becomes more important than extensification. Moreover, extensification is environmentally unsustainable in Ecuador since agriculturalists long ago occupied virtually all land in the country that lends itself well to farming; as early as 1970, total agricultural land use was equivalent to the area that the National Program for Agrarian Regionalization (PRONAREG) has identified as suitable for agriculture.

If anything, farmers and ranchers' encroachment on forests and other natural habitats has become a more predominant response to demand growth in recent times. A recent assessment of Ecuadorian agriculture for the Inter-American Development Bank reveals that extensification accounted for two-thirds of all production increases achieved between the middle 1960s and middle 1980s. But during the late 1980s and early 1990s, more than four-fifths of the additions to output related to growth in farmed area (Table 11-1).

The predominance of extensification is largely a consequence of inadequate investment of the kind that contributes to agricultural intensification. Real spending on agricultural research, for example, declined by 7.3 percent a year from 1975 through 1988, at which time research budgets amounted to 0.20 percent of agricultural GDP (Whitaker, 1990). This is a deficient level of support, even by the standards of neighboring countries. That opportunities for intensified crop production have been foregone because of a weak science and technology base is plain to see from the differences that exist between yields in Ecuador and those of adjacent nations. Even though Ecuadorian production of rice, for example, is concentrated in the Guayas River Basin, where growing conditions are ideal, average Ecuadorian yields are less than half of Colombia's or Peru's (Whitaker, 1990).

Just as too little has been spent in Ecuador on research and extension in support of crop and livestock production, human capital formation has been inadequate, especially in rural areas, which has exacerbated the pressure on natural habitats. The limited available evidence suggests that agricultural land-clearing is not financially rewarding. Annual net cash flow on recently deforested parcels in the northwestern part of the country averages $25 per hectare; of those settlers willing to name a price, the vast majority would part with their land if offered just $300 per hectare (Southgate, 1998). That the meager returns to deforestation are still attractive enough for agricultural colonists is explained, of course, by their lack of skills needed to compete for better-paying jobs. The problem of deficient skills is indicated by the finding that, along Ecuador's agricultural frontiers, average educational attainment for rural adults is just two or three years (Southgate, 1998; Pichón, 1997).

An additional impulse for deforestation relates to weak incentives for timber production – a combination of low prices and attenuated property rights, to be specific (Southgate et al., 2000). Low prices are largely a consequence of public policy. As has been done in many other Latin American countries, the Ecuadorian government long prohibited the export of unprocessed timber, not out of environmental concerns but instead to keep raw material expenses low for domestic manufacturers of wood products. Timber prices also have been held down by weak or non-existent property rights in tree-covered land. For all intents and purposes, many forests are an open-access resource, so logging takes place whenever and wherever the revenues gained by doing so are greater than or equal to the opportunity costs of labor, machinery, and other inputs to timber extraction. As a result, stumpage values hover around zero and no one has an incentive to manage forests for timber regeneration.

Table 11-1. Sources of change in agricultural production in Ecuador, 1988-89 through 1994-95

Region	Increase	Source of Increase: Area	Yield	Interaction
Coast	4,062,000 MT	82.2%	11.4%	6.4%
Highlands	365,000	67.5	6.7	25.8
Entire Country	4,428,000	81.0	11.0	8.0

Source: D. Colyer (West Virginia Univ.), personal communication, April 1996.

11. Economic Progress in the Countryside

For many years, about the only way to acquire property rights in Ecuador's tree-covered hinterlands was through agricultural land-clearing. Informal tenurial regimes in frontier settings typically stipulated that obvious and "productive" use of a newly occupied parcel (i.e., farming it) was a prerequisite for de facto ownership. This arrangement was codified in agrarian reform and colonization laws adopted throughout Latin America during the middle of the twentieth century. In Ecuador, as in other places, this arrangement stimulated deforestation (Southgate and Whitaker, 1994).

To summarize, a policy regime that favors agriculture's geographic expansion at the expense of natural habitats exists in Ecuador. Elements of this regime include:

- inadequate support for agriculture's scientific base, which hinders intensification and causes extensification to be the primary response to growing demands for crops and livestock;
- under-funding of education in the countryside, which creates a sizable segment of rural population that find the meager returns of deforestation to be attractive; as well as
- trade restrictions and weak property rights that diminish the returns to forest management and encourage agricultural land-clearing.

Because of these policies, forest loss is very high. According to the Forest Resource Assessment (FRA), carried out by the U.N. Food and Agriculture Organization (FAO), average annual deforestation during the early 1990s was approximately 189,000 hectares, which was equivalent to 1.6 percent of standing forests. The only South American country with a higher rate of relative deforestation was Paraguay, which at the time was losing more than 2.5 percent of its forests annually (Wunder, 2000).

Determining the environmental costs of high deforestation in Ecuador, where natural habitats contain a wide variety of plant and animal species is a challenge. Evidence that the costs are high is contained in a study of the value of tropical forests as a source of biological specimens for pharmaceutical research. According to that study, this value in western Ecuador, a place with a large number of endemic species and where deforestation has reached an advanced cumulative stage, is approximately $20 per hectare. This exceeds similar values calculated for threatened "hot spots" of threatened biodiversity in other parts of the world (Simpson, Sedjo, and Reid, 1996).

4. INTENSIFICATION OF CHILE'S RURAL ECONOMY

During the early 1970s, opening of oil fields in the northeastern part of Ecuador and rising fossil fuel prices allowed the country's government to assume a dominant role in the national economy – much to the detriment of tree-covered habitats, as it turns out. As the decade wore on, a very different approach was followed farther down the Pacific coast of South America. The military leaders who seized control of Chile in a 1973 coup d'etat initially adopted interventionist measures similar to those applied in Ecuador. But by the middle 1970s, the shortcomings of this approach were impossible to ignore and a switch was made to policies more conducive to free markets and investment. This switch has helped to bring about agricultural intensification as well as sizable increases in tree-covered area and timber output.

It has been argued that policies pursued during the 1960s and early 1970s – agrarian reform, in particular – helped set the stage for the agricultural development occurring after the Chilean economy was liberalized. To be specific, putting an end to the old hacienda system is said to have induced efficiency gains and brought new entrepreneurs into the agricultural sector. Of the land that was expropriated, much was not being used productively and the threat of having one's land taken away encouraged many large owners to farm more intensively or sell a portion of their holdings (Brown, 1991). One observer contends that the availability of medium sized properties, many of which were carved out of large estates, facilitated the expansion of intensive fruit production (Cruz, 1988). One could say that much of the transformation that coincided with agrarian reform was induced by fundamental economic forces and would have occurred regardless of public policy. But a number of observers would disagree, arguing that increased fruit output during the 1970s and 1980s had much to do with the prior redistribution of land (Cereceda and Dahse, 1980).

In any event, the military government made sweeping changes in rural lands policy soon after coming to power. Among these was to permit beneficiaries of agrarian reform to sell their holdings on the open market. This promoted the economically rational use of farmland. So did the transfer of state-owned properties to private parties, which also took place after 1973 (Cereceda and Dahse, 1980).

The tilt toward free-market policies has not prevented the government from sponsoring productivity-enhancing investment, which has been another ingredient of Chile's agricultural success. A national irrigation program, featuring the construction of dams, canals, and other infrastructure, was put in place in 1985 (Comisión Nacional de Riego, 2000). Eight years earlier, in 1978, the Integral Technology Transfer Program (PTTI) had been adopted to encourage the adoption by small farmers of improved seeds, fertilizers and pest control systems, as well as mechanization. This program has proved to be more effective than past approaches to agricultural extension (Bebbington and Sotomayor, 1998).

With inefficient haciendas broken up and improvements made in irrigation infrastructure as well as farming technology, agricultural yields have gone up markedly. Per-hectare output of wheat, oats, and barley have tripled and maize yields have doubled since the 1970s. In addition, Chile's agricultural economy has been able to respond in a dynamic fashion to the opportunities that exist where commodity prices are determined by competitive market forces. In particular, the adoption of a liberalized trade regime during the 1980s proved to be a major impulse for exports of wine, fruit, and other goods (Kay, 1996). Central Bank data indicate that, between 1990 and 1997, the balance of trade in agricultural products rose from $860 million to $1,542 million.

Increased agricultural productivity coincided with a reduction in agricultural land use. By the middle 1990s, the area planted by crops was 15 percent lower than what it had been earlier in the decade (Figure 11-1). Cereal area declined by more than 25 percent – from 824,000 hectares in 1989/1990 to 594,000 hectares in 1999/2000 – and land used for legume and tuber production fell by more than one-third, from well over 150,000 hectares in the early 1990s to 100,000 hectares in 1999/2000.

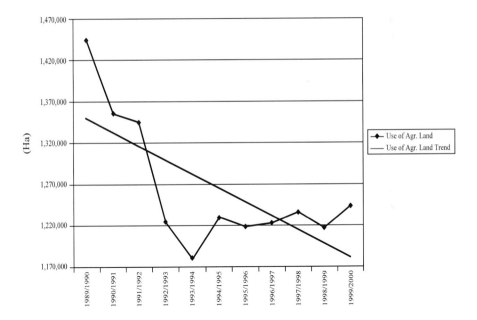

Source: ODEPA (2001).
Figure 11-1. Agricultural land use in Chile, 1989-90 through 2000-01

These reductions, which coincided with increased production of cereals, legumes, and tubers, far outweighed the modest increases in the area dedicated to fruit and vegetable production, in which Chile holds a strong comparative advantage, during the same period (ODEPA, 2001). According to data collected by the FAO, 13.8 percent of the national territory was used for intensive agricultural production in 1990; by 1998 that portion had fallen to 10.8 percent.

As agricultural land use has contracted, forestry has expanded, with tree-covered land going up from 17.8 percent of the national territory in 1990 to 24.1 percent in 1998. Between 1990 and 1997, Chile's balance of trade in timber, paper, and related products nearly doubled, from $798 million to $1,587 million. Roundwood and plywood output grew by 37 and 63 percent, respectively. Especially noteworthy has been the increased production of manufactured items. For example, output of wood panels has more than doubled.

As indicated in Figure 11-2, commercial tree plantations grew from 1,500,000 hectares during the early 1990s to 2,000,000 hectares at the end of the decade. A little less than half of this increase has to do with plantings of radiata pine, which are currently approaching 1,500,000 hectares. Practically all the rest of the increase relates to plantings of eucalyptus, which more than tripled between 1990 (102,000 hectares) and 2000 (359,000 hectares). In previous years, public subsidies provided a strong impulse for reforestation. In October 1974, the military government enacted Decree Law (DL) 701, which privatized tree plantations owned by the state and provided generous incentives for reforestation. To be sure, this law helped to prime the pump. However, Chilean forestry development has continued mainly because of free trade and the absence of conditions (e.g., tenure insecurity) that would discourage investment.

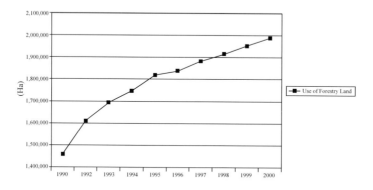

Source: ODEPA (2001).
Figure 11-2. Land used for commercial forestry in Chile, 1990 through 2000

The social and environmental impacts of commercial forestry in Chile have aroused considerable debate and concern. Most production comes from plantations, the spread of which has diminished rural employment in the southern part of the country. Moreover, land ownership is reconcentrating, which exacerbates inequality since timber production is characterized by pronounced economies of scale (Kay, 1998). Especially controversial is the displacement of primary forests by plantations, which detracts seriously from biodiversity. Although most radiata pine and eucalyptus (neither of which is a native species) has been planted on old cropland and pasture, displacement has indeed occurred. Evidence of this is contained in a report prepared for the Chilean partner of Friends of the Earth, which focuses on Empedrado – a rural community in the central part of the country. Between 1961 and 1991, tree plantations increased by 21,652 hectares. Nearly two-thirds of this area – 13,529 hectares – was formerly dedicated to crop and livestock production. In much of the remaining one-third, secondary forests were cleared to make way for stands of radiata pine. But the spread of plantations also helps to explain why the extent of species-rich primary forests has fallen below 2,000 hectares (Sepúlveda and Verscheure, no date). As in other parts of the world, production forestry has not been kind to native forests.

Granted, the intensification of economic activity in the countryside is not without adverse environmental costs. Aside from the loss of biodiversity that results when tree plantations displace native forests, agricultural intensification raises the risk of water pollution and can impair the health of farm workers. Nevertheless, the environmental benefits of economic growth in rural areas should never be ignored. Because agriculture requires less land, Chile now has more geographic space for securing the benefits of forests and other habitats. Likewise, economic growth gives the country's population the means to pay for these benefits. Thus, there is good reason to believe that Chile has passed the peak of the EKC for tree-covered land.

5. SUMMARY AND CONCLUSIONS

Occasionally forgotten in the debate over deforestation is the simple truth that agricultural development often coincides with the displacement of natural habitats and, furthermore, that this displacement is not uniformly detrimental from a social standpoint. If accessible land lends itself well to crop and livestock production, the benefits of clearing it for agriculture are apt to be considerable, perhaps or probably even exceeding whatever environmental costs are associated with the shrinkage of natural habitats. Agriculture's geographic expansion in the United States during the nineteenth century is undoubtedly a case in point. So was the displacement of tropical forests by banana plantations in coastal Ecuador after the Second World War (Wunder, 2000). But as a country's total agricultural land use approaches or exceeds its stock of prime farmland, the benefits of additional expansion decline. Meanwhile, the environmental damages associated with habitat loss mount. At some point, intensification becomes a more efficient response to demand growth.

This point was reached in Ecuador long ago. With just about all prime farmland occupied long ago, farming and ranching are going on in many places that are poorly suited to these activities, which strongly suggests that the peak of the Environmental Kuznets Curve has not been reached in the country. That extensification continues has much to do with public policy. Funding for road construction has been an important stimulus for agricultural extensification. In addition, there has been too little support both for agricultural intensification, which would have reduced the need to use more land to meet mounting commodity demands, and for human capital formation, which would have raised the opportunity cost of labor dedicated to deforestation. Furthermore, incentives for the wise use and management of forests have been weak, timber prices being low and forest ownership being attenuated.

Sorting out the specific contributions that individual elements of this policy regime have made to land use change is difficult. Undoubtedly, the regime's combined impacts, indicated by Ecuador's having one of the highest rates of deforestation in South America, cannot be cancelled out by applying what might be called fashionable approaches to tropical forest conservation. In particular, it is inconceivable that promoting nature-based tourism or the harvesting of non-timber forest products, which some argue saves trees and enhances forest dwellers' living standards, can come close to offsetting the policy-induced impulses for deforestation that have been at work in Ecuador (Southgate, 1998).

Much more effective, Chilean experience indicates, is to foster economic progress in the countryside by liberalizing trade and investing in human, social, and other non-environmental wealth. When and where sustainable intensification of agriculture, forestry, and other economic sectors does not lead directly to diminished pressure on forests and other natural habitats, demand for conserving biodiversity is bound to be enhanced, along with the means to do so. This is the key to reaching that portion of the Environnental Kuznets Curve where economic growth and environmental quality are complementary.

ACKNOWLEDGMENTS

The authors gratefully acknowledge the assistance provided by Paul Fuentes and Daniel Solis, who at the time this paper was written were research assistants in the University of Connecticut's Office of International Affairs. Horacio Cocchi, a post-doctoral fellow in that office, provided editorial assistance.

REFERENCES

Bebbington, A. and Sotomayor, O., 1998, Agricultural extension in Chile, in: *Financing the Future*, J. Beynton, ed., Oxford Policy Management, pp. 106-124.

Bohn, H. and Deacon, R., 2000, Ownership risk, investment, and the use of natural resources, *Am. Econ. R.* **90**:526-549.

Bromley, R., 1981, The colonization of humid tropical areas in Ecuador, *Singapore J. Tropical Geog.* **2**:15-26.

Brown, M., 1991, Radical reformism in Chile, in: *Searching for Agrarian Reform in Latin America*, W. Thiesenhusen, ed., Unwin Hyman, pp. 216-239.

Comisión Nacional de Riego, 2001, La política de riego nacional (11 October 2002),;http://www.chileriego.cl/cuerpo.htm.

Creceda, L. and Dahse, F., 1980, Dos décadas de cambios en el agro chileno, Instituto de Sociología, Pontificia Universidad Católica de Chile, Santiago.

Cruz, J., 1988, La fruticultura de exportación, *Colección de Estudios de CIEPLAN* **25**:79-114.

Dasgupta, S., Laplante, B., Wang, H., Wheeler, D., 2002, Confronting the environmental Kuznets curve, *J. Econ. Perspectives* **16**:147-168.

Kaimowitz, D. and Angelsen, A., 1998, *Economic Models of Tropical Deforestation*, Center for International Forestry Research, Bogor, pp. 7-14 and 92-93.

Kaimowitz, D. and Smith, J., 2001, Soybean technology and the loss of natural vegetation in Brazil and Bolivia, in: *Agricultural Technologies and Tropical Deforestation*, A. Angelsen and D. Kaimowitz, eds., CABI Publishing, Wallingford, pp. 195-211.

Kay, C., 1996, Globalización, agricultura tradicional y reconversión en Chile, *Comercio Exterior* **46**:625-631.

Kay, C., 1998, The agrarian question and the peasantry in the Chilean democratic transition, Institute of Social Studies, The Hague.

Oficina de Estudios y Políticas Agrarias (ODEPA), 2001, *Compendio Estadístico Silvoagropecuario 1999-2000*, Santiago, pp. 69 and 160.

Pichón, F., 1997, Colonist land allocation decisions, *Econ. Devt. and Cultural Change* **45**:707-744.

Sepúveda, C. and Verscheure, H., no date, Tree plantations, Comité Nacional Pro-Defensa de la Fauna y Flora and Friends of the Earth-Chile, Santiago.

Simpson, D., Sedjo, R., and Reid, J., 1996, Valuing biodiversity, *J. Pol. Econ.* **104**:163-185.

Southgate, D., 1998, *Tropical Forest Conservation*, Oxford University Press, New York, pp. 128-129, 141, and 147-149.

Southgate, D., Salazar-Canelos, C., Camacho-Saa, C. and Stewart, R., 2000, Markets, institutions, and forestry, *World Devt.* **28**:2005-2012.

Southgate, D. and Whitaker, M., 1994, *Economic Progress and the Environment*, Oxford University Press, New York, pp. 44-45.

Tweeten, L., 1998, Dodging a Malthusian Bullet in the twentieth century, *Agribusiness* **14**:15-32.

Whitaker, M., 1990, The human capital and science base, in: *Agriculture and Economic Survival*, M. Whitaker and D. Colyer, eds., Westview Press, Boulder, pp304 and 313-314.

Wunder, S., 2000,*The Economics of Deforestation*, Macmillan Press, London, pp. 95, 108-109, and 127-128.

Chapter 12

ENVIRONMENTAL IMPLICATIONS OF CUBA'S DEVELOPMENT STRATEGY DURING THE SPECIAL PERIOD

Sergio Díaz-Briquets[a] and Jorge Pérez-López[b]
[a] *Casals & Associates, 1199 N. Fairfax Street, Alexandria, Virginia, 22314,* sdiazbriquets@casals.com.;
[b] *International Economist, 5881 6th Street, Falls Church, VA 22041,* jperezlopez@cox.net.

Abstract: In August 1990, Cuba entered what has been called a 'special period in time of peace' (*período especial en tiempo de paz*), a severe economic crisis triggered by the disruptions in imports of oil and other raw materials from socialist countries. During the special period the Cuban government opened up virtually its entire economy to FDI, emphasizing a rich natural resource base, a well-trained and docile labor force, and a government partner that is willing to make deals that will result in substantial profit margins for foreign investors. This chapter examines Cuba's development strategy during the special period, focusing on case studies of joint ventures in the tourism, nickel mining, and oil production sectors. It highlights the role of foreign investment in these sectors and sets out some of the environmental implications.

Key Words: Cuba; environmental policy; pollution; foreign direct investment; tourism; nickel mining; oil production

1. INTRODUCTION

In August 1990, at a time when the socialist bloc was crumbling, President Castro declared an economic austerity program aimed at 'resisting' change and remaining in power. Cuba was said to be entering a 'special period in time of peace' (*período especial en tiempo de paz*), a severe economic crisis triggered by the disruptions in imports of oil and other raw materials from socialist countries. The strategy to deal with the special period was to conserve energy and raw materials, increase food production, secure new markets for exports, attract foreign investment, particularly in the tourism industry, and implement managerial reforms.

This strategy was woefully inadequate in stopping the economic tailspin: between 1989 and 1993, Cuba's gross domestic product (GDP) contracted by 34.8 percent and GDP per capita by 36.7 percent; gross domestic investment as a percent of GDP fell from 26.7 percent to 5.4 percent; the fiscal deficit as a percent of GDP mushroomed from 7.3 percent to 33.5 percent; merchandise exports and imports declined by 78.9 percent and 75.6 percent, respectively; and production of almost all key agricultural and industrial products plummeted, causing sharp declines in consumption and in the population's standard of living.[1]

In the summer of 1993, the Cuban government implemented measures to revitalize the economy. Among these were the legalization of the holding and use of foreign currencies, the authorization to engage in self-employment in selected occupations, and the break up of state farms into agricultural cooperatives. This was followed in 1994 by authorization for the establishment of agricultural and artisan markets, and in 1995 by a new foreign investment law that sought to make the investment climate more favorable to foreign investors. As the economy recovered in the second half of the 1990s, new reform measures slowed down and eventually came to a standstill. While Cuba has recorded seven consecutive years of positive economic growth beginning in 1994, by 2000 the nation's GDP was still 15.4 percent below its 1989 level, and GDP per capita 20.5 percent below its 1989 level.

One of the success stories of the Cuban economy during the special period has been the attraction of foreign direct investment (FDI) in the form of joint ventures with domestic state-owned enterprises. From 1989 through 2000, Cuba reportedly attracted about $5 billion in foreign investment commitments,[2] of which about one-half has been actually released by foreign investors. While this is a fairly small volume of FDI in comparison with flows to other countries in the region, it is very significant for Cuba since the island is essentially shut out from international credit markets as a result of its default on hard currency debt in 1986. In addition to its salutary effect on the balance of payments, FDI has also provided Cuba with new technology, management expertise, and access to foreign markets for its goods and services exports.

In its efforts to attract FDI, during the special period the Cuban government has opened up virtually its entire economy to foreign investors,[3] emphasizing a rich natural resource base, a well-trained and pliable labor force, and a government partner that is willing to make deals that will result in substan-

---

1. These statistics are taken from Pérez-López (2001:46).
2. Interview with Minister of Foreign Investment and Economic Cooperation Marta Lomas, in Veloz (2001).
3. The only sectors of the economy in which foreign investment is explicitly prohibited are health, education services, and national defense (other than commercial enterprises of the armed forces). See Ley No. 77 (1995), Article 10.

12. Cuba's Development Strategy

tial profit margins for the foreign investor. In particular, Cuba has aggressively sought foreign investment to expand its relatively underdeveloped tourism, mining, and oil sectors, creating the potential for severe environmental disruption unless environmental concerns are taken into consideration at all stages of project design, development, and implementation.

An equally attractive enticement for foreign investors seeking profitable opportunities in natural resource-based industries has been the absence or lax application of environmental standards in a cash-strapped Cuba willing to bend backwards to attract foreign capital. Only in 1995 did the Ministry of Science, Technology and the Environment (*Ministerio de Ciencia, Tecnología y Medio Ambiente*, CITMA) issue regulations for environmental impact assessment and the granting of environmental licenses (CITMA, 1998:26). But despite these regulations and the request and issuance of hundreds of licenses, CITMA was complaining that by 1998 few people in Cuba regarded environmental licenses as requirements to be adhered to. Further, as shown in this paper, there is evidence that environmental regulations are feebly enforced and foreign investors in priority sectors may be allowed to pass along to the Cuban State certain environmental costs.

Penalties even for flagrant environmental abuses are relatively inconsequential. A rare instance reported in the Cuban press describes the light slap managers of a beach hotel in Trinidad Province, Southern Cuba, received when it was discovered they had illegally removed several tons of sand from a nearby beach after Hurricane Michelle struck in 2001. The penalty was to return the sand and restore the beach. Only when an investigation revealed that the amount of sand removed far exceeded what the culprits had admitted, were the Cuban hotel managers expelled from Cuba's Communist Party, and the foreign investor received only a verbal reprimand. In addition, the journalist reporting the story found that in this particular region, environmental standards were virtually ignored. Violations included building tourism facilities without environmental licenses, disrupting sand dunes, and using motor vehicles on beaches (Peláez, 2002).

Not to be ignored either is the adverse environmental impact that the pervasiveness of corruption in Cuba may have. In Cuba's centralized economy of scarcity, administrative corruption is rampant.[4] Officials empowered with discretionary power to apply laws and regulations may be willing to look the other way for a few dollars, or in exchange for consumer goods. Corruption is more likely to be prevalent in the tourism sector, where Cubans and foreigners interact on a daily basis and where market exchanges and a culture of service are tolerated. Environmental damage may be caused, for example, by corrupt

---

4. For a treatment of administrative corruption in socialist Cuba see, e.g., Pérez-López (1995a).

inspectors who knowingly overlook a normative violation, or by a hotel worker who sells an endangered species to a foreign tourist.

Our central argument *is not* that FDI has located in Cuba because of weak environmental standards and lax implementation.[5] Rather, we argue that Cuba's socialist development model is responsible for environmental disruption in the island. However, to the extent that it has accelerated the growth of natural resource-based industries, FDI has compounded the environmental stresses associated with socialist development. Our hypothesis is that had Cuba implemented these development projects without foreign investment, the environmental results would have been similar.

This chapter begins with a brief discussion of Cuba's development model and its impact on the environment. This is followed by case studies of the tourism, nickel mining and oil production sectors that highlight the role of foreign investment and establish some of the environmental implications. It concludes with observations regarding environmental disruption during Cuba's special period and suggestions for further research.

2. CUBA'S DEVELOPMENT MODEL AND THE ENVIRONMENT

In the early 1960s, Cuba's revolutionary government adopted socialism and began to replicate the institutions and practices that directed the economy in the former Soviet Union and the countries of Eastern Europe. The essential elements of this model were: (1) government ownership over the means of production; (2) centralized decision making; (3) allocation of resources via a central plan; (4) state monopoly of foreign trade; and (5) a development strategy that emphasized rapid industrialization.

Although advocates of socialism argued on theoretical grounds that environmental disruption could not occur in a socialist society, the reality was very different. Writing in the early 1990s, a scholar put it as follows: "One of the ironies of the former centrally planned economies, we have come to learn, is how little they cared about protecting their environment. Such rapacious behavior should not be so prevalent in societies whose purported objectives were defined in terms of the social rather than the private good. Nevertheless, virtually every one of the countries of Central and Eastern Europe is confront-

---

5. The 'pollution haven' hypothesis refers to the possibility that multinational firms, particularly those engaged in highly-polluting activities, relocate from developed countries with strong environmental standards to developing countries with weaker standards. Despite the popularity of this hypothesis, there is little evidence in the literature to support it. See, e.g., Smarzynska and Wei (1991).

ing a Herculean task in slowing down the rate of pollution and cleaning up decades of environmental neglect" (Bohi, 1994:vii). As Feshbach and Friendly (1992:1) described it with regard to the Soviet Union:

When historians finally conduct an autopsy of the Soviet Union and Soviet Communism, they may reach the verdict of death by ecocide. ... No other great industrial civilization so systematically and so long poisoned its land, air, water, and people. None so loudly proclaiming its efforts to improve public health and protect nature so degraded both. And no advanced society faced such a bleak political and economic reckoning with so few resources to invest toward recovery.

Although Cuba eagerly implemented the socialist model, the environmental consequences of socialism in Cuba through the late 1980s differ in several respects from those of the former Soviet Union and Eastern European socialist countries. In socialist Cuba, industrial pollution was a relatively minor problem compared to the environmental disruptions that occurred in the agricultural sector. The principal factor contributing to this divergence in environmental paths was Cuba's initial weak industrial base and its assigned role within the socialist division of labor: a producer of primary goods for the more industrialized economies within the socialist bloc.

The specific features of the socialist development model varied from country to country, according to political, cultural, and national circumstances, but the basic blueprint was inspired by the Soviet Union's historical experience. The essential characteristics of the agricultural organizational model that emerged in the Soviet Union and later adopted by other socialist countries were: (1) large-scale production units, what some analysts referred to as 'farm gigantism' (Volin, 1962:254); (2) extensive cultivation, which required bringing additional land under cultivation in order to increase production (Willett, 1962:101); (3) mechanization; (4) technological interventions to solve bottlenecks in the agricultural sector (Timoshenko 1953:254; Akiner, 1993:256); and (5) heavy use of chemical inputs, especially fertilizers, pesticides and herbicides.

Cuba's application of the socialist capital-intensive agricultural development model, with its very large farms and heavy reliance on heavy equipment to work them, has led to soil compaction, and the excessive application of chemical inputs has contributed to a litany of maladies, ranging from contamination of soils and water bodies to problems with secondary pest infestations. There are alarming reports of widespread erosion, but assessing its extent, severity, and consequences must wait for carefully conducted studies of soil conditions in different regions.

The attempt to conquer nature and increase agricultural yields drove an ambitious water development program, based on the construction of hundreds of large and small dams for irrigation, that has contributed to the salinization of many of the country's soils and underground water resources. Major culprits were inattention to proper drainage of irrigated fields, saltwater intrusions due to the excessive extraction of subterranean waters from aquifers, and tampering with the natural flow of rivers and streams. In some of Cuba's rivers, seawater reaches the walls of inland dams. Fourteen rivers in the southern plain of Pinar del Río province had their sources of water affected by the construction of 18 dams. Such projects, coupled with excessive withdrawal of underground waters and loss of forest cover, have contributed to the desertification of the plain, with desertification reportedly advancing eastwardly in Pinar del Río province by 11 kilometers annually. The municipality of Sandino has been particularly affected, with sand overtaking fields that were formerly used for tobacco production and many native plants disappearing because of the excessive acidity of the soil (Pedraza Linares, 1997). Desertification is also reported in the Cauto River basin and other Cuban regions (Alfonso, 2001).

Surface waters have been contaminated by the runoff of chemicals associated with the use of fertilizers, pesticides and herbicides. There is also concern about the long-term consequences of some of the water development projects since it has been posited that they could accentuate the destructive effects of natural disasters, such as hurricanes.

A question mark remains regarding the potentially adverse health effects that socialist development policies, particularly the use of chemical inputs in agriculture, may have had in contaminating the national water supply, and thus on the nation's health. Recent health statistics and epidemiological studies are not available to assess potentially negative trends (although some limited data sets suggest a deterioration of health standards). However, there are numerous press reports suggesting that high water pollution levels may be having adverse impacts on the health of the population (e.g., Orrio, 1993; León Moya, 2002; Hernández, 2002; Gálvez, 2002).

Many of the adverse environmental impacts associated with the application in Cuba of the socialist agricultural model have come to a head in a particular region of eastern Cuba, the Cauto River basin. This region covers about 8 percent of the country's territory and accounts for 9.3 percent of agricultural land; it grows 25 percent of the rice and 10 percent of the sugar produced in the country. According to a scientist who studied the environmental problems of the basin, "the depletion of the environmental quality of Cauto River basin over the last three or four decades is a serious national problem. The damage covers a broad range of ecological issues. One third of the basin suffers from severe erosion; salt water intrusions have spoiled most of the groundwater reservoirs; the natural runoff has been reduced by 60 percent in recent decades; the forested areas have been nearly wiped out" (Portela, 1997:11). In addition, water in the basin is heavily polluted by discharges from 652 sources, including farms and sugar mills, in the provinces of Santiago de Cuba, Holguín, and Las Tunas (Cepero, 2001a).

3. FOREIGN INVESTMENT IN NATURAL RESOURCE-BASED INDUSTRIES

With the fall of socialism in Eastern Europe and the dissolution of the former Soviet Union, in the early 1990s Cuba lost its main trading partners and was forced to change drastically its development model, which relied heavily on subsidies and economic aid from the former socialist countries. One of the changes that the Cuban leadership introduced was an opening to foreign investment. Available information suggests that three natural resource-based industries—tourism, nickel mining, and oil exploration and production—have been quite successful in attracting foreign investment during the special period.

According to official statistics, there were a total of 392 joint ventures with foreign investors active at the end of 2000.

- Thirty of these joint ventures were in the development and construction of hotels (with capital from Spain, Canada, France and Mexico, among other countries) and of port facilities for cruise ships catering to international tourists. In addition, 17 foreign corporations managed 50 international tourism hotels (Banco Central, 2001:37).
- There are no recent statistics on the number of joint ventures in mining, but according to the Ministry of Foreign Investment and Cooperation (*Ministerio para la Inversión y la Cooperación Extranjera*, MINVEC), at the end of 2000 there were 92 joint ventures in the heavy industry sector (which includes mining as well as oil exploration and production).[6] In 1994 a Canadian company and a Cuban enterprise entered into a joint venture to mine and process nickel at Moa, in eastern Cuba (Pérez Villanueva, 2001:19); other foreign companies have also considered investing in the Cuban nickel industry, but these investment plans apparently have not materialized (Pérez-López, 1995b).
- At the end of 2000, there were 20 joint ventures between foreign and Cuban enterprises in oil exploration and extraction (Banco Central, 2001:37; Oramas, 2001).

---

6. Cited in Spadoni (2001:22).

3.1 Cuba's Environmental Regulatory Framework

On paper, Cuba has a well-developed legal framework of environmental protection. Cuba's socialist Constitution, as amended in 1992, recognizes the duty of the state and of all citizens to protect the environment and incorporates the concept of sustainable development.[7] Cuba first adopted a broad and general environmental law in 1981. In 1997, Cuba enacted Law No. 81, a statute that emphasizes prevention, promotes participation by the community and all citizens in environmental actions, and emphasizes sustainable development. The statute also contains other innovations, such as the use of economic tools (such as taxes, fees, and differential pricing) to promote environmental objectives, and a broad range of sanctions against polluters. Cuba has also ratified over two dozen international agreements, conventions, or protocols dealing with wildlife and habitat, oceans, atmosphere, hazardous substances, occupational safety and health, and regional environmental concerns.

In practice, environmental protection in Cuba is weak. Cuban environmental laws are general and difficult to enforce. Following the practice of other (former) socialist countries, the Cuban economy is controlled by sectoral ministries that have both promotional and regulatory functions. When conflicts arise over environmental matters, the feeble environmental protection institutions are routinely overruled by sectoral ministries that are able to appeal to the imperative of promoting economic objectives. This has been particularly the case during the special period, when the economic crisis threatened the very survival of the regime. Finally, Cuba has the dubious achievement among socialist countries of not permitting independent environmental organizations to operate openly. The lack of legal independent environmental organizations, coupled with governmental monopoly over the mass media, means that Cuban citizens are very limited in their ability to influence actions that affect the local and national environment.

When economic and environmental priorities are placed on the balance, the scale tips in favor of the former. When decisions had to be made regarding the construction of causeways linking Cuba to Cayo Coco to facilitate the development of tourism (more on this below), Cuban scientists proposed that the road link be constructed using bridges spanning gaps between naturally-existing landmasses, thereby permitting water circulation vital to the survival of marine species. The description of what transpired is instructive:

> The construction of these bridges was deemed too expensive by the Cuban government. Because an agreement could not be reached within COMARNA's [Comisión Nacional de Protección del Medio Ambiente

---

7. For description and analysis of the legal framework of environmental protection in socialist Cuba see Díaz-Briquets and Pérez-López (2000), especially Chapter 3.

y del Uso Racional de los Recursos Naturales] meeting, the decision was deferred to the Council of Ministers. An important official then intervened by pointing out that the planned road/bridge structure did not follow a direct route. He then proceeded to take out his pen and draw a straight line from Cayo Coco to the nearest point on the mainland.

The design was adapted, and a straight, bermed road with intermittent underwater tunnels was constructed. Scientists argued that there were too few underwater tunnels to maintain natural water flows. Through negotiations, they were able to double the number of passages, a small victory considering their original opposition to the plan.

This example illustrates that while COMARNA was able to settle minor issues itself, it did not posses the authority to make a final decision regarding controversial matters. When a project was deemed highly attractive, the environmental protection system could be manipulated to serve a more important agenda. In this case, development and the need to attract foreign investment prevailed (Collis, 1995:2).

3.2 Tourism

In the early years of the revolution the Cuban regime shunned international tourism for ideological reasons. As an expert analyst puts it,

Tourism was perceived as too closely associated with the capitalist evils of prostitution, drugs, gambling and organized crime. The revolutionary government discounted tourism as a vehicle for economic growth and development. During the 1960s and 1970s, no major investment in tourism was undertaken. The vast tourism infrastructure built up during the pre-revolutionary years was left for the use of Cuban citizens and international guests from socialist and other friendly countries or simply abandoned. Some sixteen hotels were shut down and hotel capacity declined by 50 percent (Espino 1994:148).

In the second half of the 1980s, Cuba began to develop its neglected international tourism industry focusing on participation of foreign firms, establishing two corporations, Cubanacán S.A. and Gaviota S.A. (the latter reported to have connections with the Cuban armed forces), to develop joint ventures with foreign investors. In the 1990s, Cuba intensified its efforts to attract foreign investment in the tourism industry.

Cuba has been quite successful in developing an international tourism industry. International tourist arrivals nearly quadrupled between 1984 and 1994 (from 158,000 to 619,000 arrivals). For most years in the 1990s (see Table 12-1), tourist arrivals grew at double-digit rates, with growth rates exceeding 20 percent in several years. Cuba achieved 1 million foreign visi-

Table 12-1. International Tourism Industry Indicators, 1990-2000

	1990	1991	1992	1993	1994	1995	1996	1997	1998	1999	2000
Tourist arrivals (000)	340	424	461	546	619	745	1004	1170	1416	1603	1774
Growth rate (%)	--	24.8	8.7	18.4	13.4	20.4	34.8	16.5	21.0	13.2	10.7
Convertible currency revenue (million USD)	243	402	550	720	850	1100	1333	1515	1759	1901	1948
Growth rate (%)	--	65.4	36.8	30.9	18.1	29.4	21.2	13.7	16.1	8.1	2.5
Occupancy rate (%)	NA	NA	NA	57.9	59.1	62.9	64.9	75.4	76.1	71.7	74.2
Employment (000)	39.7	43.0	41.9	43.9	46.1	52.3	56.2	53.6	62.8	66.0	68.0
Share of total employment (%)	1.1	1.1	1.1	1.2	1.2	1.5	1.6	1.5	1.7	1.7	1.7
No. of hotels and motels	328	347	346	356	331	348	375	400	412	424	431
Growth rate (%)	--	5.8	0.0	2.9	-7.0	5.1	7.8	6.7	3.0	2.9	1.7
No. of rooms in hotels and motels	18565	20816	23221	24262	24884	27236	29161	1043	34891	36252	37178
Growth rate (%)	--	12.1	11.6	4.5	2.6	9.5	7.1	6.5	12.3	3.9	2.6
Daily lodging capacity in hotels and motels	37740	1633	44956	49433	50410	50562	59350	62674	70222	74620	75869
Growth rate (%)	--	10.3	8.0	10.0	2.0	0.3	17.4	5.6	12.0	6.3	1.7

Note: Employment statistics are from the Comisión Económica para América Latina y el Caribe (CEPAL).
Sources: 1990-92: Statistical Annex, CEPAL (2000); 1993-2000: ONE (2001).

12. Cuba's Development Strategy

tors in 1996 and was poised to double that number (to 2 million visitors) by 2001; in 2000, nearly 1.8 million visitors arrived in the island. As a result of the contraction in international travel following September 11, the 2 million visitors target was not reached, with the number of arrivals in 2001 about the same as in 2000. In 1998, Cuba was fourth in the Caribbean with regard to international tourist arrivals, behind Puerto Rico, the Dominican Republic, and the Bahamas, three nations that have been promoting international tourism for a much longer period than Cuba. Revenue generated by tourism in Cuba grew by leaps and bounds during the 1990s, from $340 million in 1990 to $1.9 billion in 2000. In 1993-94, revenue from international tourism first exceeded sugar export revenues; this trend has continued, with the gap between the two growing over time.

The tourism infrastructure expanded greatly in the 1990s. The number of hotels and motels grew from 328 in 1990 to 431 in 2000, or by 31 percent. Meanwhile, the number of rooms in these facilities more than doubled, from 18,565 in 1990 to 37,178 in 2000. In 1998, Cuba's stock of hotel and motel accommodation facilities was surpassed in the Caribbean region only by the Dominican Republic.[8] As has been mentioned above, joint ventures with foreign investors have been a key source of capital, management expertise, and markets for the international tourism industry.

The adverse environmental impact on the Caribbean region of unbridled mass tourism is well documented. An analyst who has studied the tourism industry in the region, with its focus on increasing tourist arrivals, describes the environmental impact as follows:

> The catalogue of environmental destruction directly attributed to the growth of the tourist industry [in the Caribbean] is long. It includes the erosion of beaches, the breakdown of coral reefs, marine and coastal pollution from watersports, the dumping of waste and the non-treatment of sewage, sand mining and the destruction of wetlands and salt ponds. In many cases, the impact is interrelated, locked into a chain of tourist development where short-term gain takes precedence over long-term protection. For example, a hotel cuts down coastal trees to improve the view from its bedrooms; this accelerates coastal erosion and sand loss; then, when a jetty is built for a new dive shop even more sand is lost because sand from the newly shaped beach is washed on to the coral reef. The result is two-fold: the sandy beach has become smaller and the marine environment has been spoiled (Pattullo, 1996:105-106).

The Cuban government claims to be sensitive to the fact that growth in the tourist industry depends not only on the availability of more and better facili-

---

8. Based on statistics of the World Tourism Organization, as cited by CEPAL (2000:513).

ties, but also on preserving the country's natural tourist attractions. However, in expanding its tourism infrastructure—hotels, recreation facilities, roads, airports, and so on—Cuba has often emphasized speed and low cost to the detriment of the environment. Cuban environmental official Helenio Ferrer has been quoted as saying that "tourism, like nuclear power, is like chemotherapy for us. It is something we need, yet it is something that is bad for us." Ferrer went on to say to a foreign visitor that "we built a seaside resort and the beach disappeared," reinforcing the dependence of tourism on the preservation of natural resources and illustrating a common problem in the Caribbean region when tourism industry developers tamper with the environment.[9]

With the emphasis on sun, sea, and sand tourism, in the 1990s Cuba has developed several of the country's pristine outer cays and islets. The northern cays of Ciego de Avila province, for example, did not have a single hotel room in 1993. By 1997, they had over 1,500 rooms, an airport, more than 200 kilometers of roads, electricity generation plants, and 100 kilometers of pipes to supply waters from the mainland; 6,000 rooms were expected by 2000 (Bohemia article, 1997). Promotional materials from the Cuban government advertise tourism facilities at Cayo Coco and Cayo Guillermo, in the Sabana-Camagüey archipelago, as the largest tourism complex in the Caribbean, with over 1,700 hotel rooms, managed through joint ventures between Spanish corporations Tryp Hotels, Sol–Meliá, and RIU, the Italian corporation Viaggi del Ventaglio, and Cuban enterprises Cubanacán S.A. and Gran Caribe (Visión real, 1998).

To allow access by tourists to beaches in the numerous pristine small cays that surround the island, particularly in the northern coast, Cuban tourism authorities have constructed causeways (stone embankments), called *pedraplenes*, bridging barrier islands to the mainland and to each other. These *pedraplenes* block the movement of water in the intracoastal waters, exacerbating contamination and destroying coastal and marine habitats (Espino, 1992:335). Many of these semi-enclosed water bodies were already subject to weak circulation regimes and high organic matter contents (Alcolado, 1991). The disregard for the environmental consequences of the *pedraplenes* is exemplified by the words of President Fidel Castro, who is quoted as instructing construction workers: "Here the task is to dump stones and not worry about it."[10]

In all, the *pedraplenes* affect over 1,760 square kilometers of fragile ecosystems of the Sabana-Camagüey archipelago, contributing to the destruction of natural fauna and flora and the creation of lagoons of stagnant, polluted

---

9. Both quotes are from Kaufman (1993:33).
10. The quote in Spanish is "aquí hay que tirar piedras y no mirar para adelante." Cited in Cepero (2001:93).

water devoid of wildlife (Wotzkow, 1998:138-144). Among the *pedraplenes* that have caused substantial harm to the environment and marine resources are those joining the islands of Turiguanó and Cayo Coco and the northern region of Ciego de Avila province (Solano, 1995).

The Sabana-Camagüey archipelago ecosystem is a conservation priority region due to its biodiversity and economic importance as Cuba's second richest fishing ground (Guevara et al., 1998:48). The extensive environmental damage the archipelago has experienced as a result of the crash development of the tourism infrastructure is officially acknowledged (CITMA, 1999:48). The conclusions of studies conducted by Cuban scientists under the project 'Biodiversity Protection and Sustainable Development in the Sabana-Camagüey Archipelago Marine Ecosystem,' sponsored by the United Nations Development Program, are worrisome. One such study, a longitudinal environmental assessment of the San Juan de los Remedios and Buenavista bays, provides conclusive evidence that the Caibarién-Cayo Santa María *pedraplén* severely disrupted the marine ecosystem. It caused significant hydrological changes, including a major increase in water salinity that could potentially lead to the disappearance of 60 marine species, and was also responsible for accelerating water flows, the speed of marine currents tripling (Pérez Santos et al., 1998:60-61). The scientists recommend that to mitigate the damage, several new bridges be built to facilitate water flow.

Similar conclusions emerged from detailed studies of the environmental impact of the 26.1-kilometer *pedraplén* joining the Isla de Turiguanó to Cayo Coco, also in the Sabana-Camagüey archipelago. In Bahía de los Perros, a body of water split in half by the *pedraplén*, where water salinity was high (54 percent) even before the structure was built, salinity has increased to as much as 80 percent. Because of high salinity and an associated increase in the concentration of suspended solid particles, the sea bottom has deteriorated perceptibly. Surrounding waters can hardly sustain marine life, as the natural food chain has been affected by the loss of algae and other life forms. The study concludes that the "passing of the years has once again shown that arbitrary natural world changes can be catastrophic over the long and the short term" (Romero Ochoa et al., 1998:107).

Another study, based on ten oceanographic expeditions conducted between 1991 and 1997 to assess the hydro-chemical characteristics of the archipelago's waters, including dissolved oxygen content, pH levels, and nutrient load, found that environmental damage in some areas was substantial (Penié Rodríguez et al., 1998). While natural conditions (e.g., low water flow in coastal lagoons) contribute to less than optimal biochemical conditions, the presence of *pedraplenes* aggravates this condition, already made more precarious by low coastal fresh water flows arising from the damming of rivers on Cuba's mainland facing the archipelago. If proper restoration and conservation measures are not implemented, the study warned, what was then localized damage could become generalized, leading to "an ecological catastrophe of great magnitude." By the late 1990s, the zones in the archipelago with the most adversely affected biochemistry were the Cárdenas, Jigüey, los Perros, and Nuevitas bays; both extremes of Buenavista bay; and the eastern portions of Fragoso key. Also damaged were the Caibarién coast; the waters adjacent to the town of Camacho in San Juan de los Remedios bay; and the North side of Sabinal, Guajaba, Coco, Guillermo, and Francés cays.

Damage to the extensive mangroves typical of the regional ecosystem is also reported to have been severe, with one leading student of Cuba's environment claiming that as a result of *pedraplén* construction, more than 5,000 hectares of mangroves were lost (Cepero, 2001b:93). Cayo Coco, site of several hotels and tourism facilities developed with foreign investment, is the most significant breeding ground of the roseate spoonbill and the greater flamingo, species that have been severely affected by human activity (Silva Lee, 1996). Reportedly, several colonies of flamingos that used to nest in the Sabana-Camagüey archipelago have left this area because of the destruction of their habitat associated with tourism development and the construction of *pedraplenes* and settled in the Bahamas, much to the delight of the tourism industry of that country (Wotzkow, 1998:144). The famed corals of Northern Cuba are also under increasing stress—most in a critical state—due to global climatic changes and man-made causes, some of them perhaps associated with the rapid rise of Cuba's archipelagos as international tourist destinations (Peláez, 2001a).

Other reports indicate that in several of the tourist enclaves established in the Sabana-Camagüey archipelago, particularly in Cayo Coco, beach erosion is a major concern, with a 1999 report suggesting that in Playa Las Coloradas, the beach practically disappeared following a storm. An evaluation by Cuba's *Centro de Investigación de Ecosistemas Costeros* determined that natural erosion processes were aggravated by the construction of hotels and other tourism infrastructure on coastal dunes (Cubaeco, 2001a). To arrest the beach's decline, in late 2001 it was announced that a barrier would be built under the sea to minimize the erosion effect of sea waves (Cepero, 2001c). An indication that these practices are commonplace, despite official claims that environmental standards are being enforced, was the partial collapse in mid-2000 of the five-star hotel 'Las Terrazas,' operated by the Spanish firm Sol–Meliá, in Cayo Guillermo. The site for this hotel was partly reclaimed from a swampy beach area that was filled with earth removed from the key's interior (Hotel, 2001).

Unsustainable exploitation practices along Cuba's beaches are nothing new, however. Several reports describe how in the 1990s, Cuba's tourism authorities were forced to allocate scarce financial resources to restore beaches that in earlier decades were severely damaged by shortsighted sand mining practices to supply the construction industry. Even the famed Varadero beach was subjected to this practice, as between 1968 and 1978 more than one million cubic meters of sand were removed from the insular platform facing the resort (Peláez, 2001b). Similar restoration efforts have been reported in two beaches (Pesquero Viejo and Don Lino) in Holguín province slated for tourism development (Cubaeco, 2001b). Also, during the construction of the tourist infrastructure, the cays' natural forest vegetation has been disturbed and numerous pits dug for construction materials. The latter, plus inattention to preservation of natural geographic features, road building, and construction of excessively large and heavy tourist facilities, are believed to be interfering with the region's surface water flows (Cubaeco, 1999).

Human pressure on the fragile archipelago ecosystem will continue to intensify as new tourism development poles come on line. In late 2000, a feasibility study was underway to begin developing Cayo Paredón Grande, also in the Sabana-Camagüey archipelago (González Martínez, 2001). With the completion in Cayo Coco in early 2002 of an international airport capable of handling one million passengers a year, coupled with a modern barge port completed two years earlier to serve the tourist industry, the number of visitors will increase many-fold (González Martínez, 2002). One irony related to the port is that it was built to relieve transportation pressures on the *pedraplenes* that were constructed, to begin with, to facilitate access to the new tourist resorts. There is fear that the operation of this port and its future expansion may further harm the regional ecosystem (Cosano Alén, 2000).

New resorts are under construction or sites are being evaluated for possible tourism development that impinge on other relatively untouched Cuban ecosystems, such as several cays in Cuba's southern coast, beaches along the northeastern coast of Holguín province, and in the Guanahacabibes peninsula in westernmost Cuba. This peninsula—sections of which have been off limits to the average Cuban for years—was declared a Biosphere Reserve in 1987, and as such received some protection (Arroyo, 2002). Five of its 22 pristine beaches have been slated for development, current plans calling for the construction of between 1,500 and 1,600 hotel rooms in the area (Suárez Ramos, 2001).

Another concern of independent Cuban environmentalists regarding the Guanahacabibes peninsula is the construction of access roads and the establishment of a service marina to cater to the more than 12,000 pleasure and commercial vessels that cross the eastern edge of the Yucatán Strait annually. While the government claims that information gained during more than ten years of studies will be used to preserve the ecosystem, there is justified room for skepticism.

3.3 Nickel

Cuba's nickel reserves are the world's fourth largest, and its reserve base the largest. In 1999, Cuba was the world's sixth largest producer of mined nickel (Torres, 1999). Cuba's nickel deposits, concentrated in the Nipe Mountain Range, on the north coast of Holguín province, occur in the form of laterites. Laterites, which typically contain nickel mixed in complex fashion with other minerals, tend to occur near the land surface and are amenable to strip mining. Cuban laterites contain about 1.3 percent of nickel as well as minor concentrations of other metals, particularly cobalt. To obtain refined nickel, laterites must be treated either with pyrometallurgical (smelting) or hydrometallurgical (leaching) processes.

The development of Cuba's nickel mining and processing industry was closely connected with the strategic defense needs of the United States during World War II. The first processing plant, at Nicaro, began operations in 1943. It was built by agencies of the U.S. Government and operated by a subsidiary of the Freeport Sulphur Company. The production process used in the plant—ammonia leaching—permitted the recovery only of nickel and some cobalt, in the form of nickel-cobalt oxides and sinter, with iron and chromium disposed of as waste products. The Nicaro plant ceased operations in 1947, as demand for nickel subsided, and was not reopened until 1952 as a joint venture between the National Lead Company and Cuban investors. This plant is currently called 'Comandante René Ramos Latour.' In 1953 the Freeport Sulphur Company began construction of a new nickel production plant at Moa Bay, east of Nicaro on the north coast. The Moa plant relied on a different technology—sulphuric acid leaching—to process the nickel ore. The plant began to undergo production testing in 1959; it was nationalized by the Cuban government in 1960 and began operations in 1961. This plant is currently called 'Pedro Sotto Alba.' Cuba's revolutionary government began construction of two new nickel plants in the 1980s in the same geographic area, one at Punta Gorda (the 'Comandante Ernesto Che Guevara' plant) and the other at Las Camariocas (the 'CAME-I' plant). Both of these plants relied on the ammonia leaching process used at Nicaro. The Punta Gorda plant began commercial production in 1987.

Canadian nickel-cobalt producer Sherritt, Inc., began to purchase nickel and cobalt sulfides produced by the Moa plant in 1991 for further processing at its refinery in Fort Saskatchewan, Alberta, Canada. In 1994, Sherritt and Cuba's *Compañía General del Níquel*, the state-owned entity that manages the Cuban nickel industry, created a joint venture to develop and market Cuban nickel. According to the agreement, both the Moa Bay nickel production plant and the Fort Saskatchewan refinery are owned by the joint venture. The Cuban government granted the joint venture mining concessions to sup-

12. Cuba's Development Strategy

ply the Moa Bay plant for 25 years. Reportedly, Western Mining Corporation of Australia and the Cuban enterprise Commercial Caribbean Nickel S.A. signed a letter of intent in 1994 to form a joint venture to assess developing a nickel deposit at Pinares de Mayarí with the intention of building a new production plant at that location if the feasibility justified such an investment. Apparently the results of the feasibility study were not positive and the joint venture did not materialize.

Spearheaded by the joint venture with Sherritt, the Cuban nickel industry has performed well in the 1990s. Nickel production fell from 46,600 metric tons in 1989 to 27,000 metric tons in 1994, consistent with the sharp drop in output during the special period. By 1995 production was close to the level in 1989 and has expanded steadily since then, reaching 71,400 metric tons in 2000, and a record 75,000 metric tons in 2001.

As mentioned above, Cuban laterite deposits occur very close to the surface and are mined using strip mining techniques. Typically, a thin strip of subsoil is removed from the mountainous areas where the deposits are located, and then dragline shovels and heavy mining equipment extract huge volumes of raw materials that are transported to the processing plants via trucks and/or conveyor belts. Areas that have been mined have the appearance of 'lunar landscapes,' checkered with craterlike depressions where the mining equipment has dug deep (Núñez Jiménez, 1980).

One of the most visible environmental disruptions caused by strip mining can be mitigated by reclamation, that is, returning the land as closely as possible to the original condition, by leveling the area, covering it with topsoil, and planting a vegetable cover. Although there is evidence of reclamation activities in certain strip-mined areas, it appears that this practice is not generalized. According to Cuban officials, with regard to the Moa Bay region, reclamation is not a priority "since there is still usable mineral, although its exploitation will require the construction of a plant with a different technological process" (Pozo, 1980:8). Environmental official Helenio Ferrer has stated that "some reforestation has been done, but at a slower pace than the mining itself. Reforestation is more complex [in strip mined areas] ... since it involves bringing back the soil and guaranteeing proper fertilization as well as planting" (Reed, 1993:53).

When exposed to moisture, the strip mined areas tend to give off acids that are carried by rainwater and pollute nearby streams and rivers. Reportedly, heavy erosion from surface mining is filling Moa Bay with earth (Knox, 1995). Ian Delaney, chairman of Sherritt, Inc., has stated that Moa Bay's worst pollution comes from erosion after ore is stripped from the hill sides: "[Cubans] have not reforested to the extent that they should have" (Knox, 1995).

Cuba's nickel plants are significant sources of air, land, and water pollution (see Table 12-2). The ammonia-leaching technology used at Nicaro (and replicated without improvement at Punta Gorda and Las Camariocas) "proved to be one of the most polluting processes that ever went into operation due to excessive dust emission" (Habashi, 1993:1173). The towns of Nicaro and Moa are often engulfed in dust and smoke clouds. The situation in Nicaro is so bad that "many residents of the area suffer from acute respiratory illnesses" (García, 2001). Several independent reports affirm that some physicians have recommended to patients with respiratory problems, of which there are many, to move away from the region (Garcell, 2001).

The Moa Bay plant is prone to leaks of toxic hydrogen sulfide into the atmosphere (Habashi, 1993:1173). A worker at the plant reported that "the rain in Moa is acidic enough to produce *picazón* (a rash) when it falls on bare skin. ... He and other residents said children tend to suffer from respiratory problems such as asthma and persistent coughing" (Knox, 1995). A report by CITMA (1999:3-4) states that a trend toward increase in acid rain has been observed in several parts of the country during the 1990s. Most seriously affected are Cuba's major urban agglomerations and the mining regions of Holguín province. In order to achieve "natural" rainfall acid levels, it would be necessary to reduce SO_2 and NO_2 emissions by anywhere between 30 and 80-100 percent, the latest values applying to the Mariel and La Habana urban areas, as well as the mining regions of eastern Cuba (García, 2001; Garcell, 2001).

Residues from the ammonia leaching process, in the form of slurry, are disposed of in holding ponds. Because the bottom of the ponds is not impermeable, water quickly drains underground, leaving a solidified mass that amounts to 2 million tons of material per year for each plant. The water that seeps below the surface contaminates underground water deposits (Habashi, 1993:1177). Other unofficial reports speak of poorly constructed and managed oxidation and retention pools that, because of overflowing, discharge heavy metals into the sea and pollute the surrounding coastline (Wotzkow and Casañas, 2002).

The nickel plants also discharge large quantities of pollutants into streams, rivers, bays, and, ultimately, the ocean. In a study conducted by Cuban scientists in 1985-86 of a 72.5-square kilometer marine area in the eastern edge of Holguín province to assess sedimentation rates, contamination levels were alarmingly high, with acid and metallic levels exceeding tolerable limits for release into the environment. Concern was expressed about potential contamination of the food chain. Moa Cay bay was most impacted, but high levels of contamination were also reported in Yaguasey and Yagrumaje (Martínez Canal et al., 1998). There is no evidence of the implementation of measures to limit heavy metal discharges into the sea since the study was completed, and

12. Cuba's Development Strategy

Table 12-2. Main Sources of Pollution of Cuban Nickel Processing Plants

Facility	Range of ecological impact of pollution	Type and magnitude of pollution
'Pedro Sotto Alba' Moa Bay, Holguín Province	Moa area, national, and potentially international impact. The amount of H_2S launched into the atmosphere is capable of generating acid rain in the eastern part of Cuba and its surroundings and spilling into Moa Bay. The coral reef barrier has suffered heavy damage and will be completely destroyed in about 10 years.	Great quantities of H_2S launched into the atmosphere. A huge amount of dust. Nearly 2 million tons of waste per year.
'Comandante René Ramos Latour' Nicaro, Holguín Province	Nicaro area and national impact. Significant damage to the towns of Nicaro, Felton, and Levisa. Waste in the marine shelf is affecting the environment from the mouth of the Mayarí river to Tánamo Bay.	Colossal quantities of smoke, soot, NH_3, noise, and vibrations. A huge amount of dust. Two million tons of technological waste per year in an open pond. Potential NH_3 discharge.
'Comandante Ernesto Che Guevara' Punta Gorda, Holguín Province	Local area and national impact. Significant damage to the town of Punta Gorda. Muck runs through the Punta Gorda river and out to sea.	Great amount of smoke and soot, NH_3, noise, and vibrations. A huge amount of dust. Two and one half million tons of technological waste per year. Potential NH_3 discharge.
'CAME-I' Plant (Not in operation) Las Camariocas, Holguín Provice	Most likely the same impact as the 'Comandante Ernesto Che Guevara' plant.	Most likely the same impact as the 'Comandante Ernesto Che Guevara' plant.

Source: Culled from Oro (1992:80-82).

therefore the environmental situation could only have worsened with the increase in nickel production.

A Cuban journalist reported in 1994 that Cuban scientists were working on a method to recover some of the minerals and chemicals produced as waste by the Moa Bay plant. According to this report, liquid waste produced by the plant "flows at the rate of 12,000 cubic meters a day and carries into the sea a wide range of light and heavy metals, such as sulfates and great amounts of sulfuric acid, which is used in the lixivating process of the laterictic mineral.

Every day 72 tons of aluminum, 48 tons of chrome, 15 tons of magnesium, and 30 tons of dangerous sulfuric acid get dumped into the sea. This harms the marine flora and fauna and, in the long term, could cause irreversible damage" (Nickel Plant, 1994).

A Soviet shipping and marine environment specialist described the situation in Moa Bay in the 1980s as follows: "Each day, more than 450 cubic meters of waste products from the nickel enrichment process are dumped into the ocean. When a ship arrives at that port, it is as if it were subject to a galvanic bath, with the hull being cleaned from any adhesions. For this reason, it is imperative to do something to preserve the nature at Moa" (Hernández, 1982).

The expanded nickel-mining operations of the joint venture with Sherritt, Inc. has put additional pressure on an already serious environmental situation. According to a Canadian journalist that visited the joint venture, "because of leaky equipment and other factors, the sulphur compounds used in the process pollute the air and water, producing what residents say is acid rain. Heavy erosion from surface mining is also filling Moa Bay with earth" (Knox, 1995). Residents of Moa told the same journalist that they took it for granted that "one of the reasons a foreign mining company would be interested in operating in Cuba was that environmental standards would be lower" (Knox, 1995).

3.4 Oil Exploration and Production

Commercial oil production in Cuba began in 1915 with the discovery of the Bacuranao field; another commercial field was discovered at Jarahueca in 1943. Oil production from these two fields was small, averaging about 4,000 metric tons per annum during 1950-54. The discovery of the important Jatibonico field in 1954 shut up production to an average of about 30,000 metric tons per annum during 1955-58 and gave rise to a flurry of concession applications and exploratory drilling activities by domestic and foreign companies. Small fields were subsequently discovered at Catalina, Cristales, and Guanabo. However, as most exploratory wells either turned up dry or found oil in quantities too small or of too low quality to justify commercial exploitation, the exploration boom subsided.

With financial and technical assistance from the Soviet Union and Romania, Cuba undertook an ambitious program aimed at boosting oil production. Output for 1960-67 averaged about 50,000 tons per annum, rose to over 200,000 tons per annum in 1968-69 when production peaked at the Guanabo field, and steadied at about 140,000 tons in 1970-74 as production declined at mature fields (such as Cristales and Jatibonico). Production from new fields east of La Habana (Boca Jaruco and Varadero), pushed output above 250,000 tons per annum in 1975-78.

12. Cuba's Development Strategy

In the 1980s and 1990s, domestic oil production increased greatly as newly-discovered fields were brought under production. Output for the first time reached 1 million tons in 1993 and exceeded 2 million tons in 1999. In 2000, oil production reached a record 2.7 million tons. The very rapid increase in domestic production has reduced Cuba's reliance on imported oil and oil products. Data in Table 12-3 indicate that domestic production as a percentage of apparent oil and oil products supply rose from 5-6 percent in the second half of the 1980s to the 15-20 percent range in the first half of the 1990s, and to a record 31 percent in 2000.

Since 1991, Cuba has permitted foreign companies to participate in oil exploration and production activities. With the assistance of these foreign companies, Cuba has conducted seismic work to identify potential oil deposits and drilled many exploratory wells. Forty-five shallow blocks along the north coast of Cuba were offered for exploration; foreign companies won the right to explore in 22 of them (CEPAL, 2000:470-471). Reportedly, foreign investors from Canada, France, the United Kingdom, Sweden, Spain, and Brazil invested over $600 million in oil exploration and production over the period 1991-2000 (CEPAL, 2001:7). President Castro has invited U.S. oil companies to explore and produce oil in Cuba under the same terms and conditions as companies from other countries (Invitan, 2001); U.S. companies are prohibited from doing so by the U.S. embargo on Cuba.

According to a Cuban analyst, out of the 14,000 square kilometers identified as most likely to contain oil deposits, 50 percent are located in the marine platform (Figueras, 1994:26). Exploration by foreign companies has led to the discovery of new fields in the northern basin, particularly at Varadero and Río Escondido; this basin contains high-density oil with a high sulphur content. Several exploratory wells have been drilled by foreign oil companies in the southern basin, without commercial success to date (CEPAL, 2000:471).

The drive to increase crude production has meant that drilling and production have been permitted in certain areas that were formerly considered environmentally fragile, such as coastal areas. In early 2000, Cuba offered 59 offshore blocks in its deepwater economic exclusive zone in the Gulf of Mexico for oil exploration by international oil companies.

The joint ventures approved with foreign firms to explore offshore risk the possibility of an oil spill that could harm the marine environment in Cuba and Florida (Rosendahl, 1991). This concern was made more immediate by the announcement by the joint venture of Spain's Repsol YPF and Cupet, its Cuban partner, that seismic studies would get under way in 2002. The area to be studied is located in the Gulf of Mexico between La Habana and Cárdenas, too close for comfort to the Florida keys, where there are several environmentally protected areas and tourism is a leading economic sector (Cepero, 2002). Oil released by an accidental spill from a drilling platform in this area could

Table 12-3. Apparent Supply of Oil and Oil Products (million metric tons)

	Domestic Production	Imports	Apparent Supply	Domestic Production/ Apparent Supply (%)
1985	0.9	13.3	14.2	6.3
1986	0.9	12.9	13.8	6.5
1987	0.9	13.3	14.2	6.3
1988	0.7	13.1	13.8	5.1
1989	0.7	13.1	13.8	5.1
1990	0.7	10.2	10.9	6.4
1991	0.5	8.1	8.7	5.7
1992	0.9	6.0	6.9	13.0
1993	1.1	5.5	6.6	16.7
1994	1.3	5.7	7.0	18.6
1995	1.5	6.2	7.6	19.7
1996	1.5	6.6	8.1	18.5
1997	1.5	7.1	8.6	17.4
1998	1.7	6.6	8.2	20.7
1999	2.1	6.0	8.1	25.9
2000	2.7	6.0	8.7	31.0

Source: Imports, domestic production and apparent supply: ECLAC (2001:25). Domestic production/apparent supply: Calculated from columns 1 and 3.

be rapidly transported to the U.S. mainland by the current that swiftly flows from north of Cuba to southeastern Florida (Agencia Ambiental, 2002). It goes without saying that such an oil spill could also put a severe damper on Cuba's tourist industry, where the major beach resorts are found in Varadero and the Sabana-Camagüey archipelago.

The location of Cuba's oil deposits—in the northern basin near areas that are suitable for tourism—has caused conflicts within the Cuban leadership. According to President Castro, he had to intervene personally to stop efforts to drill for oil in the Hicacos Peninsula (the seat of Varadero Beach), an activ-

12. Cuba's Development Strategy

ity that would have been harmful to the tourism industry. As President Castro said in 1990:

> Of course we need oil! But when we did the analysis and the calculations, no matter how much oil might be deposited under the [Hicacos] peninsula, it would not generate as much revenue for the country as tourism. I personally had to do the calculations, showing them the value of the 100,000 tons of oil that they planned to extract from this area. Moreover, oil is an exhaustible resource while resources such as sun, air, and sea are inexhaustible. Finally, the oil [under the Hicacos Peninsula] is a heavy oil, with high sulphur content, and a low value in the world market (Castro, 1992:10).

The financial pinch during the special period has reduced Cuba's ability to import high quality oil and has prompted the island to use its domestically-produced crude. According to the Minister of Basic Industries, in 2000, 50 percent of the electricity generated used domestically-produced crude oil; this share had risen to 70 percent by the first quarter of 2001 and was expected to reach 90 percent by the end of 2001 (Molina, 2001). This is not good news for the environment, as the high-density and high-sulphur Cuban crude is a heavy air polluter.

The decision to begin drilling for offshore oil deposits so close to Cuba's main sun, beach, and sand tourism destinations suggests that once again economic imperatives are taking precedence over environmental concerns. With the severe economic contraction that followed September 11—hard currency revenues contracted sharply as the number of tourist arrivals fell off, emigrant remittances took a dip, and international prices for Cuba's principal commodity exports (sugar and nickel) weakened—it is clear that the pressure to increase oil production at any cost will only intensify.

4. CONCLUDING OBSERVATIONS

Three foreign-invested sectors, tourism, nickel, and oil, have been the star economic performers during Cuba's special period. According to the United Nations Economic Commission for Latin America, in 2000 tourist arrivals grew by 10 percent, nickel production by 7.4 percent, and oil production by 26.2 percent (CEPAL, 2001:7). Rapid development of these sectors, intensified by foreign investment, has carried a substantial environmental price tag, with some regional ecosystems appearing to have been damaged.

Regarding tourism, some of Cuba's previously untouched ecosystems are being overwhelmed by poorly managed growth and lax implementation of environmental safeguards. The best documented instance of environmental deterioration is in the Sabana-Camagüey archipelago, where the health of beaches and surrounding waters is being tested by hotels sited in vulnerable locations and poorly designed *pedraplenes*. Recognition of the harm caused by these projects and flourishing speeches about the need to preserve the environment notwithstanding, Cuban authorities continue developing new tourism resorts with foreign participation.

In northern Holguín province, site of Cuba's nickel industry, ever increasing ore extraction operations have not been accompanied by the requisite investments to minimize air, water, and land pollution, and to reclaim strip-mined areas. A growing number of reports speak about contamination of water bodies and poor air quality that are associated with respiratory and skin ailments.

Of potentially serious consequences is the emphasis on oil exploration and production. Oil wells in Cuban waters must be drilled at considerable depths, in the neighborhood of the heavily-transited Florida Straits, and in treacherous seas. Oil spills could seriously harm tourism, the mainstay of the economies of Florida and Cuba. Moreover, the oil found in Cuba thus far is dense and high in sulphur content and its use (in electricity generation plants) worsens air pollution in La Habana and elsewhere in the nation.

The Cuban government's primary objective during the special period has been survival. In this pursuit, it has been eager to develop natural resource-based industries in a political and economic environment where the checks and balances of public participation, independent advocacy groups, and a free press are lacking. Except for scattered reports from some dissident scientists in the island, all information on the Cuban environment comes from government sources. Researchers can play an important role in monitoring and reporting on the environmental effects of investment projects with a view to providing an objective assessment of the environmental situation in the island.

REFERENCES

Agencia Ambiental Entorno Cubano, 2002, Planes petrolíferos cubanos en el Golfo de México amenazan el norte caribeño, *Encuentro en la Red* (January 15), http://www.cubaencuentro.com.

Akiner, S., 1993, Environmental Degradation in Central Asia, in: *Economic Development in Cooperative Partner Countries from a Sectoral Perspective*, NATO Economics Directorate, Brussels.

Alcolado, P. M., 1991, Ecological Assessment of Semi-enclosed Marine Water Bodies of the Archipelago Sabana-Camagüey (Cuba) Prior to Tourism Development Projects, *Marine Pollution Bulletin* **23**.

Alfonso, P., 2001, Una nueva reyerta diplomática en La Habana, *El Nuevo Herald* (February 21), http://www.elherald.com.

Arroyo, V. R., 2002, Guanahacabibes: Tesoro en peligro, (January 28), http://www.cubanet.org.

12. Cuba's Development Strategy

Banco Central de Cuba, 2001, *Informe económico 2000*, La Habana.

Bohemia Article on Effect of Tourism in Key Areas, 1997, Havana Prensa Latina (April 30), in: Foreign Broadcast Information Service, *FBIS-LAT-97-125* (May 5).

Bohi, D. R., 1994, Foreword, in: *Pollution Abatement Strategies in Central and Eastern Europe*, M. E. Toman, editor, Resources for the Future, Washington, pp. ii-vii.

Castro, F., 1992, *Ecología y Desarrollo: Selección Temática 1963-1992*, Editora Política, La Habana.

Cepero, E., 2001a, La destrucción del Cauto, *Encuentro de la Cultura Cubana* **21/22** (Summer/Fall): 91-93.

Cepero, E., 2001b, Dolor en los cayos, *Encuentro de la Cultura Cubana* **21/22** (Summer/Fall): 93-96.

Cepero, E., 2001c, Michelle, Castro y las playas, *El Nuevo Herald Digital* (November 9) http://www.elherald.com.

Cepero, E., 2002, El jinetero ambiental del Caribe, *Encuentro en la Red* (February 11), http://www.cubaencuentro.com.

Collis, D. S., 1995, *Environmental Implications of Cuba's Economic Crisis*, Cuba Briefing Paper No. 5, Georgetown University, Washington.

Comisión Económica para América Latina y el Caribe (CEPAL), 2000, *La economía cubana: Reformas estructurales y desempeño en los noventa*, Fondo de Cultura Económica, Mexico.

Comisión Económica para América Latina y el Caribe (CEPAL), 2001, *Cuba: Evolución económica durante 2000*, (May), CEPAL, Mexico.

Cosano Alén, R., 2000, Puerto en Cayo Coco refuta necesidad de construcción de pedraplenes, (April 19), http://www.cubanet.org.

Cubaeco, 1999, Impacto ambiental en Cayo Guillermo y Cayo Coco, al Norte de Cuba, (July 1), http://www.cubanet.org.

Cubaeco, 2001a, Intensa erosión deteriora las playas del polo turístico Cayo Coco, *Encuentro en la Red* (October 23), http://www.cubaencuentro.com.

Cubaeco, 2001b, Reconoce el régimen destrucción de al menos dos playas en la costa de Holguín, (December 24), http://www.cubaencuentro.com.

Díaz-Briquets, S., and J. Pérez-López, 2000, *Conquering Nature: The Environmental Legacy of Socialism in Cuba*, University of Pittsburgh Press, Pittsburgh.

Espino, M. D., 1992, Environmental Deterioration and Protection in Socialist Cuba, in: *Cuba in Transition—Volume 2*, Association for the Study of the Cuban Economy, Washington, pp. 327-342.

Espino, M. D., 1994, Tourism in Cuba: A Development Strategy for the 1990s?, in: *Cuba at a Crossroads*, Jorge F. Pérez-López, editor, University Press of Florida, Gainesville, pp. 147-166.

Feshback, M., and A. Friendly, 1992, *Ecocide in the USSR*, Basic Books, New York.

Figueras, M. A., 1994, *Aspectos estructurales de la economía cubana*, Editorial de Ciencias Sociales, La Habana.

Gálvez, J. C., 2002, Contaminación de la presa Zaza provoca intoxicaciones, *Encuentro en la red* (July 26), http://www.cubaencuentro.com.

Garcell, J.C., 2001, Contaminación ambiental pone en peligro la salud de residentes de Moa, (November 27), http://www.cubanet.org.

Garcell, J. C., 2002, Contaminación ambiental daña la salud de trabajadores de planta de níquel, (February 4), http://www.cubanet.org.

García, E. J., 2001, Planta de níquel cubana reduce plantilla y contamina el entorno, (September 21), http://www.cubanet.org.

González Martinez, O., 2001, Avanza estudio conjunto para el desarrollo de Cayo Paredón Grande, *Granma Digital Edition* (November 14), http://www.granma.cubaweb.cu.

González Martínez, O., 2002, Comienza operaciones aeropuerto internacional en Cayo Coco, *Granma Digital Edition* (January 15), http://granma.cubaweb.cu.

Guevara, V., A. Rivero, and others, 1998, El clima y la protección de la biodiversidad en el ecosistema Sabana-Camagüey para el establecimiento de un desarrollo sustentable, in: Instituto Superior de Ciencias y Tecnología Nucleares, Cátedra de Medio Ambiente, *Contribución a la educación y la protección ambiental*, Editorial Academia, La Habana, pp. 48-51.

Habashi, F., 1993, Nickel in Cuba, in: *Extractive Metallurgy of Copper, Nickel and Cobalt, Vol. I: Fundamental Aspects*, Summary of the Paul E Queneau International Symposium, Minerals, Metals, and Materials Society, Warrendale, Pennsylvania, pp. 1165-1178.

Hernández, G., 1982, Detener la contaminación de nuestras aguas marinas, *Bohemia* 74:18 (April 30):28-31.

Hernández, N., 2002, Agua de consumo doméstico contaminada con albañales, CPIC (June 12), http://www.cubanet.org.

Hotel de la Sol-Meliá se desploma parcialmente antes de ser inaugurado, 2000, (July 5), http://www.cubanet.org.

Invitan a firmas de EU a buscar petróleo en la isla, 2001, *El Nuevo Herald* (December 21):29A.

Kaufman, H., 1993, From Red to Green: Cuba Forced to Conserve Due to Economic Crisis, *Agriculture and Human Values* **10**(3):31-34.

Knox, P., 1995, Sherritt Breathes Life into Cuban Mine, *Globe and Mail* (Toronto) (July 31).

León Moya, H., 2002, Contaminación del Guaso pudo ser evitado, *Granma* (February 28), http://www.granma.cubaweb.cu.

Ley No. 77: Ley de inversion extranjera, 1995, *Gaceta Oficial* (September 6):5-12.

Martínez Canal, M., R. Pérez Díaz, A. Rodríguez Vargas, and Y. Lorente Pérez, 1998, Nivel de contaminación de los sedimentos de fondo de algunas zonas de la plataforma insular cubana, in: Instituto Superior de Ciencias y Tecnología Nucleares, Cátedra de Medio Ambiente, *Contribución a la educación y la protección ambiental*, Editorial Academia, La Habana, pp. 52-55.

Ministerio de Ciencia, Tecnología y Medio Ambiente (CITMA), 1999, *Situación ambiental cubana 1998*, La Habana.

Molina, G., 2001, A fines de este año el país producirá el 90% de su electricidad, *Granma Internacional Digital* (April 27), http://www.granma.cu/espanol/abri4/16basica-e.html.

Nickel Plant Said to Pollute Sea with Metals, Acids, 1994, Prensa Latina News Service (December 9), Foreign Broadcast Information Service, *FBIS-LAT-94-238* (December 12):15.

Núñez Jiménez, A., 1980, Defensa de la naturaleza cubana en la Asamblea Nacional, *Granma* (July 4): 4.

Oficina Nacional de Estadísticas (ONE), 2001, *Anuario estadístico de Cuba 2000*, La Habana.

Oramas, J., 2001, Se elevó en seis veces la producción nacional con una inversión de USD 600 millones, *Granma Internacional Digital* (January 15), http://www.granma.cubaweb.cu/.

Oro, J. R., 1992, *The Poisoning of Paradise: The Environmental Crisis in Cuba*, Endowment for Cuban American Studies, Miami.

Orrio, M. D., 1997, Acusan incremento de enfermedades de declaración obligatoria, (April 29), http://www.cubanet.org.

Pattullo, P., 1996, *Last Resorts: The Cost of Tourism in the Caribbean*, Latin American Bureau, London.

Pedraza Linares, H., 1997, Alarmante situación ambiental en Pinar del Río, Habana Press (June 18), http://www.cubanet.org.

Peláez, O., 2001a, Arrecifes coralinos bajo amenaza, *Granma Digital Edition* (December 25), http://www.granma.cubaweb.cu.

Peláez, O., 2001b, Jaque a la erosión en Varadero, *Granma Digital Edition* (May 31), http://www.granma.cubaweb.cu.

Pelaez, O., 2002, Piratas de la naturaleza, *Granma Digital Edition* (February 2), http://www.granma.cubaweb.cu.

Penié Rodríguez, I., and I. de los A. García Ramil, 1998, Hidroquímica y calidad ambiental del archipiélago Sabana-Camagüey, in: Instituto Superior de Ciencias y Tecnología Nucleares, Cátedra de Medio Ambiente, *Contribución a la educación y la protección ambiental*, Editorial Academia, La Habana, pp. 155-159.

Pérez-López, J. F., 1995a, *Cuba's Second Economy*, Transaction Publishers, New Brunswick.

Pérez-López, J. F., 1995b, *Odd Couples: Joint Ventures Between Foreign Capitalists and Cuban Socialists*, Agenda Paper No. 16, North-South Center, University of Miami, Coral Gables, Florida.

Pérez-López, J. F., 2001, Waiting for Godot: Cuba's Stalled Reforms and Continuing Economic Crisis, *Problems of Communism* **48**(6):43-55.

Pérez Santos, I., A. Morales Abreu, and N. Capetillo Pinar, 1998, Afectaciones producidas en el acuatorio marino por la construcción del pedraplén Caibarién-Cayo Santa María, in: Instituto Superior de Ciencias y Tecnología Nucleares, Cátedra de Medio Ambiente, *Contribución a la educación y la protección ambiental*, Editorial Academia, La Habana, pp. 60-63.

Pérez Villanueva, O. E., La inversion extranjera directa en Cuba: Evolución y perspectiva, paper presented at the 2001 meeting of the Latin American Studies Association, Washington, September.

Portela, A., 1997, Cuba Assesses Damage, Finds it Excessive, *Cuba News* 5(7):11.

Pozo, A., 1989, ¡Anduvo, anda y andará!, *Bohemia* **72**:1 (January 4):16-23.

Reed, G., 1993, Protecting the Environment During the Special Period, *Cuba Update* 29 (September-October):29, 31-33.

Romero Ochoa, N., D. López García, and others, 1998, Un caso de la afectación de los cambios antrópicos al medio en las aguas de una bahía del archipiélago Sabana-Camagüey, in: Instituto Superior de Ciencias y Tecnología Nucleares, Cátedra de Medio Ambiente, *Contribución a la educación y la protección ambiental*, Editorial Academia, La Habana, pp. 106-110.

Rosendahl, B., 1991, Cuban Oil Drilling Could Spoil the Keys, *The Miami Herald* (May 5):1C.

Silva Lee, A., 1996, *Natural Cuba*, Pangae, Saint Paul, Minnesota.

Smarzynska, B. K, and S. Wei, 1991, *Pollution Heavens and Foreign Direct Investment: Dirty Secret or Popular Myth?*, World Bank, Washington.

Solano, R., 1995, Catástrofe del ecosistema cubano, *El Nuevo Herald* (April 22):15A.

Spadoni, P., 2001, The Impact of the Helms-Burton Legislation on Foreign Investment in Cuba, in: *Cuba in Transition—Volume 11*, Association for the Study of the Cuban Economy, Washington, pp. 18-36.

Suárez Ramos, R., 2001, Marcha plan de desarrollo de Guanahacabibes, *Granma Digital Edition* (September 27), http://www.granma.cubaweb.cu.

Timoshenko, V.P., 1953, Agricultural Resources, in: *Soviet Economic Growth*, A. Bergson, editor, Row, Peterson and Company, Evanston, Illinois, pp. 246-271.

Torres, I. E., 1999, The Mineral Industry of Cuba—1999, in: *Minerals Yearbook—1999*, U.S. Geological Survey, Washington, http://minerals.usgs.gov/minerals/pubs/country/9509099.pdf.

Veloz, M., 2001, Inversión extranjera: El justo lugar, *Opciones* (January), http://www.elecono-mista.cubaweb.cu/2000/2001/nro127_127169.html.

Visión real de lo maravilloso, 1998, *Granma Digital Edition* (April 2), http://www.granma.cubaweb.cu.

Volin, L., 1962, Agricultural Policy in the Soviet Union, in: *The Soviet Economy: A Book of Readings*, M. Bornstein and D. Fusfeld, editors, Richard D. Irwin, Homewood, Illionois, pp. 243-276.

Willett, J. H., 1962, The Recent Record in Agricultural Production, in: *Dimensions of Soviet Economic Power*, U.S. Congress, Joint Economic Committee, U.S. Government Printing Office, Washington, pp. 91-93.

Wotzkow, C., 1998, *Natumaleza cubana*, Ediciones Universal, Miami.

Wotzkow, C., and E. Casañas, 2002, Un desierto submarino, *Encuentro en la Red* (January 15), http://www.cubaencuentro.com.

Index

Abatement, 122-128, 132, 134, 137, 138, 141-142
Agrarian reform, 61, 103, 259-260
Agricultural extensification, 253-264, 271
Agricultural intensification, 58, 178, 191, 253-264
Agricultural policy, 75-92, 100, 253-264
Agricultural productivity, 80, 81, 253-264, 271-272
Agriculture, 5, 6, 12, 17, 19, 57-58, 67, 75-92, 96, 99-100, 110, 112, 253-264
Agroecology, 175-198
Air pollution, 96, 130-131, 135, 137, 139-140, 142-143, 215, 284
Air quality, 99, 130, 143, 210, 214, 215
Andes, 55-56, 65
Arawaks, 5-7, 9, 11
Argentina, 208
Australia, 46-47
Authoritarianism, 97, 106, 208

Barbados, 8, 17-19
Belize, 142, 160-161
Biodiversity, 55-68, 83, 84, 111, 125, 150, 175-198, 253, 259, 263-264, 279
Biological diversity, *see* Biodiversity
Birds, 17, 37, 42, 56, 175-198
Bolivia, 55-68, 142, 189, 254
Brazil, 6, 62, 140-141, 189, 207-228, 254

Caribs, 5-6, 11, 14, 19
Causeways (Cuba), 275-282
Cedros Island (Mexico), 31-50
Central America, 161-165, 168-169, 188
Certification, 95-116, 175-198
Chapter 11 (of NAFTA), 107
Chile, 32, 46, 62, 140-143, 173, 208, 253-264
Citizen participation, 95-116, 231-248
Civil society, 61, 231-248
Coffee, 84, 88-90, 175-198
Colombia, 140-141, 154, 160-

165, 169, 191, 208, 257
Command-and-control policy, *see* Regulations
Commission for Environmental Cooperation (CEC), 107, 197, 233-235, 239-248
Common property, 95-116, 149-173
Commons
  international, 149-173
  tragedy of the, 97-98, 113-114, 149-151, 153, 167
Community-based forest management, 95-116
Comparative advantage, 255
Conservation
  of biodiversity, 55-68, 175-198, 264
  ex situ, 47
  of forests, 31-50, 55-68, 95-116, 175-198, 253, 254, 264
  institutions, 59, 60, 65
  in situ, 47
Conservation genetics, 31-50
Constitution
  of Brazil, 207-228
  of other countries, 208
Cooperative, 34, 45, 75, 77, 88-92, 179, 198, 268
Coral reefs, 5, 277, 280, 285
Corruption, 97, 100, 105, 110-111, 131, 223, 224, 226, 269-270
Costa Rica, 142, 151, 154, 158, 160-162, 164-167, 169, 171-174, 189, 191, 208
Cost-effectiveness, 122-123, 134, 144
Cuba, 189, 267-291
Culture, 3-5, 14, 27, 227
Currency over-valuation, 256

Deforestation, 55, 57, 64-66, 77, 86, 99-101, 104, 110-112, 135, 177-178, 212, 214-215, 253-264
Democracy, 95-116, 223-224, 231
Deposit-refund system, 132, 138-139, 142
Devaluation, 256-257
Distributional concerns, 130, 144
Dolphins, 3-7, 9-10, 14, 17, 19-29, 153-153, 158-159, 166-167, 171

Eco-labeling, 153, 166-167, 175-198
Ecological diversity, *see* Biodiversity
Ecology, 40, 47, 175-198
Ecosystem, 36-37, 41, 44, 46, 55-58, 63-34, 67, 98-99, 103, 115, 166, 171, 175-198, 279-281, 288, 290
Ecotourism, 43-44, 50, 105, 264
Ecuador, 141, 151, 154, 160-165, 168-169, 189, 208, 253-264
El Salvador, 154, 160-163, 172, 189
Endangered species, 33, 40, 42-44, 49, 50, 107, 182
Enforcement, 10, 12, 26, 111, 115, 128-132, 141, 143-144, 154, 157, 165, 181, 207-228, 235, 260-270, 274
Environmental awareness, 95, 232, 235, 243, 245-246
Environmental law, 10, 19, 24, 26, 106-107, 111, 115, 207-228

Environmental justice, 98
Environmental Kuznets Curve (EKC), 256, 263
Environmental non-governmental institutions (ENGOs), 64, 65, 231-248
Environmental protection (institutions), 103, 131-132, 135, 144, 207-228, 238, 274-275
Environmental standards, see Regulations
Erosion, 59, 81, 86-87, 96, 99, 105, 112, 135, 138, 214, 271-272, 277, 280-281
European environmental policies, 136-140

Fertilizer, 85-89, 216, 260, 271
Firewood, 43, 58, 65
Fishery, 3-27, 149-173
Foreign direct investment, 163, 267-291
Forests, 33, 44-45, 47, 55-68, 95-116, 253-264
Forest certification, 95, 109, 113-114
Forest Resource Assessment, 259
Forest trade, 95, 98
Forestry, 59, 61, 67, 95-116, 262-263

Gender, 114
Genetic diversity, 36, 38, 46, 47
Genetic forest resources, 42, 43, 46, 48, 50
Genetic restoration, 32
Governance, 68. 95-116
Green Revolution, 75-92, 255

Grenada, 8, 14-17, 19, 22-27
Guadalupe Island (Mexico), 31-50
Guatemala, 154, 160-163, 172, 189, 190, 192
Guyana, 16

Honduras, 142, 160-161, 189
Hybrid instruments, 127-128, 144

Imperfect information, 134
Indigenous communities, 6-7, 9, 22-23, 95-116, 176
Information
dissemination, 79, 231-248
as regulation, 128, 140, 143-144, 231-248
Infrastructure (roads), 42, 56, 64, 66, 257, 264
Inter-American Development Bank (IADB), 167, 257
Inter-American Tropical Tuna Commission, 149-173
Integrated pest management, 179
Intercropping, 80-81
International collaboration, 47-51, 149-173
International Ecosystem Management, 175-198

Justice, 207-228

Kyoto Protocol, 65, 142

Law, 106-107, 111, 115, 207-228
Lawsuit, public interest, 207-228
Lumber, 46, 95-96, 108, 110, 114

Maize, 75-92
Manatees, 4-7, 9, 12, 15, 17, 20, 22-23
Marine mammals, 3-27, 152
Market-based policies, 121-144, 175-198
Meso-Indians, 5, 9
Mexico, 31-52, 75-92, 95-116, 132, 140-143, 179-180, 186, 188-192, 197-198, 231-248
Mining, 12, 35, 40, 273, 282-286
Monopoly, 133
Monterey Pine, 31-52

National parks, 55-68
National Pollutant Release Inventory (Canada), 232, 238
Nicaragua, 142, 154, 160-163, 189, 208
Nickel mining, 282-286
Non-governmental institutions (NGOs), *see also* Environmental NGOs, 49, 60-61, 64-65, 68, 231-248
North American Commission on Environmental Cooperation (CEC), 107, 197, 233-235, 239-248
North American Free Trade Agreement (NAFTA), 95-116, 197, 238

Oil, 55, 63-65, 67-68, 214, 216, 218, 257, 260, 286-289
Open access fishery, 149-173

Paleo-Indians, 5
Panama, 142, 154, 160, 162-164, 166, 189, 194

Paraguay, 189, 208, 259
Partido Revolucionario Institucional (PRI), 76
Peasants and political change, 75-92, 105
Pedraplenes (Cuba), 275-282
Permits, tradable, 124, 126-127, 134, 139-140, 142-144
Peru, 62, 142, 160-164, 167, 173, 189, 194
Pesticides, 84-85, 179, 181, 191, 271
Pinus radiata, 31-52
Pollutant Release and Transfer Registry (Mexico), 231-248
Pollution, air, *see* Air pollution
Pollution, water, *see* Water pollution
Pollution haven hypothesis, 270
Poverty, 55-57, 65-66, 99, 101, 110, 256
Privatization, 97-98, 103, 106, 108, 135, 262
Productivity, agricultural, 253-264
Prosecution, 207-228
Protected areas, 34, 36, 40-43, 49, 55, 58-67, 96
Protected status, 31, 49, 51
Protection, wildlife, 10, 26, 55, 58
Public education, 31, 235
Public interest lawsuit, 207-228
Purse-seine, 149-173

Radiata pine, 31-52
Regulations, 61, 65, 102-104, 107, 121-124, 131, 135-137, 140, 142, 144, 167, 172, 212, 215, 235

Index

Roads, *see* Infrastructure

St. Vincent and the Grenadines, 4-5, 19-22, 23-27
Seeds, 43-44, 46-51, 102, 260
SEMARNAT (Mexico), 31-34, 36, 42, 103, 111, 233, 243, 245
Socialism, 267-291
Solid waste, 137, 212, 214, 215
Special period in time of peace (Cuba), 267-291
Species loss (*see* Biodiversity)
Standards, *see* Regulations
Subsidies, 124-125, 137-138, 141-142, 262
Sustainable development, 59, 61, 64, 103, 171, 175-198, 274,
Sustainable forestry, 95-116, 264

Tax and subsidy combinations, 132, 138-139, 142
Taxes, 121-144, 274
Technology
 and agriculture, 57, 75-92, 253-264
 and forestry, 46, 101
Timber, 60, 101-104, 108, 110-113, 176, 258, 260, 262-264
Tourism, *see also* Ecotourism, 14, 17, 19, 43-45, 237, 264, 275-282
Toxics Release Inventory (United States), 140, 232, 238, 240, 243, 246
Tradable pollution permits, *see* Permits, tradable
Tragedy of the commons, 97-98, 113-114, 149-151, 153, 167
Trinidad and Tobago, 4-6, 8, 11-14, 21, 23-26, 189
Tropical ecology, 175-198
Tuna, 149-173,

Uncertainty, 127, 133-134, 139
United Nations, 48, 51-52, 237, 239
United Nations Convention on Biodiversity, 55

Venezuela, 4-11, 22-26, 155, 160, 162-165, 168-169, 172, 189, 207
Voluntary instruments, 142, 161, 231-248

Water pollution, 99, 130, 135, 137, 139-141, 214, 215, 272, 284
Whaling, 3-27
Wildlife protection, 10, 26, 55, 58